T0396724

Proceedings of the NATO Advanced Research Workshop on
Climate Change, Human Health and National Security
Dubrovnik, Croatia
28–30 April 2011

Library of Congress Control Number: 2011940792

ISBN 978-94-007-2500-3 (PB)
ISBN 978-94-007-2429-7 (HB)
ISBN 978-94-007-2430-3 (e-book)
DOI 10.1007/978-94-007-2430-3

Published by Springer,
P.O. Box 17, 3300 AA Dordrecht, The Netherlands.

www.springer.com

Printed on acid-free paper

National Security and Human Health Implications of Climate Change

edited by

H.J.S. Fernando
University of Notre Dame
Notre Dame, IN, USA

Z.B. Klaić
University of Zagreb
Zagreb, Croatia

and

J.L. McCulley
Arizona State University
Tempe, AZ, USA

 Springer

Published in Cooperation with NATO Emerging Security Challenges Division

NATO Science for Peace and Security Series

This Series presents the results of scientific meetings supported under the NATO Programme: Science for Peace and Security (SPS).

The NATO SPS Programme supports meetings in the following Key Priority areas: (1) Defence Against Terrorism; (2) Countering other Threats to Security and (3) NATO, Partner and Mediterranean Dialogue Country Priorities. The types of meeting supported are generally "Advanced Study Institutes" and "Advanced Research Workshops". The NATO SPS Series collects together the results of these meetings. The meetings are co-organized by scientists from NATO countries and scientists from NATO's "Partner" or "Mediterranean Dialogue" countries. The observations and recommendations made at the meetings, as well as the contents of the volumes in the Series, reflect those of participants and contributors only; they should not necessarily be regarded as reflecting NATO views or policy.

Advanced Study Institutes (ASI) are high-level tutorial courses to convey the latest developments in a subject to an advanced-level audience

Advanced Research Workshops (ARW) are expert meetings where an intense but informal exchange of views at the frontiers of a subject aims at identifying directions for future action

Following a transformation of the programme in 2006 the Series has been re-named and re-organised. Recent volumes on topics not related to security, which result from meetings supported under the programme earlier, may be found in the NATO Science Series.

The Series is published by IOS Press, Amsterdam, and Springer, Dordrecht, in conjunction with the NATO Emerging Security Challenges Division.

Sub-Series

A.	Chemistry and Biology	Springer
B.	Physics and Biophysics	Springer
C.	Environmental Security	Springer
D.	Information and Communication Security	IOS Press
E.	Human and Societal Dynamics	IOS Press

http://www.nato.int/science
http://www.springer.com
http://www.iospress.nl

Series C: Environmental Security

National Security and Human Health Implications of Climate Change

Preface

This NATO Advanced Research Workshop was aimed at examining *the relationship between Climate Change, Human Health and (Inter) National Security*. The subject is widely discussed internationally both at the military and at the civilian level. A plethora of movie fictions have illustrated a great variety of possible scenarios. The direct impact of climate change on health has been shown and accepted. The CO_2 levels are now the highest of the last 500,000 years, the global temperature is clearly on the rise, glaciers melt at the poles, but also at the continental level, and extreme events are on the rise. In our cities, we face a level of pollution that is increasingly relevant in the pathogenesis of human and animal diseases. It is time to very seriously evaluate these new or newly arisen threats, which are at levels higher than the "watch" or "guard" levels of the last century's environmental conditions. Unless these threats are appropriately studied, carefully assessed and prevented, they can, at least in their more direct impacts, have devastating effects on our health, social organization and, thus, on our security.

The questions we should ask ourselves are therefore:

1. Which climactic changes can represent a threat to our security and why?
2. Which elements increase the effects of climate change on health?
3. Which actions we must undertake?

Political and military leaders of the major countries asked themselves the same questions. The UN Security Council decided to tackle the problem and even the U.S. Central Intelligence Agency decided to open a center devoted to "Climate Change and National Security." For the next few decades the forecasted effects of climate change are primarily the extreme events: typhoons, floods, rising sea levels, reduction of polar ice, peaks of extreme heat, and conditions which favour the spreading of disease, such as malaria, dengue fever, schistosomiasis as well as increasing the risk of water-borne diseases.

These events, both at the national and international level can cause migrations of individuals or entire populations, but also situations favouring internal conflict and can create political instability and humanitarian disasters. Regional impacts of climate change include the following.

Africa: Increased political instability, reduced agricultural productivity, famines, civil wars, which favour terrorism: Darfur/Ethiopia. Eritrea, Somalia, Angola, Nigeria, Cameroon, Western Sahara are clear examples.

Asia: The forecast is for a warming of the Asia/Pacific region where hundreds of millions people are at risk because of the melting of the Tibetan glaciers.

Middle East: In this region water is crucial and the situation can be summarized by *"ABUNDANT OIL, SCARCE WATER AND INTERNATIONAL CONFLICT"*

The Western Hemisphere (US): The major risks for the American continent are cyclones, fires, whether naturally occurring or by arson, at times of huge dimension and duration, and tropical storms (Katrina, etc.) which pose a major challenge for the social infrastructures and the organization of the alarm and support systems.

The Western Hemisphere (EEC): Europe faces a warming phenomenon (unfortunately at this point we cannot speak of a trend) so that for some areas we speak now of desertification. Only in the last few decades have we witnessed the phenomena of coastal erosion, rivers overflowing, abnormal heat waves, and torrential rains responsible for landslides and snowslides. In 2003 a single heat wave alone has caused over 35,000 deaths. Unfortunately, not all countries have an efficient system of civil protection. While industrial countries may have effective social infrastructure to adapt to modified climatic conditions effectively, it is much lower in the less developed countries such as The Balkans, Moldova and the Caucasian regions.

What Can We Do?

Climatic changes at the international level must fit in to a global geo-political strategy, which must take into account existing resources and structures. We need programs of specific information targeted to policymakers like the scientific publication entitled: *Research on Environmental Management in a Coastal Industrial Area: new indicators and tools for air quality and river investigations ISBN 9788860818997* performed by ENEA (MC. Mammarella et al) with the scientific support of American, European and Russian research groups leading at environmental level. It is imperative to devote resources to specific research, information, and training of civil and military personnel by a qualified international task force. It is important to develop equipment, strategies and preventive measures, creating infrastructures and networks, both national and international levels, which are capable of responding quickly and effectively in emergency situations. We should also support the weaker governments and help them to achieve the ability to implement all the preventive measures to face the effects of climatic changes on population. The climate change can be conquered, but quick action is needed.

President   Vincenzo Costigliola MD
European Medical Association
Bruxelles

Contents

Introduction: Climate Change, Human Health and National Security

Prime facie, the title of this volume appears as three timely topics, disconnected yet juxtaposed, but a closer look indicates that they are indeed interconnected through the fabric of *quality of life*. The latter is defined in terms of ensuring safe, healthy and equitable existence for every human, with access to adequate resources at present and in the future. As depicted in Fig. 1, however, climate change has threatened human health and security through numerous manifestations. To understand the tripartite interplay between human health, climate change and security of nations and citizens, a workshop was held in Dubrovnik, Croatia, during 28–30 April 2011, with sponsorship from the NATO Science for Peace and Security Program. Entitled 'Climate Change, Human Health and National Security,' the workshop was intended to facilitate discussions on each of the three themes, their interconnectedness and ensuing feedbacks. Thirty-two attendees from 17 countries were invited. The highlight was the multidisciplinary inclusiveness, where leading modelers, natural, political and social scientists, engineers, politicians, military experts, urban planners, industry analysts, epidemiologists and healthcare professionals parsed the topic on a common platform. The papers presented at the workshop are included in this volume.

Climate change impacts on humans are numerous, and at times can be pernicious, encompassing human comfort to food, energy and water shortages to armed conflicts. Human security implies freedom from the risk of loss of damage to attributes that are important for survival and well-being (Matthew et al. 2010). National security is the component of human security that deals with safety against armed conflicts and terrorism (hard security). As Kjeld Rasmussen[*] pointed out, while most discussions tend to be centered on hard security, soft security that deals with individuals is equally important in the current geopolitical atmosphere. Since the end of cold war, traditional definitions have been expanded to include additional threats such as social and political instability and ethnic rivalries. Because of the complexity of the problem and the formidable number of governing factors involved, addressing climate related issues requires a system approach, noted Julian Hunt[*].

---

[*] quotations made during the meeting

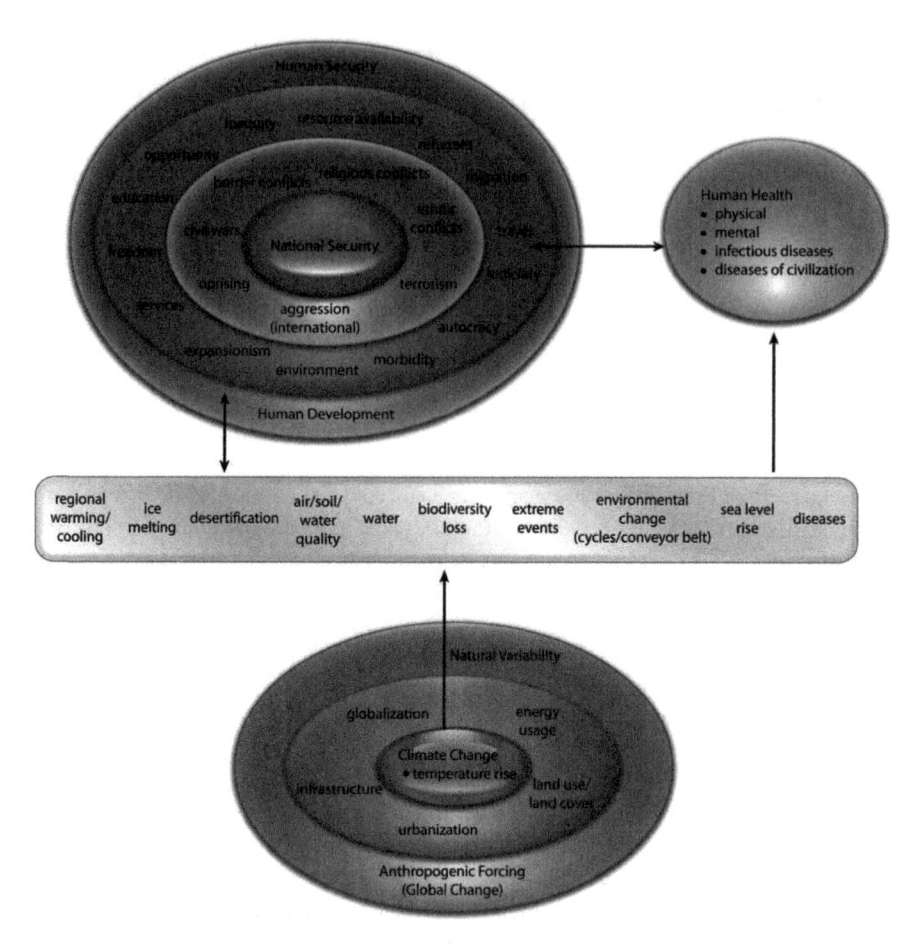

Fig. 1 Interplay between climate change, human health and national security

Through social reforms, 'out of the box' thinking and state-of-the-art technology utilization, it is possible to assess and fight off many negative impacts of climate change. "We need to convert conflict to cooperation," noted Jacques Ganoulis[*]. "Introduce therapy as soon as symptoms come out," added Vincenzo Costigliola[*].

Societal and ecosystem impacts of climate change are pervasive. Vector borne diseases will appear or reappear, and vectors will expand their poleward operating range as the temperature increases (Alebić-Juretić, Tourre, Paz)[†]. High temperatures may increase heat-stress related illnesses such as heat strokes and dehydration, which may increase the mortality rate (Peretz[†]). Weather variability is expected to produce high pressure regions conducive for heat waves and air pollution episodes (Kambezidis[†]), and teleconnections between different regions may cause climatic interdependences (Herceg Bulić[†]). According to Anne-Lise Beaulant[*], "several

[†] see the paper of this author in this volume

yearly episodes of the ilk of [the] Paris-2003 heat wave are possible toward the year 2100" (Barriopedro et al. 2011).

Resource shortages due to climate change, especially the reduction of water and food supply, may spark conflict for resources. Extreme events triggered by climate variability such as intense hurricanes, heat waves and desertification may lead to human catastrophes, thus impacting human security. Those affected will aggressively search for means of adaptation and/or resettlement, leading to mass migration. Uncontrolled influx of migrants (climate refugees) sparks conflict between nations, in addition to intra-nation social and economic segregation. "Climate change is a threat multiplier," argued Marcus King[*]. Feeling the sense of injustice is an acute cause of uprising, added Lukas Rüttinger[*]. Governments may have to divert significant energies and resources, which could have been otherwise used for productive means, to help those afflicted and to quell uprisings. Of those, the most affected are the poor and vulnerable as well as smaller nations and islands, stoking issues of equity and justice, and hence political and social instabilities (Radović[†]). "For Pacific states, climate change is our main security concern," pointed out Nancy Lewis[*].

In addition to indirect influence, climate change may directly affect the military enterprise, for example, through physical damage to military installations caused by extreme events, opening of sea-lanes due to ice melting, health impacts on warfighters caused by vector born diseases and poor air quality, and political instability of nations that house military assets. Conversely, military machinery can help ameliorate climate impacts on humans by providing physical, material and psychological humanitarian assistance, including mass evacuations, food distribution and emergency medical services. "It will be necessary to assess the current assets, their vulnerabilities as well as future requirements," contended Marcus King[*]. For example, fewer ice breakers and more hospital ships will be required in the future due to ice melting and increased disease and humanitarian assistance needs. Sound socio-economic analyses as well as cutting-edge resilience and risk assessment models can help conducting such assessments.

Urban areas are the centers of greenhouse gas emissions, and indications are that they will bear the brunt of climate change given their concentrated populations and intense on-going land use changes. Climate change may exacerbate the urban heat island and may cause a marked decrease of the diurnal temperature range in urban areas, thus affecting both human and ecosystem health. "[The] atmospheric boundary layer in which ecosystems are immersed is most sensitive to climate change," pointed out Sergej Zilitinkevich[*]. Changes to it will have consequences in pollution distribution, and hence to human health (Jeričević, Klaić, Fernando).[†] Also affected will be wind patterns and speeds, which will impact civil infrastructure and wind energy availability (Kozmar[†]).

"Tipping" between climatic states is another issue of interest. While the IPCC 4th Assessment Report discounts the possibility of strong nonlinearities (or catastrophic shifts), regime shifts are possible over regional and local (urban) scales, driven by positive feedbacks amongst processes (Rasmussen, Fernando).[†] The workshop attendees call for physical understanding of phenomena and mechanisms of local climatic tipping, which is imperative in preparing for local climate variability.

The attendees also commented on the lukewarm response of regional and local governments to climate change challenges. "Global climate strongly interacts with local climate – some for good and some for bad," said Robert Bornstein[*]. Governments tend to work with 4–5 year time scales, and hence pay lesser attention to 10-year averages. Obviously local climate adaptation should not rely on IPCC predictions, as local responses can be markedly different. Sea Breeze, land use change, rainfall redistribution - all influence the local climate. New models, measurement tools and information technologies are necessary for rapid dissemination of climate and environmental risk information to stakeholders (Baklanov, Mammarella, Costigliola).[†]

Interdisciplinary, multi-scale and collaborative approaches are imperative in handling critical trans-boundary issues of climate change (NAS 2005). The workshop attendees identified possible ways to break communication barriers within multidisciplinary audiences, foster harmony within climate science enterprise and turn climate woes into opportunities. They noted that "water can be a powerful source to foster peace," since nations are unwilling to deprive others of accessing water resources lest grave humanitarian crises arise (Rüttinger[†]). "Water issues cannot be looked at in isolation; food, water and energy are all interrelated, and all underpin ecosystem services," noted Roger Falconer[*]. Many countries, especially those in Middle East, Africa and Asia, will be impacted by the changes to hydrological cycle (Elsaeed, Oroud),[†] but special attention should be paid to local water resources, the climatic response of which is largely unknown. "We know that we don't know about [it]," remarked Jacques Ganoulis[*]. Accounting for water is not a straight jacket issue, as 'virtual water' (used in the production of goods or services) needs reckoning, Roger Falconer[*] added. Extreme rainfall over narrow land areas as well as shifting of rain over to oceans may leave some catchment areas devoid of rain (Alpert[†]).

How sensitively the earth system responds to climate mitigation strategies depends on the resilience of large water bodies, such as oceans and inland seas, to environmental change (Zavialov[†]). Even if there is no further release of anthropogenic CO_2 to the atmosphere, because of the slow response of oceans, climate warming will continue to occur over the twenty-first century albeit at a slower place, increasing by several tenths of a degree over the century (Royal Society 2010). About half of the CO_2 released since the industrial revolution has been absorbed by the oceans, which has been a source of ocean acidification and coral bleaching. A change of CO_2 injection permeates to the ocean very slowly, and hence greenhouse gas mitigation strategies only sluggishly come into effect.

Ecosystems response to climate change can be diverse, and include loss of biodiversity and indigenous species as well as arrival of invasive species (Bashmakova, Vardanian).[†] Landscape planning, ecosystem health and air quality are effective platforms for climate mitigation and adaptation discourse at the local level. For example, management of parks, deltas, rivers and wetlands require melding of social, political, economic and ecological teams. Residents pay attention to ecosystems, air pollution, visibility and aesthetics, and are eager to see that local governments ensure a healthy environment sooner than later, said Adnan Kaplan[*]. In this context, naturally, climate change becomes a part of the consideration.

A bane for the progress of climate science is the paucity of data. Only some 1,400 data stations are being used for global averaged temperature, and some of them have now become urban over time, introducing biases. Change of flow patterns can also introduce unrepresentative trends. "Sound physics-based protocols must be developed for data processing and rejection, rather than relying on preconceived trends," noted Robert Bornstein[*]. Satellites are stepping up to the challenge of global temperature monitoring and provide extensive spatial coverage. More representative data stations are needed, with frequent evaluation of their suitability for climate research. Data should be transparent and easily available, with metadata, to all researchers. "Governments and international organizations such as WMO, WHO and the UN ought to develop data exchange, reposting and cataloging plans" proposed Julian Hunt[*]. Voluntary data also can be used after proper quality control procedures.

The workshop was a resounding success in bringing scientists with a myriad of different backgrounds together to communicate on how climate change can trigger health and security concerns. The seeds of the conference were germinated by Dr. Vincenzo Costigliola, former Medical Chief of NATO and the President of the European Medical Association. The workshop could not come to light without painstaking contributions of many colleagues, co-workers and students. Jennifer McCulley, Arizona State University, acted as the conference coordinator, Stipo Sentic, Scott Coppersmith, Melissa Unruh and Marie Villarreal, University of Notre Dame, helped with fine tuning of logistics and maintaining the website and Sahan Fernando, Gonzaga University, helped in preparing this ARW volume. Both University of Notre Dame and Faculty of Science, University of Zagreb, provided generous support in numerous ways, including financial contributions, for which we wish to express sincere gratitude. The enthusiastic participation of conference attendees and their willingness to exchange information made the conference a memorable event that is bound to spark future workshops of this ilk. We are grateful to the NATO for financial support through grant # EAP.ARW.984000.

Harindra Joseph Fernando

College of Engineering, University of Notre Dame, Indiana, USA

Zvjezdana Bentić Klaić

Faculty of Science, University of Zagreb, Zagreb, Croatia

References

Matthew RA, Barnett J, McDonald B, O'Brien KL (eds) (2010) Global environmental change and human security. MIT Press, Cambridge

Barriopedro D, Fischer EM, Trigo RM, Garcia-Herrera R (2011) The hot summer of 2010: redrawing the temperature record map of Europe. Science 32:220–224

NAS, US National Academy of Sciences (2005) Facilitating interdisciplinary research. National Academy Press, Washington, DC

Royal Society (2010) Climate change: a summary of the science, The Royal Society, London

Chapter 1
National Security and Human Health Implications of Climate Change

Marcus DuBois King

Abstract The first section of the paper presents key findings from the 2007 report, *National Security and the Threat of Climate Change* by the CNA Corporation, including that projected climate change: (1) Poses a serious threat to U.S. National Security; (2) Acts as a threat multiplier for instability in some of the most volatile regions in the world and; (3) Adds tensions even in stable regions of the world. In the second section I summarize work conducted by myself and Dr. Ralph Espach at CNA that identifies exactly which countries are most relevant to the CNA Military Advisory Board's original findings. By compiling data from a variety of sources, we identify the states most exposed to the impacts of climate change both in the short and long term. The next section introduces estimates of the resilience of these countries, and combines our evaluation of country exposure and expected resilience to create a 3-tiered ranking of countries most vulnerable to political and/or humanitarian crises as a result of climate impacts.

Keywords Climate change • Stability • Resilience • Exposure

1.1 Introduction

This paper will build upon the findings of the CNA Military Advisory Board (CNA MAB) study, *National Security and the Threat of Climate Change* published in 2007 and subsequent research our group has performed on climate change and state

M.D. King, Ph.D. (✉)
CNA Corporation, 4825 Mark Center Drive, Alexandria, VA 22311, USA
e-mail: mdking@gwu.edu

H.J.S. Fernando et al. (eds.), *National Security and Human Health Implications of Climate Change*, NATO Science for Peace and Security Series C: Environmental Security, DOI 10.1007/978-94-007-2430-3_1, © Springer Science+Business Media B.V. 2012

stability. The CNA MAB is an elite group of retired three-and four-star flag and general officers from the U.S. Army, Navy, Air Force and Marine Corps that studies pressing energy and environmental issues of the day to assess their impact on America's national security.

The mention of this CNA study in the website material providing justification for this conference is a testament to its continued relevance. It is therefore worth reviewing key findings of the study in some detail.

1.2   Findings of the CNA MAB

Finding 1: Projected climate change poses a serious threat to America's national security

The CNA MAB found that potential threats to U.S. national security require careful study and prudent planning—to counter and mitigate potential detrimental outcomes. Based on the evidence presented, the CNA MAB concluded that it is appropriate to focus on the serious consequences to our national security that likely stem from unmitigated climate change. In already-weakened states, extreme weather events, drought, flooding, sea level rise, retreating glaciers, and the rapid spread of life-threatening diseases will themselves have likely effects. The effects may include increased migrations, further weakened and failed states, expanded ungoverned spaces, exacerbated underlying conditions that terrorist groups seek to exploit, and increased internal conflicts. In developed countries, these conditions threaten to disrupt economic trade and introduce new security challenges, such as increased spread of infectious disease and increased immigration.

Overall, the study found that climate change has the potential to disrupt our way of life and force changes in how we keep ourselves safe and secure by adding a new hostile and stressing factors into the national and international security environment.

Finding 2: Climate change acts as a threat multiplier for instability in some of the most volatile regions of the world

The CNA MAB found that many governments in Asia, Africa, and the Middle East are already on edge in terms of their ability to provide basic needs: food, water, shelter and stability. Projected climate change will exacerbate the problems in these regions and likely add to the problems of effective governance. Unlike most conventional security threats that involve a single entity acting in specific ways at different points in time, climate change has the potential to result in multiple chronic conditions, occurring globally within the same time frame. Economic and environmental conditions in these already fragile areas will further erode as food production declines, diseases increase, clean water becomes increasingly scarce, and populations migrate in search of resources. Weakened and failing governments, with an already thin margin for survival, foster the conditions for internal conflict, extremism, and movement toward increased authoritarianism and radical ideologies. The U.S. or its allies may be drawn more frequently into these situations to help to provide relief, rescue, and logistics, or to stabilize conditions before conflicts arise.

Because climate change also has the potential to create natural and humanitarian disasters on a large scale its consequences will likely foster political instability where societal demands exceed the capacity of governments to cope. As a result, the U.S. or its allies may also be called upon to undertake stability and reconstruction efforts once a conflict has begun.

Finding 3: Projected climate change will add to tensions even in stable regions of the world

The CNA MAB report found that developed nations, including the U.S. and Europe, may experience increases in immigration and refugees as drought increases and food production declines in Africa and Latin America. Pandemic disease caused by the spread of vectors and extreme weather events and natural disasters may lead to increased domestic missions for US military personnel—lowering troop availability for other missions and putting further stress on its already stretched military, including National Guard and Reserve forces [1].

1.3 Analysis of Global Climate Change and State Stability

In 2008, CNA took a deeper, more analytical look to determine which parts of the world the second and third conclusions of the CNA MAB study best applied to. The question in this further research undertaken by myself and Dr. Ralph Espach was to determine exactly which states (strong or weak) were most exposed to the impacts of climate change and what sort of resilience these countries might have? [1] This research was undertaken at the request of the U.S. National Intelligence Council (NIC), a center for midterm and long-term strategic thinking within the U.S. Intelligence Community. The NIC asked CNA to examine countries that could become unstable from climate change in the near (2020–2025) and long (2040–2045) terms. Specifically, we were asked to:

- Identify those countries that are most exposed to climate impacts (water scarcity, agricultural degradation, sea level rise, and extreme weather events) both in the short term and the long term;
- Assess these countries' resilience to the impacts of climate change; and
- Discuss the implications of these findings for the security interests of the United States.

CNA was not asked to conduct any original data collection for this study, but instead to base our analysis on existing research from reputable sources [2].

1.3.1 Methodology

We derived a list of the countries most exposed to climate impacts in the short term (2020–2025) based on the historical record of frequency and intensity of droughts, severe weather events, agricultural degradation, and sea-level rise. To this data, which we obtained from the International Development Association (IDA) of the World Bank, we added an additional risk category of water scarcity related to a country's geography and precipitation rates.

To estimate long-term (2040–2045) exposure, we assumed the continuation of current and short-term trends in climate effects, but added consideration of new effects caused by glacial melt and other causes of water scarcity.

We estimated resilience of these countries based on analysis from the Joint Global Change Research Institute (JGCRI) at the University of Maryland, USA, which assigns resilience scores based on quantitative indicators of economic and social factors. We then combined our categorization of countries most exposed to climate effects in the short term with resilience scores to create a 3-tier ranking of countries most susceptible to political and humanitarian crises as a result of future climate change effects.

1.3.2 States' Exposure to Climate Change

1.3.2.1 Short Term (2020–2025)

Most models of climate change effects such as those of the Intergovernmental Panel on Climate Change (IPCC) 2007 [3], estimate effects decades into the future. To assess which countries are most vulnerable to climate change in the shorter term, we obtained the index of countries "most at risk from climate-related threats" from the World Bank International Development Association (IDA) [4].

This list was drawn from all IDA-eligible countries, and is based on the 25-year record of naturally-caused crises in these countries and their economic and human costs. This list is especially suitable for our study because it disaggregates countries' exposure to climate change to specific types of risk, including drought, flood, severe weather events, sea-level rise, and agricultural degradation.

In our short-term projection, the assumption implicit in the IDA list is that, generally speaking, regions and countries that have suffered the most from naturally caused events in the recent past are those most likely to suffer from similar events in the near future. This assumption—that the short-term future will most likely be similar to today, but with current trends continuing—is supported by virtually all models of future climate change effects, including the United Nations IPCC 2007 report.

The IDA report lists the countries expected to be most affected by all the types of future risk relevant to our study except one: water scarcity. Several countries and regions of the world are threatened already by a shortage of water for human and agricultural use, and are exhausting aquifers faster than they can be replenished. Again, in our analysis the implicit assumption was that these countries that are already experiencing water scarcity are likely to be those to experience it the most in the future, largely because the replenishment of many of these fresh water sources is impossible and we expected the human demand for these resources only to increase over the next 15 years.

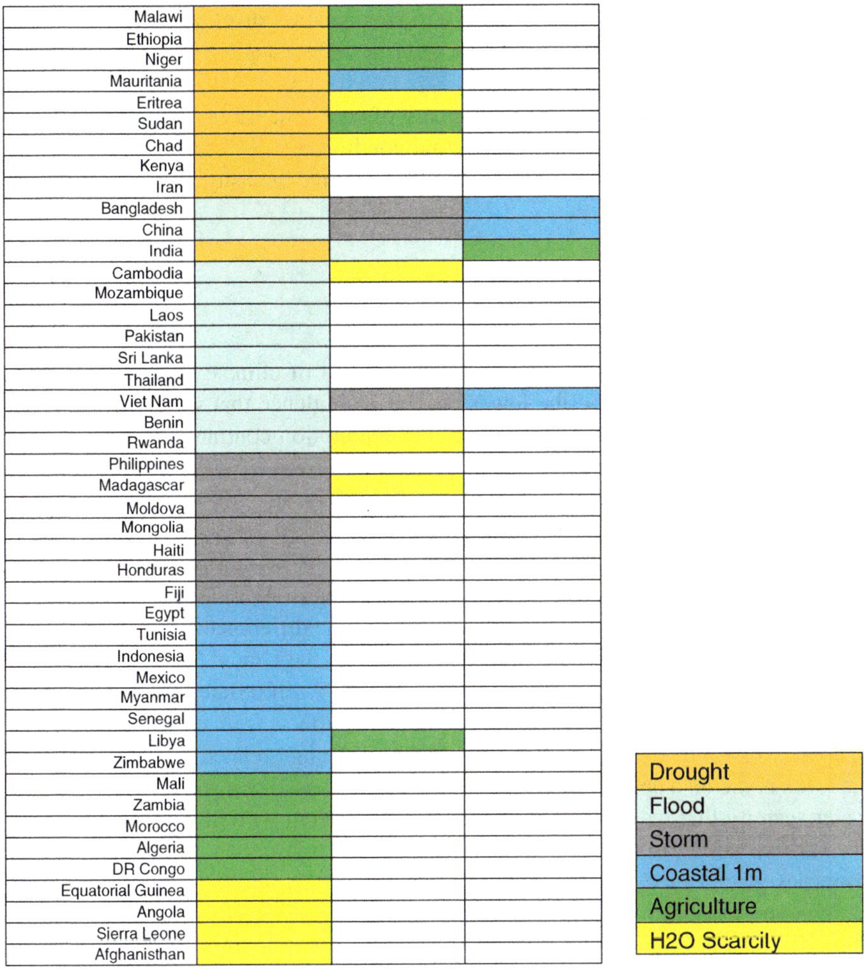

Malawi, Ethiopia, Niger, Mauritania, Eritrea, Sudan, Chad, Kenya, Iran, Bangladesh, China, India, Cambodia, Mozambique, Laos, Pakistan, Sri Lanka, Thailand, Viet Nam, Benin, Rwanda, Philippines, Madagascar, Moldova, Mongolia, Haiti, Honduras, Fiji, Egypt, Tunisia, Indonesia, Mexico, Myanmar, Senegal, Libya, Zimbabwe, Mali, Zambia, Morocco, Algeria, DR Congo, Equatorial Guinea, Angola, Sierra Leone, Afghanisthan

Legend: Drought, Flood, Storm, Coastal 1m, Agriculture, H2O Scarcity

Fig. 1.1 Climate change exposure in selected countries

Our estimate of the likely risk, for all IDA-eligible countries, of water scarcity, is based on a joint publication of the United Nations Environment Programme (UNEP) and its collaborating centre UNEP/GRID-Arendal in Norway [5]. The 12 countries listed as the most threatened by water scarcity are: Morocco, Mauritania, Tunisia, Algeria, Kenya, Rwanda, Burundi, Burkina Faso, Yemen, Eritrea, Egypt, and Djibouti.

Figure 1.1 presents the 44 countries identified by the IDA and the UNEP/GRID-Arendal report to represent overall exposure to the impacts of climate change. For each country we show the categories of threat to which it is exposed. We assumed that the more categories to which the country is exposed, the greater will be its overall exposure to negative effects of climate change. Several of the nations shown face only one of the six categories of exposure; no nation displays more than three categories. The countries that show exposure in three categories are: Bangladesh, China, India, Mauritania, and Vietnam.

This index does not consider the intensity of each risk factor. Because intensity is related to panoply of national factors and conditions, it was difficult to evaluate with confidence. We return to considerations of relative intensity in the next section of the paper.

1.3.2.2 Long Term (2040–2045)

The farther into the future we estimated the effects of climate change, the more variables were involved and the lower was our confidence that we could make an accurate and comprehensive prediction. Our assumption continued to be that currently observed trends would generally continue; however, over the longer term (2040–2045) other trends would also emerge. Therefore, all of the countries expected to be highly exposed to climate change effects in the next 20 years were also those most vulnerable 30–40 years from now, though some additional countries were expected to face new, severe risks particularly relating to water scarcity. In terms of the evolving effects of climate change, a 10-year time difference is generally insignificant. It is partly for this reason, for example, that the United Nations' latest climate change report extends its projections to the 2077–2100 time frame, in order to capture the longer-term, more dramatic anticipated effects from climate change.

One potentially severe type of future climate change-related risk that was not captured in the data upon which our country list was based was glacial melt, and the water scarcity that may result. Glacial melt is occurring with increasing rapidness around the world. In the Andean countries of South America and in Himalayan countries glacial melt poses a high risk of future water shortage. The capitals of Bolivia and Ecuador, for example, draw most of their water from sources fed by shrinking glaciers, as do various agricultural zones. Nepal is similarly threatened, as are major river valleys in China and India.

In addition, sea level rise poses a serious threat to low-lying island states such as the Maldives or the Pacific microstates of Kiribati, Nauru, and Vanuatu. Over the long term, particularly 100 years or more into the future, sea-level rise could threaten the existence of these nations. However, these small island nations were not included on our list because we judged that these expected effects from sea-level rise were not likely to occur within the timeframe of our study. Subsequent research may indicate that we will have to revisit these findings.

1.3.2.3 Summary

By comparing countries' degrees of exposure to six potential impacts of climate change, we identified six countries that are most vulnerable. These six countries are likely to be significantly affected by at least three of the six identified risk factors.

With the exception of China, these countries are located in South Asia or Africa. Considering the severe human and economic damage that can result from any one of these factors, any combination of such factors poses a tremendous danger to these populations and, potentially, to their economic and political systems. We then turned to the question of what capabilities these countries, and others, are likely to have in responding to these dangers.

1.3.3   Comparing States' Exposure and Resilience

1.3.3.1   Resilience Index

We next addressed the expected resilience of the countries that we determined to be most significantly exposed to climate change.

We obtained resilience data from the Vulnerability-Resilience Indicators Model (VRIM) developed by JGCRI, a joint research program between the University of Maryland and the Pacific Northwest National Laboratory [6]. The VRIM model allows for resilience score comparison between countries based on a combination of social, economic and environmental factors.[1] In the model, resilience is defined as the ability to cope with or recover from exposure to climate change induced shocks. The model calculates resilience scores per country on a scale of 1–100.

The VRIM does not include political risk or governance factors in the calculation of the resilience scores. We address this limitation below.

1.4   Results

1.4.1   Base Case

The figure below combines the short-term exposure of the 44 countries from Fig. 1.1 with their resilience scores according to the VRIM. We categorized the countries into three tiers. The Tier-1 countries have high exposure but low resilience. These 14 countries are shown in the top left section in Fig. 1.1. Tier-2 countries, shown in the middle section, have either high exposure and high resilience or low exposure and low resilience. Tier-3 countries, in the bottom right, show low exposure and high resilience.

---

[1] These factors include settlement/infrastructure sensitivity; food security; ecosystem sensitivity; human health sensitivity; water resource sensitivity; economic capacity; human and civic resources and environmental capacity.

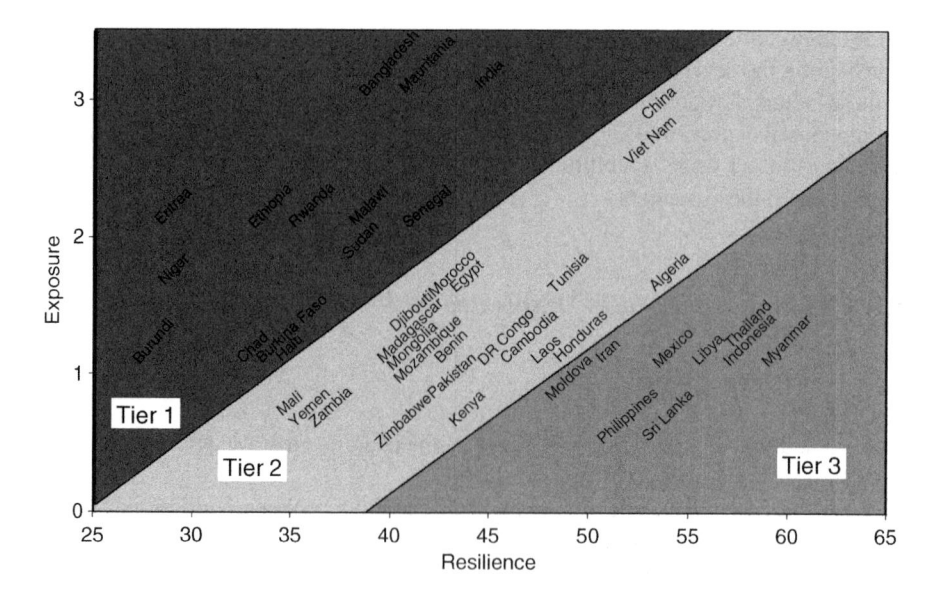

1.4.2 Summary of Findings

In the short term (2020–2025), 11 of the 14 of the countries we identified as most vulnerable to political and/or humanitarian crises as a result of climate impacts (i.e., our Tier 1) were located in Africa, and 9 of the 14 are in north or central Africa. In the long term (2040–2045) these same countries remain highly vulnerable; however, due to decreasing water supplies caused by glacial melt Bolivia, Ecuador, and Nepal also merit special attention, and crises in China and India are increasingly likely. Our study suggested that climate change poses the greatest threat in terms of political and social upheaval to the regions of north and central Africa, and south Asia.

1.4.3 Further Considerations

We found that the VRIM is extremely useful as a quantitative estimate of national resilience. However, it is not comprehensive. There are several additional factors, or country characteristics, that must also be considered when evaluating a country's resilience to climate change effects. Most importantly, the VRIM does not include assessments of governance or the capacity of a government to provide effective security and public services. Although this is captured to some extent in data on economic growth, infrastructure, and human capital, other factors are excluded including levels of corruption, bureaucratic efficiency, and political instability. Other factors that must be considered are a country's degree of integration into the

Table 1.1 Governance scores in percentile bins

Percentile bin			
0–10%	10–25%	25–50%	50–75%
Eritrea	Ethiopia	Madagascar	Tunisia
Sudan	Niger	Moldova	Morocco
Chad	Iran	Mongolia	
Myanmar	Bangladesh	Honduras	
Zimbabwe	Cambodia	Fiji	
DR Congo	Laos	Egypt	
Afghanistan	Pakistan	Indonesia	
	Haiti	Mexico	
	Libya	Senegal	
	Angola	Mali	
	Sierra Leone	Zambia	
		Algeria	
		Philippines	
		Rwanda	
		Malawi	
		Mauritania	
		Kenya	
		China	
		India	
		Mozambique	
		Sri Lanka	
		Thailand	
		Vietnam	
		Benin	

global economy, rates of population growth, and potable water provision. Recalculation of the formal model behind the VRIM was outside of the scope of our work. Instead, we added these factors qualitatively, on a case-by-case basis, with considerations of how they affect our assumption that short-term trends generally continue into the longer-term future.

1.4.3.1 Governance

We first examined the World Bank Governance Indicators obtained from the organization's web site [7]. The World Bank Governance Indicators reflected the statistical compilation of responses on the quality of governance given by a large number of survey respondents in industrial and developing countries as reported by a number of survey institutes, think tanks, non-governmental organizations, and international organizations. These indicators measure the following aspects of a national government's operation: Voice and Accountability; Political Stability and Absence of Violence; Government Effectiveness; Regulatory Quality; Rule of Law; and Control of Corruption. Table 1.1 presents the governance scores for the countries listed in Fig. 1.1.

We found that there was a general relationship between governance and resilience. Of the 44 countries in Fig. 1.1, only Tunisia and Morocco have governance scores above the 50th percentile. It is important to consider that the VRIM scores did not include a comprehensive measure for governance. Therefore, our resilience estimates are likely to be overly optimistic.

1.4.3.2 Globalization, Global Inequality

We utilized a report from the Development Concepts and Doctrine Center (DCDC) of the UK Ministry of Defence that projects strategic trends out to 2036 [8]. It identifies three key issues: globalization, climate change, and global inequality, and associates specific risks associated with each of these issues.

DCDC suggests that Africa is likely to be the hardest-hit region when taking these factors into account, and notes that Latin America could face challenges as well. DCDC identifies Colombia, Peru and Mexico as three countries whose profile could make them candidates for our Tier-1 by 2040. However, these projections are based on the assumption that current levels of insurgency in these countries are likely to continue or increase to 2040, which may not occur. DCDC estimates that Haiti is likely to suffer ongoing political instability partly as the result of climate change effects. In Haiti, a combination of environmental and man-made stressors will likely continue to produce requirements for massive humanitarian assistance.

1.4.3.3 Population Growth to 2040

We found that countries that are expected to experience explosive population growth by 2040 are also likely to show less resilience toward climate change. Overpopulation will affect indictors used to calculate resilience in the VRIM model such as food security, economic capacity and human and civic resources. The International Institute for Applied Systems Analysis (IIASA) has formulated probabilistic population growth estimates to 2040[9]. We made the following observations based on their estimates:

- World population as a whole is expected to rise by 21%; however significant regional variance is predicted.
- Sub-Saharan Africa's population is expected to nearly double, from its current level of approximately 740 million assuming that no new major disease pandemics occur. Rapid population growth in Africa could generally lower resilience scores.
- The countries of Benin, Zimbabwe, Democratic Republic of the Congo, Kenya, Mozambique and Angola display low exposure and low resilience scores on Fig. 1.1. Rapid population growth may be an additional factor pushing these countries toward instability.
- A 27% increase in population is expected for South Asia. This change will likely have an adverse impact on the resilience of India and Bangladesh. India could

shift to the left on Fig. 1.1 as population growth erodes its resilience. A rise in population will likely dampen the prospects for successful adaptation to the effects of climate change in Bangladesh.

- Declining population may improve China's resilience capacity.

Pacific Asia will experience a 22% increase in population. This trend, compounded with sea level rise, could contribute to instability as coastal regions and small island states are increasingly threatened.

In sum, by 2040 a general rise in population will put pressure on resilience capacity in much of Asia, the Middle East and Latin America, but no region will experience these effects more than Sub-Saharan Africa.

1.4.3.4   Water Scarcity in 2040

In addition to population growth, water scarcity is likely to have a profound effect on many of the countries by 2040. Water scarcity as a geographic issue was included in our identification of countries most exposed to climate change effects in the short term. However, in regard to the longer term, the capability of a country to access and provide potable water to its population under conditions of duress is an extremely important element of its resilience. The distinction is that water scarcity as an element of exposure is related to a country's geography; water scarcity as an element of resilience is related to the country's capacity for providing water to meet the needs of its people. For example, a wealthy country facing extreme water scarcity could afford to import water, while a poor country in the same situation may face large-scale emigration and/or collapse.

According to the United Nations World Water Assessment Program [10], the 12 most water stressed countries are: Afghanistan, Ethiopia, Chad, Cambodia, Sierra Leone, Angola, Mauritania, Rwanda, Equatorial Guinea, Democratic Republic of Congo, Eritrea, and Madagascar. In these cases the disappearance of naturally occurring water sources combines with poor government capacity for water provision, suggesting that long-term impact from water scarcity could be extremely severe. In other parts of central Africa, however, precipitation is expected to increase.

In the Middle East, it is possible that some of the least resilient Arab countries such as Iraq could rise to a higher level of concern by 2040 despite their rather high resilience scores. One reason for this is their reliance on water desalinization units that could be subject to malfunction or sabotage.

1.4.4   Summary

The inclusion of factors such as governance, globalization, inequality, population trends, and water availability suggests that some countries not at risk in the short term might be in the 2040–2045 time period. Specifically, trends in water availability and population growth suggest that India, Bangladesh and Thailand may be worse off in the 2040–2045 timeframe than is reflected by their current position. In Latin America, Bolivia's and Peru's resilience could decrease from their current moderate levels.

Rapid population growth, water scarcity and in some cases poor governance are likely to be key drivers that will cause several Middle Eastern countries to become at risk. The effect of these factors on Africa is likely to be profound. While water scarcity's effect on Africa will be mixed, population growth, a recent history of conflict, and poor governance indicate that by 2040, resilience will continue to diminish in this region.

These results support the CNA MAB's finding that Africa is a region that will be heavily affected by climate change but least able to cope. The results support the CNA MAB's recommendation that the U.S. should commit to global partnerships that help African nations build the capacity and resilience to better manage climate impacts. The establishment of U.S. Africa Command (AFRICOM), with a directorate that includes officials from the development community is a positive step toward ensuring that the national security community can play a positive role in this endeavor.

References

1. CNA (2007) National security and the threat of climate change. CNA, Washington, DC, November 2007
2. King MD, Espach RH (2009) Global climate change and state stability (U). CNA Research Memorandum D0020868.A1, July 2009
3. IPCC (2007) Summary for policymakers. In: Solomon S, Qin D, Manning M, Chen Z, Marquis M, Averyt KB, Tignor M, Miller HL (eds) Climate change 2007: the physical science basis. Contribution of working group I to the fourth assessment report of the Intergovernmental Panel on Climate Change. Cambridge University Press, Cambridge/New York
4. International Development Association (2009) IDA and climate change: making climate action work for development sustainable development network, October 2007. Accessed 30 June 2009 at: http://siteresources.worldbank.org/IDA/Resources/Seminar%20PDFs/73449-1172525976405/3492866-1175095887430/IDAClimateChange.pdf.
5. UNEP Vital Water Graphics (2008) An overview of the state of the world's fresh and marine waters. Accessed at: http://www.unep.org/dewa/vitalwater/article192.html.
6. Elizabeth LM, Antoinette B (2008) Vulnerability, sensitivity, and coping/adaptive capacity worldwide (Chapter 3). In: Ruth M, Ibarraran M (eds) The distributional effects of climate change: social and economic implications. Edward Elgar, Cheltenham, April 2008
7. World Bank (2009) Governace matters 2009, worldwide governance indicators 1996–2008. Accessed 15 June 2009 at: http://info.worldbank.org/governance/wgi/index.asp.
8. UK Development, Doctrine and Concepts Center (2009) Strategic trends programme 2007–2036. Accessed 1 June 2009 at: http://www.mod.uk/NR/rdonlyres/94A1F45E-A830-49DB-B319-DF68C28D561D/0/strat_trends_17mar07.pdf.
9. Lutz W, Sanderson W, Scherbov S (2009) IIASA's 2007 probabilistic world population projections, IIASA world population program online data base of results 2008. Accessed 20 June 2009 at: http://www.iiasa.ac.at/Research/POP/proj07/index.html?sb=5.
10. World Water Assessment Program (2009) Water in a changing world: the United Nations world water and development report 3, facts and figures. Accessed 1 June 2009 at: http://www.unesco.org/water/wwap/wwdr/wwdr3/pdf/WWDR3_Facts_and_Figures.pdf.

Chapter 2
Islands in a Sea of Change: Climate Change, Health and Human Security in Small Island States*

Nancy Lewis

Abstract Small island states are often seen as the *cause célèbre* of climate change, although the total population at risk in small island states is substantially less than the dense populations at risk in low lying coastal areas globally. Nonetheless, Islands remain particularly vulnerable to climate change and climate variability. Viewing only the vulnerability of islands, however, limits the scope of island adaptation and denies island peoples agency. Human security, the relationship between environmental degradation, resource scarcity and conflict, and the use of the concept "climate refugees" are briefly discussed. The relationship between climate change and health is small islands is then explored using examples from the extreme ENSO event of 1997–1998. An argument is made for robust, multisectoral, stakeholder based approaches to climate change adaptation in islands. New paradigms including transdisciplinary climate change science must be embraced.

Keywords Climate change • Health • Human security • Small island states • Vulnerability • Adaptation • Climate refugees • 1997–1998 ENSO

Small Island States, particularly Small Island Developing States (SIDS), are heralded as the *cause célèbre* in discussions of climate change. Relative to the number of people at risk, the populations of the SIDS are dwarfed by the dense human populations at risk, many in rapidly growing cities where planning is often inadequate or poorly implemented, on the coasts and in the expansive river deltas of Asia – coastal Vietnam, the Indo-Gangetic Plain, or in Jakarta – or, for that matter, in the Netherlands

---

*The author acknowledges Epeli Hau'ofa [15] for his influential conceptualization of "our sea of islands."

N. Lewis (✉)
Research Program, East-West Center, Honolulu, HI, USA
e-mail: lewisn@eastwestcenter.org

H.J.S. Fernando et al. (eds.), *National Security and Human Health Implications of Climate Change*, NATO Science for Peace and Security Series C: Environmental Security, DOI 10.1007/978-94-007-2430-3_2, © Springer Science+Business Media B.V. 2012

or on the east coast of the United States. But island states and their populations, their health, their ecosystems, their cultures, and their future generations, and some would argue, their sovereignty are, along with peoples in the Arctic, particularly vulnerable to climate change, as well as to climate variability, a phenomenon to which island peoples have been adapting since they first arrived on the shores of the islands which they inhabit.

An important caveat must be stressed at the beginning of this discussion. In a recent volume, Barnett and Campbell [3] argue that the concept of "island vulnerability" cannot be sustained empirically. While islands are assumed to be among the most vulnerable locations in the world, they are some of the locations for which we have the least empirical knowledge of climate vulnerability and adaptation [3, 22, 23].[1] These authors further argue that the hegemonic focus on island vulnerability limits the scope of island adaptation and denies island peoples agency. The hegemony of "Big Science" and the IPCC, along with modeling, and climate impacts and assessment research, marginalizes local knowledge. The application of climate science also has limitations in island locations. For example, in Hawaii (as for high islands in general) the location of the islands, their scale and the highly varied topography currently limits the utility of downscaled climate forecasts. In Pacific Island cultures the land, *'aina* in Hawaiian, cannot be separated from those who belong to it. Western interpretations of climate change that do not take this relationship into consideration separate people from their land and increase their vulnerability by doing so. Islands have in fact shown amazing resilience to change over the centuries [5, 28]. We do need more empirical data, and more is becoming available, but the small size of most islands, their isolation, and the extent of their coasts relative to their land mass, as well as their economies and their histories, including their colonial histories, render them vulnerable.[2] I continue the discussion with this caveat and related considerations in mind.

2.1 Island States

The examples for the paper are drawn primarily from the small island nations and states of the Pacific Ocean (Table 2.1). They span a third of the world's surface, largely but not exclusively south of the equator. These states, especially the

---

[1]Mimura et al. [22] note that there were relatively fewer specific island studies between IPCC's Third Assessment Report (TAR 2001) and the Fourth Assessment Report AR4 (2007) than there had between the Second Assessment Report (1995) and the TAR.

[2]IPCC's Fourth Assessment Report (AR4) reports with very high or high confidence that for islands sea levels are expected to rise, exacerbating coastal hazards, that water resources on small islands are likely to be seriously compromised, that there will be heavy impacts on coral reefs and other marine based resources, that subsistence and commercial agriculture on small islands is likely to be adversely affected, that there will be direct and indirect effects on tourism that will be largely negative and, with medium confidence, that health will be impacted primarily adversely [22]. There may be more intense tropical storms with changes in extremes and return times leading to both droughts and floods.

Table 2.1 Pacific Island states and territories

	Pop (000s)	Land area (km²)	Pop growth rate	Net migrants	Urban pop %	IMR	Life expec	GDP (USD per capita)
Melanesia	**8,642**	**542,377**	**2.0**	**−5,316**				
Fiji Islands	848	18,273	0.5	−6,489	51	17.0	65.4	3,499
New Caledonia	254	18,576	1.5	1,173	63	6.1	75.9	37,993
Papua New Guinea	6,745	462,840	2.1	0	13	56.7	54.2	897
Solomon Islands	550	30,407	2.7	0	16	24.3	61.1	1,014
Vanuatu	245	12,281	2.5	0	24	25.0	67.3	2,218
Micronesia	**547**	**3,156**	**1.5**	**−1,333**				
Fed States of Micronesia	111	701	0.4	−1,633	22	37.5	67.7	2,183
Guam	187	541	2.7	2,400	93	11.7	73.6	22,661
Kiribati	101	811	1.8	−100	44	52.0	61.0	1,490
Marshall Islands	54	181	0.7	−1,000	65	21.0	67.5	2,851
Nauru	10	21	2.1	0	100	45.8	56.2	2,071
Northern Mariana Islands	63	457	−0.1	−1,000	90	4.9	75.3	12,638
Palau	21	444	0.6	0	77	20.1	69.0	8,423
Polynesia	**664**	**7,986**	**0.7**	**−5,658**				
American Samoa	66	199	1.2	−465	50	11.3	72.5	9,041
Cook Islands	16	237	0.3	−98	72	11.6	72.8	10,875
French Polynesia	269	3,521	1.2	0	51	5.8	74.1	21,071
Niue	1	259	−2.3	−42	36	7.8	71.6	9,618
Pitcairn Islands	–	5	–	–	–	–	–	–
Samoa	184	2,935	0.3	−3,050	21	20.4	72.8	2,672
Tokelau	1	12	−0.2	−19	0	31.3	69.1	–
Tonga	103	650	0.3	−1,711	23	19.0	70.2	2,629
Tuvalu	11	26	0.5	98	47	17.3	63.6	1,831
Wallis & Futuna	13	142	−0.6	−175	0	5.2	74.3	–

Summary information from the Secretariat of the Pacific Community website (accessed 17 February, 2011). Italics indicate AOSIS member state. A full listing of AOSIS member states can be found at http://www.sidsnet.org/aosis/members.html

exclusively atoll states of Kiribati, the Marshalls and Tuvalu, (along with the Maldives in the Indian Ocean) are often represented as the "poster children" of climate change. The nations of the Pacific are representative of other islands, island territories of metropolitan powers, and to varying degrees, of archipelagic nations, coastal communities and other small states.

Excluding Hawaii and New Zealand, the combined population of the Pacific States is 9.5 million. In comparison, the population of Cuba is 11 million. The independent nations of the Pacific are members the Alliance of Small Island States

(AOSIS), a coalition formed in 1990, of countries with similar environmental interests. AOSIS has 39 members in the Pacific, Caribbean, Indian, Mediterranean, and Atlantic Oceans.[3]

Island states individually and collectively, have been highly visible in climate change debates, in part due to their previously mentioned iconic status but also because of their representation in the United Nations and in the UNFCC. Some suggest that islands have a 'moral argument' with respect to climate change and even that islands are the "conscience" of climate change, perhaps an ironically heavy burden given their very small contribution to greenhouse gas emissions. Although islands are highly visible, the lack of both the human and financial resources to support large diplomatic missions or delegations, limits the real power that they have in climate politics [3].

2.2   Human Security

The NATO workshop that this paper was prepared for was "Climate Change, Human Health and National Security", an encompassing, important and complex theme. Reflecting on the theme with small island states in mind and reviewing the descriptions in the Venn diagram provided as background material for the meeting, suggested that the security concerns to be addressed in this conference were broader than national security as conventionally defined in international relations and security studies, and I have adopted the expanded theme of human security for this paper.

This is not the place for an exploration of the relationship between the understandings of human security as they have evolved, and to a very limited degree converged, within international relations and security studies on the one hand and development studies on the other [24]. Gasper [11] explores human security as an intellectual framework and the epistemic communities that do and do not embrace the concept. Connecting environmental change to human security provides new possibilities for linkages between the research and policy communities concerned with international relations, those concerned with development, and those concerned with environmental change and sustainability [4].

Acknowledging the important contributions to our understandings of human security made by the UNDP in its *Human Development Report* [30] and the Human Security Commission [25], the definition of human security adopted here is that of the Global Environmental Change and Human Security (GECHS) project[4] which "considers human security to be a state that is achieved when and where individuals and communities have the options necessary to end, mitigate or adapt to threats to their human, environmental and social rights; have the capacity and freedom to exercise these options; and actively participate in pursuing these options" (http://www.gechs.org/human-security/ accessed on February 6, 2011).

---

[3]AOSIS is supported by the Foundation for International Law and Development (FIELD) in international climate deliberations.

[4]Initiated in 1999 this was a core project of the International Human Dimensions Programme (IHDP) that concluded in June of 2010.

In spite of the work of GECHS and the efforts of the UN and others, there has not sufficient been interaction between the Global Environmental Change and Human Security communities. Meetings such as this workshop provide a forum for furthering that interaction.

2.3 Climate Change and Security

Before turning specifically to climate change and health in small island states let me briefly address two prominent and related themes in the climate change and security debate. The first, environmental degradation, resource scarcity and conflict [2, 16] and the second, environmentally induced migration and 'climate refugees'. With respect to the first, and acknowledging the extreme Pacific island example of Rapa Nui (Easter Island) and more contemporary internal conflicts in New Caledonia, Vanuatu, Bouganville, the Solomon Islands and Fiji, I will only note that the circumstances surrounding conflict in environmentally stressed locations are typically highly complex and deeply intertwined with contemporary and historical socioeconomic and political circumstances. While environmental challenges can exacerbate inequalities and tensions, there are also examples of societies that have developed coping skills to address these challenges [10].

With respect to migration and 'climate refugees', the Asian Development Bank is scheduled to release its report , *Climate Change and Migration in Asia and the Pacific* later this spring. This is one product of an ADB project "Policy Options to Support Climate-induced Migration" aimed at increasing our understanding of climate-induced migration and stimulating the policy debate regarding how to address anticipated population movement, both gradual and abrupt, in response to a changing climate and associated extreme events. While there is relatively little empirical research on the relationship between environmental change and migration for Asia and the Pacific [1], ADB anticipates the movement of millions of people due to changing weather patterns and notes that the Pacific is at the epicenter of weather disasters. Given the publicity surrounding the extreme events of the past several years, this focus is not surprising. And, the ADB report is only one of a growing number of projects and publications on climate change and migration. It is often unclear how many migrants are crossing international borders and how many may be temporary rather than permanent migrants. Other sources suggest that most people who move in response to climate disasters, move within nations, often from one rural location to another rural locations (rather than to cities), and further, that when the movements are across national boundaries they tend to be one nation in the global South to another. Another very recent report from the International Institute for Environment and Development (IIED) suggests that predictions of millions of people being forced to migrate across international borders are alarmist. They argue that the majority of migrants are migrating for socioeconomic reasons. They argue further that migration, perhaps temporary in response to climate events, can be an adaptive response reducing migrants' dependence on natural resources. (http://www.iied.org/human-settlements/media/climate-change-governments-should-support-migration-not-fear-it, posted 2/2/2011).

The issues are complex and demand a nuanced understanding of the many contributing factors.

The term "climate refugees" is complicated and contested [14]. Some argue that it undermines the rights and status of political refugees. Burkett [6] notes that the fact that climate refugees have no legal status is an additional reason to avoid using the term. She goes on to argue that the international legal community lacks the will to address the legal implications of individuals and island communities who may be forced to abandon their islands and she also asks whether the climate refugees are stateless persons or landless citizens of a state that does not exist. With respect to the Pacific, the vast majority of island migrants and there are many, e.g., more Cook Islanders live in New Zealand than in the Cook Islands, have moved for socioeconomic, family and educational reasons. The attention in the international media to 'sinking islands' (and marooned polar bears) notwithstanding, the official response of Pacific Island nations has been largely to reject exodus as an acceptable strategy [21]. Quoting Prime Minister Apisai Ielemia of the atoll nation of Tuvalu, at COP 14 in Poznan, "We are not contemplating migration …We are a proud nation of people, we have a unique culture which cannot be relocated to somewhere else. We want to survive as a people and as a nation. And we will survive – it is our fundamental right".[5] The motivation for the often cited example of "the world's first climate refugees", residents of the Carteret Islands in Papua New Guinea, being resettled on Bouganville was not climate change but land subsidence and additional factors. The agreement signed between Tuvalu and New Zealand (for a modest 75 migrants a year) is not only a 'climate change' agreement but rather one intended to provide employment opportunities.

Beyond the Pacific, in 1989 President Maumoon Abdul Gayon of the Maldives was the first to highlight the importance of retaining territory, nationality and cultural identity. I concur with Dabelko [7] who argues against oversimplification and hyperbole in climate and security discussions and this may be particularly important with respect to islands. Dabelko suggests that climate change may be a "threat multiplier"[6] but that the link between climate change and violent conflict or terrorism should not be oversold, that ongoing natural resource and conflict problems exist, that climate change should not be assumed to drive mass migration and that climate mitigation efforts can also introduce social conflict.

2.4   In a Sea of Change

In the Pacific, as elsewhere, climate change is a major concern, but only one of the challenges that island states are facing. The islands are generally characterized by small land areas, fragmentation and distance from continental land masses,

---

[5]http://unfccc.int/files/meetings/cop_14/application/pdf/cop_14_hls_statement_ielemia.pdf.

[6]We learned from Marcus King's opening presentation that the origin of that term was the 2007 CNA report, *National Security and the Threat of Climate Change* (Washington, DC). That was an influential document in this debate.

attenuated but often unique and endangered biota, largely tropical maritime climate regimes and most often small populations, with varying population growth and migration rates (Table 2.1). They typically have fragile economies. While there are communalities, the peoples of the Pacific exhibit complex genetic, linguistic, cultural and social differences and their experiences of initial settlement, explorers, missionaries, colonization, decolonialization and most recently globalization have also varied. Their endowment of natural resources is vastly different and they have achieved different levels of 'development' or modernization [17].

Regarding migration, movement has characterized Pacific peoples from their initial settlement of their dispersed islands to the migration that is characteristic today. The complex migration of Pacific peoples, as that of others, is often a pragmatic response to multiple challenges and opportunities, including a lack of employment and development opportunities on home islands in an increasingly globally connected world. Pacific mobility has included considerable 'circular migration' and remittances from migrants contribute importantly to island economies. Mobility related to climate variability is not new in the Pacific. In Micronesia, an interisland exchange system known as *sawei* operated in the Caroline Islands between the low lying atolls and the high island of Yap. It is believed to have developed as a mechanism to maintain social relationships needed to secure aid for the low-lying islands in case of natural disasters, cyclones or severe drought [29].

2.5   Climate and Health

The attention given climate change and health escalated concomitant with the IPCC's Third Assessment Report [19, 20, 27]. Climate change can affect health both directly and indirectly (Fig. 2.1), for example, it can result in temperature and extreme event related illness and death, extreme weather-related health effects, air-pollution related health effects, water and foodborne diseases, vector-borne and rodent-borne diseases, effects of food and water shortages, and mental, nutritional, and other health effects [27, 33]. AR4 [26] confirms with a fairly high degree of confidence that climate change will be implicated in a growing burden of disease and premature deaths and that the adverse impacts far outweigh likely positive impacts on human health. It is important to note that on islands as elsewhere, inequalities exist in vulnerability, impact, resilience and ability to respond to climate variability and change [12].

2.6   Climate and Health in the Pacific

The health status of Pacific states is representative of much of the spectrum of the health transition (Table 2.1) [17]. With the adoption of Western diets of imported foodstuffs, alcohol and tobacco, and more sedentary lifestyles, dietary change and

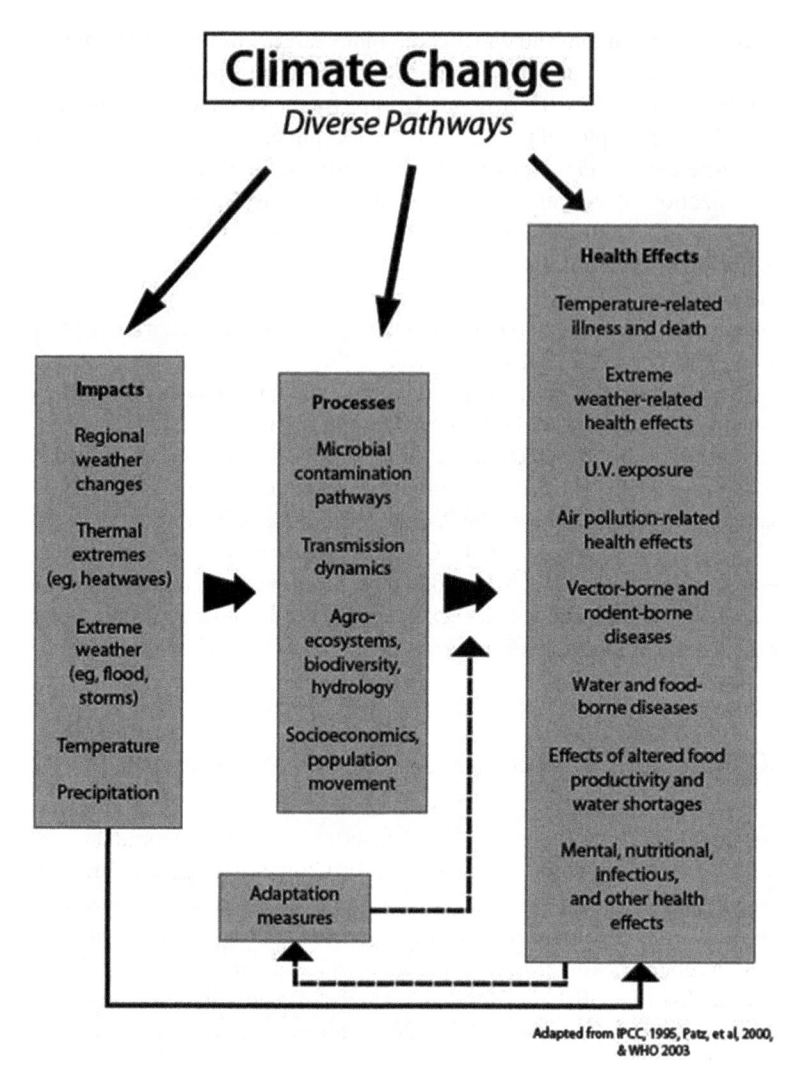

Fig. 2.1 Climate change effects

concomitant obesity have emerged and noncommunicable diseases are the leading cause of disease death in the more modernized parts of the Pacific. However infectious and environmentally based diseases including TB, dengue, filariasis, leptospirosis, malaria (in Papua New Guinea, the Solomon Islands and Vanuatu), diarrheal disease, marine toxins including ciguatera, and STDs including HIV/AIDS (most notably in Papua New Guinea) remain significant problems. Some of these health risks are climate sensitive.

Populations in the Pacific have been adapting to climate variability since the first voyagers reached island shores. Climate change is projected to increase climate variability. The globally dominant climate cycle, the El Niño Southern Oscillation or ENSO, is most strongly expressed in the Pacific Ocean. The ENSO phenomenon may provide clues to the environmental and health impacts of longer term climate change. With climate change it is expected that the prevailing conditions in the Pacific will be more "ENSO-like", generally drier in the east, wetter in the west and with shifting tropical cyclone patterns, although ENSO exhibits considerable variability from event to event and within the course of a single event.

Evidence of the health impacts of El Niño in the Pacific may be drawn from the strong the 1997–1998 El Niño, the most well studied in terms of health. Strong El Niño events can be predicted months in advance although there is much less skill in predicting weak El Niño events and La Niña events. In June 1997 the Hawaii based Pacific ENSO Applications Center (PEAC) warned governments in the American affiliated Pacific that a strong El Niño was developing. PEAC indicated that there might be changes in rainfall and storm patterns, that extreme droughts might occur and that some islands would be at unusually high risk of typhoons and hurricanes. Hurricanes hit French Polynesia and the Cook Islands. Extreme drought did occur in much of the region and on some islands, e.g., in Pohnpei in the Federated States of Micronesia and Palau at the peak of the drought, water was only available for 2 h a day. There were also substantial agricultural losses and increased financial costs of importing food and water across the North Pacific [13]. Guam and other islands experienced serious drought related wildfires. The early warning could not prevent all El Niño related consequences but in terms of public health, in spite of the water shortage in Pohnpei, fewer children than normal were admitted to hospital with severe diarrheal disease, in all likelihood due to the attention given to water quality and frequent public service announcements about water safety. South of the equator, Fiji also felt the effects of the strong El Niño. Micronutrient deficiencies were found in pregnant women, and there was a dengue outbreak,[7] attributed to the El Niño that affected 24,000 of the islands 856,000 inhabitants. [31]. To the west, Papua New Guinea, by far the largest nation in the Pacific, experienced one of the most severe droughts in 100 years and frosts in the highlands caused widespread devastation to food crops. Several hundred thousand people were at risk. It is important to stress that with respect to the application of ENSO forecasts, scale and local level variation must be taken into account [18].

The increased interest in climate change and health in the early 2000s led to a series of workshops on climate and health in small island states organized by the World Health Organization, in partnership with the World Meteorological Organization and the United Nations Environment Programme held in Samoa, Barbados and the Maldives [32, 33]. The workshops and a synthesis conference addressed the current distribution and burden of climate sensitive diseases in small island states, interventions currently used to reduce the burden of these diseases, potential future health impacts, and additional interventions that are needed to adapt to current and future health impacts [8, 9]. The workshops also highlighted the need for site specific, multi-disease hazard research, early warning systems, more efficient

---

[7]The relationships are complex and not fully understood. In a regional retrospective study dengue outbreaks were found to be more likely to occur during La Niña events [18].

approaches to health education, better surveillance and response and the overall need to strengthen health care infrastructure and sewage and solid waste practices. The participants stressed the need to explore the health impacts of climate change and climate variability in other sectors, e.g., agriculture, fisheries, coral reefs and fresh water. There was also agreement that given the limited resources available in island states, it is critical that integrated, cross-sectoral approaches be employed in adaptation planning. Key recommendations included the need to enhance awareness of the potential effects of climate variability and change on health, enhance adaptation strategies, develop policies and measures to decrease the impact of climate change, address high priority research and data needs, and develop regional forecasts.

2.7 Conclusion

Climate change is a reality for islanders in the Pacific and elsewhere, as it is for the rest of the globe. While islands are often seen as representing the iconic image of climate change, viewing them only through this lens denies island populations agency and ignores their adaptation to climate variability which has taken place over centuries. Approaches to climate change adaptation in islands must be robust, multisectoral and stakeholder based. Where science can inform decision making, now and into the future, e.g., El Niño forecasts, downscaled projections and assessments, systems need to be developed to use this information to protect human health and address negative health impacts in other sectors. 'No regrets' strategies with respect to climate change adaptation can provide ancillary benefits in terms of health. Especially in small islands where human resources are limited, communication between the meteorological units, health ministries, disaster managers and other sectors must be enhanced. Not only for islands, but more generally there is a need to create new fora to allow for enhanced dialogue among the research and policy communities addressing global environmental (importantly including climate) change, health, and human security. These challenges also demand a new way of thinking about the science of climate change and adaptation that embraces the social sciences, beyond the current 'nod' given to economics and sometimes geography, and recognizes that climate change is more than an environmental problem, but a fundamentally human problem [24]. As participants at the workshop concluded, new paradigms including truly transdisciplinary climate change science must be embraced.

References

1. Asian Development Bank (2009) Policy options to support climate-induced migration. Project number 43181-01, Research and Development Technical Assistance, Manila
2. Barnett J, Adger WN (2010) Environmental change, human security and violent conflict (Ch. 6). In: Matthew RA, Barnett J, McDonald B, O'Brien KL (eds) Global environmental change and human security. MIT Press, Cambridge, pp 119–136

3. Barnett J, Campbell J (2010) Climate change and small islands states. Earthscan, Washington, DC
4. Barnett J, Matthew RA, O'Brien KL (2010) Introduction: global environmental change and human security. In: Matthew RA, Barnett J, McDonald B, O'Brien KL (eds) Global environmental change and human security. MIT Press, Cambridge, pp 3–32
5. Bayliss-Smith T, Bedford R, Brookfield H, Latham M (1988) Islands, Islanders and the world: the colonial and post-colonial experience of Eastern Fiji. Cambridge University Press, Cambridge
6. Burkett M (2011) In search of refuge: Pacific Islands, climate-induced migration, and the legal frontier. Asia Pacific Issues, no. 98. East-West Center, Honolulu
7. Dabelko G (2009) Avoid hyperbole, oversimplification when climate and security meet. Bulletin of the Atomic Scientists. http://www.thebulletin.org/web-edition/op-eds/avoid-hyperbole-oversimplification-when-climate-and-security-meet. Accessed Feb 2011
8. Ebi KL, Lewis ND, Corvalan CF (2005) Climate variability and change and their health effects in small island states: information for adaptation planning in the health sector. World Health Organization, Geneva
9. Ebi KL, Lewis ND, Corvalan CF (2006) Climate variability and change and their potential health effects in small island states: information for adaption planning in the health sector. Environ Health Perspect 114(12):1957–1963
10. Fox J, Swamy A (2008) Introduction: natural resources and ethnic conflicts in Asia Pacific. Asia Pac Viewp 49(1):1–11
11. Gasper Des V (2010) The idea of human security. In: O'Brien K, Kristoffersen B, St. Clair AS (eds) Climate change, ethics and human security. Cambridge University Press, Cambridge, p 246
12. Gruebner O, Staffeld R, Khan MMH, Burkhart K, Krämer A, Hostert P (2011) Urban health in megacities: extending the framework for developing countries. IHDP Update 1: 42–49
13. Hamnett MP, Anderson CL, Guard CP (1999) The Pacific ENSO Applications Center and the 1997–98 ENSO warm event in the US-affiliated Micronesian Islands: minimizing impacts through rainfall forecasts and hazard mitigation. Pacific ENSO Applications Center, Honolulu
14. Hartmann B (2010) Rethinking climate refugees and climate conflict: rhetoric, reality and the politics of policy discourse. J Int Dev 22(2):233–246
15. Hau'ofa E (1994) Our sea of islands. Contemp Pac 6(1):147–161
16. Homer-Dixon TF (1999) Environment, scarcity and violence. Princeton University Press, Princeton
17. Lewis ND, Rappaport M (1995) In a sea of change: health transitions in the Pacific Island. Health Place Int J 4(1):211–226
18. Lewis ND, Hamnett M, Prasad U, Tran L, Hilton A (1998) Climate and health in the Pacific: research in progress. Pac Health Dialog 5(1):187–190
19. McCarthy JJ, Canziani OF, Leary NA, Dokken DJ, White KS (eds) (2001) Climate change 2001: impacts, adaptation and vulnerability. Contribution of working group II to the third assessment report of the Intergovernmental Panel on Climate Change. Cambridge University Press, Cambridge
20. McMichael AJ, Campbell-Lendrum DH, Corvalan CF, Ebi KL, Gotheko AK, Scheraga JD, Woodward A (eds) (2003) Climate change and human health – risks and responses. World Health Organization, Geneva
21. McNamara KE, Gibson C (2009) We do not want to leave our land: Pacific ambassadors at the United Nations resist the category of 'climate refugees'. Geoforum 40(3):475–483
22. Mimura N, Nurse L, McLean RF, Agard J, Briguglio L, Lefale P, Payet R, Sem G (2007) Small islands. In: Parry ML, Canziani OF, Palutikof JP, van der Linden PJ, Hanson CE (eds) Climate change 2007: impacts, adaptation and vulnerability. Contribution of working group II to the fourth assessment report of the Intergovernmental Panel on Climate Change. Cambridge University Press, Cambridge, pp 687–716
23. Nurse LA, Sem G, Hay JE, Suarez AG, Wong PP, Briguglio L, Ragoonaden S (2001) Small island states. In: McCarthy J, Canziani O, Leary N, Dokken D, White K (eds) Climate change 2001: impacts, adaptation and vulnerability. Contribution of working group II to the third assessment report of the Intergovernmental Panel on Climate Change. Cambridge University Press, Cambridge, pp 844–875

24. O'Brien KL, St. Clair AL, Kristoffersen B (eds) (2010) Climate change, ethics and human security. Cambridge University Press, Cambridge
25. Ogata S, Sen A (2003) Human security now. UN Commission on Human Security, New York
26. Parry M, Canziani OF, Palutikof J, van der Linden P, Hanson C (eds.) (2007) Climate change 2007: impacts, adaptation and vulnerability. Contribution of working group II to the fourth assessment report of the Intergovernmental Panel on Climate Change. Cambridge University Press, Cambridge
27. Patz JA, McGeehin MA, Bernard SM, Ebi KL, Epstein PR, Grambsch A et al (2000) The potential health impacts of climate variability and change for the United States: executive summary of the report of the health sector of the U.S. National Assessment. Environ Health Perspect 108(4):367–376
28. Sahlins M (2000) On the anthropology of modernity, or, some triumphs of culture over despondency theory. In: Hooper A (ed) Culture and sustainable development in the Pacific. Asia Pacific Press, Canberra, pp 44–61
29. Sudo Ken-ichi (1996) Rank, hierarchy and routes of migration: chieftainship in the Central Caroline Islands of Micronesia. In: Fox JJ, Sather C (eds) Origins, ancestry and alliance: explorations in Austronesian ethnography. ANU Press, Canberra, ACT, pp 55–69
30. United Nations Development Programme (1994) New dimensions of human security. In: Human development report 1994. Oxford University Press, New York
31. World Bank (2000) Cities, seas and storms: managing change in Pacific economies. World Bank, Washington, DC
32. World Health Organization (2000) Climate variability and change and their health effects in Pacific Island Countries. WHO/SDE/OEH/01.1. WHO, Geneva
33. World Health Organization (2004) Synthesis workshop on climate variability, climate change and health in small-island states. WHO/SDE/OEH/04.02. WHO, Geneva

Chapter 3
Issues of Climate Change, Health and National Security in Expanding Cities Worldwide

J.C.R. Hunt, S.E. Belcher, and Y.V. Timoshkina

Abstract The changing climate and environment in cities and their effects on human health and national security are reviewed. Science and policies need re-examining when applied to growing mega-cities as their diameters exceed 50–100 km and their populations rise beyond 30 million people. Although urban areas themselves contribute to climate change, caused by their increasing greenhouse gas (GHG) emissions associated with rapidly expanding energy use, depending on their structure and operation they may or may not contribute more per person than in rural areas. Environmental and social policies are considered for how large conurbations can be prepared for climatic and environmental hazards including health and security, and how these policies need to be coordinated with those for mitigating GHG emissions and adapting megacities to the hazards associated with climate change. Hazard refuges and urban insurance may be two techniques that need more consideration.

J.C.R. Hunt (✉)
University College London, London, UK

Center for Environmental Fluid Dynamics, Arizona State University, Tempe, AZ, USA
e-mail: jcrh@cpom.ucl.ac.uk

S.E. Belcher
Department of Meteorology, University of Reading, Reading, UK

Y.V. Timoshkina
Department of Mathematics, University College London, London, UK

Advisory Committee on Protection of the Sea, University of Cambridge, Cambridge, UK

H.J.S. Fernando et al. (eds.), *National Security and Human Health Implications of Climate Change*, NATO Science for Peace and Security Series C: Environmental Security, DOI 10.1007/978-94-007-2430-3_3, © Springer Science+Business Media B.V. 2012

3.1 Climate and Environment of Urban Areas

3.1.1 Sustainability

Since the second UN Habitat Conference in 1996 politicians, as well as environmental scientists, meteorologists and urban planners have begun to realise that the objectives of sustainability could only be achieved by understanding and dealing effectively with the interactions between cities and the wider global and regional environment ([18, 26], 2008; [17]). As the world population living in cities grows progressively, during this century from 50% to about 60–70%, their energy demands, and hence greenhouse gas emissions, are expected to increase. Ensuring their sustainability requires studying how both short and long term measures to improve their resilience against natural and other disasters need to be modified to allow for their growing size and the interactions with changes to the regional climate and to the terresbutrial, atmospheric and aqueous environments (e.g., [37]).

The high concentration of people per square meter in urban areas, about 100–1,000 times the global average, also affects their ultimate sustainability because it can make the health and security of populations more vulnerable to extreme natural hazards. In industrial countries, these dangers have been reduced by improvements in the science and technology of hazard forecasting, in disaster management, and in preventative design. Fewer lives have been lost and much less physical damage caused by hazards in major cities in developed countries compared with developing countries. However, the heat waves and floods in Europe and the USA in the past decades have shown that all societies can be vulnerable to high intensity hazards.

In Europe about 40,000 deaths were attributed to the heat wave in 2003, especially in urban areas where there was little preparation for the adverse effects on the health of elderly and vulnerable people. Large hazards like this one can sometimes be so unexpected that even the insurance industry may have overlooked them [24]. In the future, the nature and frequency of such extreme hazards are likely to change everywhere, leading to greater impacts on communities and infrastructure. New risks will arise associated with the operation of large cities in emergencies, and with changes to their local meteorology and hydrology, will interact with changes in regional climate [17, 19]. Recent developments in system dynamics provide a framework for planning and managing the increasing complexities and uncertainties of megacities now and in the future. UN organisations with responsibilities for climate change and disasters are focussing on these issues (www.ipcc-wg2/AR5/extremes and www.unisdr.org/makingcitiesresilient)

3.1.2 Changes in Climate and Environment

The scientific study of these trends needs to combine meteorological and environmental research for urban areas [5, 11–15] and their interactions on regional and global scales. The effects of climate change of climate change, seasonal variability and extreme natural hazards also have to be considered [22]. The latest projections, based on likely estimates of expanding, rather than contracting emissions of GHG by developed and developing countries, indicate that the future rise in average global temperature by the end of the century will lie between 3°C and 4°C, which are significantly greater than the 2°C discussed at the UN climate conference in Copenhagen (e.g., IEA report Nov. 2010; www.iea.org). This will hasten and amplify the impacts of climate change to a greater extent than many current plans allow for. Also, there will be an increase in the variability of weather and seasonal climate in certain parts of the world, e.g., extended periods of extreme heat and cold [9] may be long enough for diseases to develop [32]. Record extremes of high and low temperature, and drought and flood are already being recorded around the world [22].

This paper draws some conclusions from these studies about the likely changes in meteorological and hydrological hazards in urban areas both as the climate changes and as the sizes of urban areas grow. Mesoscale meteorological models used to study the climate and environment within specific local areas also depend on the influence of the weather, environment and climate outside the areas [33]. For the study of present conditions in urban areas, these 'boundary conditions' can be determined by local observational data, but for predicting future trends they are determined by the prediction of global climate models.

Observational studies of the decadal warming trend in cities worldwide show that they are greater than in the surrounding regions: in central London the urban heat island effect currently adds up to a further 5–6°C to summer night temperatures and will intensify in the future. The trapping of heat (and pollution) in the boundary layer below 200–300 m is now being monitored and studied as urban areas develop.

As the new computations in the appendix show, the temperature profiles across urban areas vary in space and time even on flat terrain; they are even more complex where there are nearby hills (e.g., [6]), coasts and local hot-spots such as airports. Over the neighbourhood scale of 1–3 km the temperatures are raised or lowered by parks, rivers, buildings, etc.

Government and communities need to understand the full range of possible hazard scenarios. In built-up areas, there are a number of critical environmental conditions which lead to primary and/or secondary dangerous impacts on the physical and social structure such as flooding, causing secondary effects of water pollution and cuts of electrical power (see Table 3.1). These different events need to be considered both individually and collectively in order to decide on long term policies and short term response. Precautionary measures may avert the necessity of drastic actions later, such as cities and communities being abandoned as they have been in past climates. Some coastal communities and island states, such as the Maldives, are preparing for the same fate during this century as sea level rises.

3.1.3 Policy Approaches

Technological development and behavioural changes in future energy use and transportation could limit GHG emissions (as in EU countries) *or* at least to limit their

Table 3.1 Changing impacts on cities of climate and environmental change and increasing scale of urbanization

CAUSES	CHANGING IMPACTS OVER TIME from climate/environmental changes	CHANGING IMPACTS OF INCREASING SCALE of urbanization L (in relationto scales L_H of hazards and L_R of meso- meteorology) $\dfrac{L}{L_H} \uparrow; \dfrac{L}{L_R} \uparrow$
PRIMARY HAZARDS		
Wind (speed U↑)	$(I_0 + \Delta I) \uparrow$ (Note(i))	$\Delta \hat{U} \uparrow \ \Delta \hat{I} \uparrow$ (Note(ii))
		As $\dfrac{L}{L_H} \uparrow, <\Delta I> \uparrow$
Flood (water level h↑). Causes: a) Local precipitation b) High river flow/ high coastal winds/ tides/ cyclones	$(I_0 + \Delta I) \uparrow$	(a) $\dfrac{L}{L_H} > 1 \Rightarrow <\Delta I> \downarrow$ (b) $\dfrac{L}{L_H} \sim 1 \Rightarrow <\Delta I> \uparrow$
Heat (T ↑)	$(I_0 + \Delta I) \uparrow$ (Note(i))	$\dfrac{L}{L_H} \sim 1 \rightarrow \Delta T \uparrow$ $\Rightarrow \Delta \hat{I} \uparrow, \ <\Delta I> \uparrow$
Pollution (C ↑)	$I_0 + \Delta I \uparrow$	$L/H \sim L \Rightarrow \Delta C \uparrow$ $\Delta \hat{I} \uparrow, \ <\Delta I> \uparrow$

SECONDARY AND GEOPHYSICAL HAZARDS

Physical/Environmental Effects	$(I_0 + \Delta I)\uparrow$ (Note(i))	$\dfrac{L}{L_H}\uparrow \Rightarrow <\Delta I>\downarrow$ (Note (i))
		if L/H $\Rightarrow \Delta \hat{I}\uparrow$ (Note (i))
Societal/Economic loss of capacity	$(I_0 + \Delta I)\uparrow$ (Note(i))	a) if $\dfrac{L}{L_H}\uparrow \Rightarrow <\Delta I>\downarrow$ (Note (i))
	$\bar{I}\uparrow (?)$	b) if $\dfrac{L}{L_H}<1, <\Delta I>\uparrow$ (Note (ii))
CONCLUSION	$<\overline{\Delta I}>\uparrow$ or $\downarrow$ (Note (i))	

Notes:

(i) $I = I_{CR}$ for $H > H_{CR}$; but $I \approx 0$ for $H < H_{CR}$, the critical hazard threshold

(ii) $\Delta \hat{I}$ is peak impact; $\bar{I}$ is time average over many events ; $<\Delta I>$ is spatial average over urban area

(iii) $\uparrow$ - implies increasing; $\downarrow$ - implies decreasing trends.

(iv) (?) means uncertain.

rate of increase as in China [20]. These developments also contribute to the other important policy objectives of reducing reliance on fossil fuels and improving energy security. But according to the UK Climate Change Committee (2009, 2010), the necessary policies for the UK are not being introduced fast enough or strongly enough to achieve the national target to reduce emissions by 80% by 2050, or to adapt to climate risks. Consequently there has been a steady rise in global temperature ΔT from its baseline in 1850 (the upward trend is more marked over land areas ΔT_L – see Hunt [19] and www.metoffice.gov.uk; Beijing Climate Centre 2008).

Governments and cities are introducing policies for adapting communities, industries and agriculture to the likely consequences of climate change [22]. The Netherlands Government, following a major review of their risks, plans to raise its dykes by several meters to allow for the eventual several meters rise in sea level [23, 29], corresponding to the worst case scenario of melting of polar ice-caps [19].

Connecting, and in some cases, integrating technical, economic and administrative policies for dealing with climatic and environmental risks (e.g., in infrastructure and operational risk management) has been shown to be more effective socially and economically [31], provided bureaucratic obstacles are avoided [28]. This approach also contributes to measures for long-term sustainability (for example, introducing renewable or high efficiency energy systems in new housing developments) as some cities and regions have already demonstrated. See Table 3.2.

3.2 Climate Change, Environmental Risks and Urbanisation

3.2.1 Factors Contributing to Risks

The main types of meteorological, hydrological, or environmental hazards (see Table 3.1) that cause damaging impacts on urban areas are the following: high wind speeds (U); raised water levels (h) caused by local precipitation, river discharges or wind induced surges and waves along coasts; high temperatures (T) associated with regional meteorology and artificial effects of produced by changes in vegetation and surface properties and by heat emissions from energy systems; high concentrations (C) of atmospheric gases and particulates arising from natural sources (e.g., wind-blown sand or noxious gases from lakes) and from artificial processes of industry, transport and agriculture, etc. Most of these kinds of hazards can occur singly or in combination, such as floods and high winds, or high temperatures and concentrations associated with forest fires around urban areas [14]. Their impact on the physical and societal structure of communities produce short or long term damage to their physical infrastructure, health and social and economic capacities. But the magnitude of the impact (denoted by I) depends on H, and also how well all these aspects of the community are adapted to reduce the impact of the hazard and to recover afterwards, i.e., its resilience or lack of vulnerability [1]. Some countries are also exposed to other kinds of equally damaging geophysical hazards such as volcanoes, earthquakes and tsunamis [21].

These hazards are firstly associated with regional weather and climate and with environmental effects that are independent of any local urban effects (even though urban areas worldwide are affecting the global climate). These non-urban impacts are denoted by I_0. Regional effects may be exacerbated by significant regional amplification of global warming mentioned above.

In urban areas the presence of large population as well as certain physical, chemical and biological processes can lead to additional hazards and impacts. In other words, urban areas are particularly vulnerable to hazards denoted by ΔI. Estimating impact risks firstly requires considering which hazards can occur simultaneously, e.g., which can happen with high winds, floods and waves. But other combinations, such as high temperature and very high synoptic scale winds, may be very unlikely, depending on the climatic region and the geography. Secondly, in urban areas different hazards can combine to enhance ΔI. This can occur with flooding and also when high urban temperatures worsen illnesses caused by high air pollution concentrations, which is documented in Africa and Asia [16].

Climate change and urban factors can also exacerbate the impacts of geophysical hazards, e.g., longer lasting stagnant atmospheric conditions following volcanic eruptions [34], tsunami impacts on coastal cities will increase with sea-level rise and may in the future occur on arctic coasts as sea ice melts [25]. Another possible geophysical hazard occurs in years of high solar activity, when ionisation of the atmosphere causes breakdown of the electrical systems that are essential for cities' infrastructure, a concern for certain Asian cities during the next sun-spot cycle in about 2012 (Lam, 2009, personal communication), or secondary hazards such as rain induced mud slides or environmental effects caused by the disruption of the cities' systems, such as overflow from drains leading to widespread water pollution [35].

Monitoring and predicting medical, social and economic impacts during and after disasters are becoming more reliable through collecting more data, aided by advances in remote sensing, and computer modelling. Predictions can be more accurate, useful and specific, through feedback from individuals and communities. By considering frequencies of hazards and climate change policy makers can assess whether communities can recover before the next hazard event occurs. Failure to do so threatens their long term viability [19] – see Table 3.1. Insurance companies now assess vulnerability risk as much in terms of the social capacity of communities as by physical impacts of extreme events and preventive measures that may have been taken.

3.2.2 Effects of Increasing Scale

As the size L of an urban area increases, its energy use and pollution emissions increase approximately in proportion to the square of L. In the evening and night time the heat transferred to the air from the buildings, is transported by the wind. Since it is confined within the urban boundary layer, the urban surface temperature increases in the downwind direction [21].

The length scale over which the urban heat island hazard is significant, denoted by L_H, therefore grows broadly in proportion to the scale L of expanding cities. But as the Paris heat wave of 2003 demonstrated, local variations in the urban environment temperatures are also very important, and even determined the local pattern of mortality [8]. In this case these hot spots effectively determine the average mortality impact per unit area over the city $<\Delta I>$. This would be expected to increase as the urban temperature increases with the scale of cities.

As air pollutants are transported across the city, some gases increase in concentration, while others undergo chemical transformations. Some can even decrease in intensity. Overall, the scale of the pollutant hazard L_H increases with L, which will increase the impact on mortality due to air pollution. In some countries, air pollution concentrations are now large enough to cause a significant loss of visibility.

Unlike heat, which diffuses to the ground, pollutants can be advected far downwind of cities. In local sea-land breeze and valley circulations, as in Los Angeles and Phoenix, pollutants are transported 30 km out of the centre and are swept back to build up the concentrations even further [12]. Such local effects are limited to central areas of the cities but become less significant as it expand. Even low hills, such as those surrounding London and Athens, confine the air flow and pollutants at night and in stable conditions.

Where there are surrounding mountains higher than the depth of the boundary layer (about 1,000 m), they have a dominant role in the local meteorology and environment even in very large cities over 100 km in diameter, e.g., in Los Angeles or some cities in China [27].

Many observational and numerical studies, e.g., Bornstein [5], have shown how over larger cities the airflow, temperature profiles and precipitation patterns differ appreciably between those in the centre and the outside the urban areas. Recent measurements over central London [4] show that the night-time depth of the mixed layer (about 200–300 m) is significantly greater than the shallow nocturnal layer (of less than 100 m) in the surrounding rural areas.

Fluid dynamical studies of perturbed stratified flows with the Coriolis effects of the Earth's rotation [30, 36] show that the changes in the direction and speed of the airflow only become substantial when the length L exceeds the 'Rossby Radius' L_R, which is about 30 km at night and up to 100 km by day. This is the distance over which the urban area affects the meteorology upwind and around the urban area. Because of the greater heat in the centre of the city, there are also significant variations over this distance within the urban area, especially at night-time.

As the scale L of 'mega' cities of the future exceeds L_R the characteristic wind, temperature and precipitation patterns will change. Typically in the day time with a steady wind blowing towards the city, zones of increased wind speed form around the periphery and extend far downwind. These are also associated with areas of marked surface convergence and divergence, which effect changes in the patterns of precipitation. The significant consequences for urban climate and environment of cities when they reach this mega scale is being studied in the EU MEGAPOLI project (see http://megapoli.dmi.dk//; [3]).

Another long term hazard in growing urbanised areas in dry regions of the world is a reduced water supply either caused by reduced precipitation locally or regionally and by depleting the water table. Water conservation and water harvesting (a growing practice in cities and rural areas of India) will have to be supplemented by desalination for coastal cities or transport of water over long distances [10]. The hydraulic energy requirement (per year) for water transport into cities (which for example is already a substantial fraction of California's energy [2]) grows more rapidly than for any other use, i.e., as L^6, (because it is proportional to Q^3, where Q is the volume flow which increases as the population, which, in turn, varies as L^2). With longer and hotter periods of drought, the risks of increased forest fires around cities, with an associated rise in air pollution, are also growing, which is particularly marked where the cities are surrounded by mountains, as Santiago and Athens have experienced.

3.2.3 Effects of Scale on the Functioning of Urban Areas

It is equally important to consider how the increasing scale of cities affects the operational, social and economic capacities to deal with hazards. Some hazards, both climatic and environmental, as well as those caused by industrial accidents and by malefactors, tend to be localised over hazard length scales L_H, which generally do not depend on the overall size of the city L. But note that L_H increases with the magnitude of the hazard. In the largest cities L generally exceeds L_H for mild and frequent hazards, such as precipitation, local wind storms, which provides these communities with relatively greater resilience. But in other cases the hazards extend across the city (where $L_H \geq L$), either over the short term such as with fluvial flooding, large earthquakes, heat waves or tropical cyclones, or over days and weeks such as with heat waves or very large fluvial flooding. Extended periods of air pollution can be fatal for vulnerable groups in the population.

When hazards are above a critical level (Table 3.2) large cities can be dangerous because damage can spread (like water-borne or air-borne debris), and because people cannot leave the endangered areas of the city within the period of short term hazard warning T_H (which varies from minutes to days depending on the hazard). Also, they may not be able to leave their homes, businesses, or aged relatives as in the Japanese earthquake and Tsunami in 2011. These considerations should influence strategic policies about increasing the size of such cities and practical policies about managing them, including evacuation plans and investments in security measures, e.g., secure refuges for people above the level of any likely floods, and secure against the danger of flying debris in high winds [7]. In the largest evacuation plan anywhere, a second city is available for people escaping from Naples in the event of a major eruption of Mount Vesuvius (S. Edwards, personal communication 2011). Multi-disciplinary modelling of these scenarios is an urgent research priority everywhere [19]. WMO has a special programme for multi-hazards early warning system for Shanghai [38].

Table 3.2 Policies and actions for climate and environmental hazards in urban areas advantages for different objectives

TYPES OF ACTION	Advantages for Mitigation of GHG Emissions	Advantages for Adaptation to climate change	
		Short-term (+Resilience)	Long-term benefits (+Sustainability)
Energy and Resource Efficiency	Short time scale (for decisions and action) Local →	Health Security ($\Delta T \downarrow$) ($\Delta C \downarrow$)	↑ ↗
New Power Sources (± Networks)	Renewable (F) ↑ Nuclear ↑	Resilience Security	↑ ↗ (?)
Land use, buildings, Planning	Bio/Ag/For/Urban →	Reduce Hazards	↗
Less Travel & more tele-communication	Non-fossil efficiency/use	$\Delta C \downarrow$ Hazard Security	Env and Economic sustainability

Notes:

(i) Actions for climate change also contribute to and link with resilience, resources, security, economics, and social capacity (also affected by integration);

(ii) Varying time scales for different actions;

(iii) Continual information and warnings needed for all policy objectives;

(iv) Response and recovery systems needed for all hazards-can be common for most cases;

(v) (?) means uncertain.

3.3 Conclusions

This paper shows that as urban areas grow to an unprecedented scale in many parts of the world there are significant risks for their populations, which differ considerably depending on geographical and climatic factors. In general, these can be greatly reduced through short and long term measures specific for these areas, both for protecting people and infrastructure. Developments in science, technology and institutional organisations are transforming warning and disaster response systems, through better understanding of the linkage between geophysical processes and detection technology and improving the education of communities that are particularly vulnerable to risk [21].

Policies in each urban area need to be coordinated so as to minimise the impacts of the likely range of hazards affecting that area. Also, these measures should be coordinated with those for reducing greenhouse gases, and adaptation to long-term climate change (see Table 3.2). and policies for sustainable development and national security. Policies should be reconsidered when these areas become so large that they adversely affect the local climate and environment, and add to natural risks, such as heat waves, flooding, extreme wind damage, air pollution events, *etc.*, as explained in this paper. Their size will mean that people cannot escape in the event of extreme hazards, as recent hurricanes and tsunamis in the USA and Indonesia have shown. Indeed where attempts have been made to evacuate multi-million populations, lives have actually been lost in the transport systems as they seized up.

Communities should be informed and consulted about the most appropriate measures in the local context, such as the dependence on natural conditions, the planning of the city and the sustainable use of energy. As is done in some hazard prone cities like Hong Kong, communities should be prepared for appropriate actions for the different kinds of hazards in their locality. Structural engineers, planners and social scientists need to consider more urgently the design of appropriate shelters in urban and also in rural areas-parks and open areas may also act as refuges. Technical solutions also have to be supported by communities – which has not always been the case. In certain developing countries, for example where preservation of livestock has been neglected in emergency measures. Estimates of the likelihood and impacts of extreme events in growing conurbations are needed to plan and justify the investment needed for these precautionary measures.

This paper has shown some aspects of the complexity of establishing optimum policies for dealing with climate change in expanding urban areas. Since this and other research is indicating how the likely degree of future climate change impacts will require qualitative changes in the planning and operation of cities depending on the area of the world concerned, environmental, engineering and societal research needs to address these problems more fully than they are at present. With a better understanding and monitoring of climate and different environmental risks in urban areas, appropriate insurance may become available to cities as well as to individuals, businesses, and agriculture.

Acknowledgments We are grateful for support from UK, French, EU and US agencies , and for conversations with Steadman J.P., Fernando J., Chun S., Masson V., Cassou C, Carruthers D.J., Bornstein R., Sabatino, S., Kareem A.

An early version of this paper was presented at American Meteorological Society meeting in Phoenix Arizona in January 2009.We acknowledge support from EPSRC Lucid project at UCL, London, GSDP EU project at UCL, and Arizona State University.

References

1. Adger W (2006) Vulnerability. Resilience, vulnerability, and adaptation: a cross-cutting theme of the international human dimensions programme on global environmental change. Glob Environ Change 16(3):268–281
2. Andrew J (2009) Adapt, flee, or perish: responses to climate change for California's water sector. Presentation for 'Ice, Snow and Water' workshop, University of California, San Diego. Retrieved Apr 2010, from http://esi.ucsd.edu/gwi/index.php?option=com_content&task=view &id=14&Itemid=26
3. Baklanov A (2011) Megacities; urban environment, air pollution, climate change and human health interactions. In: Fernando H, Klaić ZB, McCulley JL (eds) National security and human health implications of climate change. NATO advanced research workshop proceedings, Dubrovnik, 28–30 Apr 2011. Springer, Dordrecht

4. Barlow JF, Dunbar TM, Nemitz EG, Wood CR, Gallagher MW, Davies F, O'Connor E, Harrison RM (2010) Boundary layer dynamics over London, UK as observed using Doppler lidar. Atmos Chem Phys 10:19901–19938, see http://www.atmos-chem-phys-discuss.net/10/19901/2010/acpd-10-19901-2010.html

5. Bornstein R (1987) Mean diurnal circulation and thermodynamic evolution of urban boundary layers. Modeling the urban boundary layer, AMS/EPA conference, Baltimore, 1987

6. Brazel A, Fernando H, Hunt JCR, Selover N, Hedquist B, Pardyak E (2005) Evening transition observations in Phoenix, Arizona, U.S.A. J Appl Meteorol 44:99–112

7. Brewick P, Divel L, Butler K, Bashor R, Kareem A (2009) Consequences of urban aerodynamics and debris impacts in extreme wind events. In: Proceedings of the 11th Americas conference on wind engineering, San Juan, 22–26 June 2009

8. Canoui-Poitrine F, Cadot E, Spira A (2006) Excess deaths during the August 2003 heat wave in Paris, France (in English). Rev Epidemiol Sante Publique 54(2):127–136

9. Cassou C, Guilyardi E (2007) Modes de variabilite et changement climatique. La Meteorologie 59:22–30

10. Fenner R (2009) Briefing: ice, snow and water – impacts of climate change on California and Himalayan Asia. Proc ICE Eng Sustain 162(3):123–126

11. Fernando H (2008) Polimetrics: the quantitative study of urban systems (and its applications to atmospheric and hydro environments). Environ Fluid Mech 8(5–6):397–409

12. Fernando H, Lee S, Anderson J, Princevac M, Pardyjak E, Grossman-Clarke S (2001) Urban fluid mechanics: air circulation and contaminant dispersion in cities. J Environ Fluid Mech 1(1):107–164

13. Gayev Y, Hunt JCR (eds) (2007) Flow and transport processes with complex obstructions. Springer, Dordrecht

14. Gorbachev MS (2010) Opening speech at the second annual world forum on enterprise and the environment, Oxford, UK

15. Head P (2008) Entering the ecological age: the engineer's role, Brunel lecture series. Institute of Civil Engineering, London

16. HEI (2004) Health effects of outdoor air pollution in developing countries of Asia: a literature review. Health Impact Institute, Boston. Retrieved Aug 2009, from http://pubs.healtheffects.org/view.php?id=3

17. Hunt JCR (ed) (2005) London's environment: prospects for a sustainable world city. Imperial College Press, London

18. Hunt JCR, Maslin M, Killeen T, Backlund P, Schellnhuber H (2007) Introduction. Climate change and urban areas: research dialogue in a policy framework. Philos Trans R Soc A 365:2615–2629

19. Hunt JCR (2009) Integrated policies for environmental resilience and sustainability. Proc Inst Civil Eng Eng Sustain 162(3):155–167

20. Hunt JCR (2009) Why China needs help cutting its emissions. New Sci 2720:22–23

21. Hunt JCR, Kopec G, Aplin K (2010) Tsunamis and geophysical warnings. Astronomy Geophy 51(5):5.37–5.38

22. IPCC (2007) Pachauri R and Reisinger A (eds) Climate change 2007: synthesis report. Contribution of working groups I, II and III to the fourth assessment report of the Intergovernmental Panel on Climate Change. IPCC, Geneva

23. Kabat P, Fresco L, Marcel S, Veerman C, van Alphen J, Parmet B, Hazeleger W, Katsman C (2009) Dutch coasts in transition. Nat Geosci 2:450–452

24. Kalkstein A, Sheridan S (2007) The social impacts of the heat–health watch/warning system in Phoenix, Arizona: assessing the perceived risk and response of the public. Int J Biometeorol 52(1):43–55

25. Klettner CA, Balasubramanian S, Hunt JCR, Fernando HJS, Voropayev SI, Eames I (2010) Evolution of tsunami waves of depression and elevation. Proceedings European conference on mathematics and industry, (Ed, J.Norbury et al) Springer

26. Lee K (2007) An Urbanizing World. In State of the World. World Watch Institute, Washington, DC. see http://www.worldwatch.org/files/pdf/State%20of%20the%20World%202007.pdf

27. Lu X, Chow K, Yao T, Lau A, Fung J (2009) Effects of urbanization on the land sea breeze circulation over the Pearl River Delta region in winter. Int J Climatol 30(7):1089–1104
28. Mawson A (2008) The social entrepreneur: making communities work. London
29. Netherlands Government (2007) Reclaiming the Netherlands from the future. Retrieved Aug 2009, from http://www.verkeerenwaterstaat.nl/english/topics/water/water_and_the_future/water_vision/
30. Orr A, Hunt JCR, Capon R, Sommeria J, Cresswell D, Owinoh A (2005) Coriolis effects on wind jets and cloudiness along coasts. Weather 60(October):291–299
31. Parker D, Penning-Rowsell E (2005) Dealing with disasters. In: Hunt JCR (ed) London's environment: prospects for a sustainable world city. Imperial College Press, London
32. Paz S (2011) West Nile virus eruptions in summer 2010-what is possible linkage with climate change. In: Fernando H, Klaić ZB, McCulley JL (eds) National security and human health implications of climate change. NATO advanced research workshop proceedings, Dubrovnik, 28–30 Apr 2011. Springer, Dordrecht
33. Porson A, Harman I, Bohnenstengel S, Belcher S (2009) How many facets are needed to represent the surface energy balance in urban areas? Bound-Layer Meteorol 132:107–128
34. Ravilious K. (2010) *Blame the Volcano Trouble on the Sun and Global Warming*. See http://www.newscientist.com/article/dn18794-blame-the-volcano-trouble-on-sun-and-global-warming.html. Accessed 18 Mar 2011
35. Ristenpart E (2003) European approaches against diffuse water pollution xaused by urban drainage. Diffuse pollution conference, Dublin, 2003
36. Rotunno R (1983) On the linear theory of the land- and sea-breeze. J Atmos Sci 41: 1999–2009
37. Schellnhuber H-J, Cramer W, Nakićenović N, Wigley T, Yohe G (2006) Avoiding dangerous climate change. Cambridge University Press, Cambridge, 392 pp
38. Tang X (2009) Shanghai multi-hazard early warning system demonstration project. Shanghai 2009. Retrieved 21 Mar 2010, from http://www.wmo.ch/pages/prog/drr/events/documents/20070518_MHEWS_Shanghai.pdf
39. Yang X, Hou Y, Chen B (2011) Observed surface warming induced by urbanization in east China. J Geophys Res 116 D14113, 12 PP. see http://www.agu.org/pubs/crossref/2011/2010JD015452.shtml

Chapter 4
Climate Change, Tipping Elements and Security

Kjeld Rasmussen and Thomas Birk

Abstract Climate change is increasingly being described as a threat to international (as well as 'human') security. We examine the claim that it is the so-called 'tipping elements' of the Earth System which constitute the most important threats. Three examples of suggested tipping elements, (1) de-stabilization of the West-Antarctic Ice Cap, (2) acidification of the upper layers of the ocean and (3) die-back of the Amazon rain forest, are used to illustrate the ways in which tipping elements may cause insecurity, in various meanings of the term. Further, the use of the tipping element/point metaphor as a means of communicating the risks and uncertainties associated with climate change is discussed, and it is compared to the alternative terminology used by IPCC. Subsequently, we discuss the extent to which the use of the tipping element/point metaphor constitutes 'securitization' of climate change, and whether or not such securitization, in 'hard' or 'soft' versions, is desirable. It is concluded that while 'hard securitization', presumably involving use of force, is unlikely to be relevant, 'soft securitization' may be realistic – and even necessary - in order to mobilize the reform of international political institutions required to deal efficiently with climate change in general and tipping elements specifically.

4.1 Introduction

Climate change (CC) through the twenty-first century, as depicted in IPCC's Fourth Assessment Report (IPCC 4AR) [13], involves gradual and relatively slow changes in climatic variables, such as average temperatures or average annual precipitation,

K. Rasmussen, Ph.D. (✉) • T. Birk, MSc
Department of Geography and Geology & Centre for Advanced Security Theory,
University of Copenhagen, Copenhagen, Denmark
e-mail: kr@geo.ku.dk

H.J.S. Fernando et al. (eds.), *National Security and Human Health Implications of Climate Change*, NATO Science for Peace and Security Series C: Environmental Security, DOI 10.1007/978-94-007-2430-3_4, © Springer Science+Business Media B.V. 2012

as well as 'impact-variables' such as mean sea level. In most cases the changes in average values projected until 2100 do not exceed the current variability. In addition, IPCC predicts increased variability, involving greater probabilities of extreme temperatures, rainfall events, drought periods and high wind speeds. It may be argued that this may represent greater risks to society than changes in average values: The large damages and great numbers of deaths attributed to climate change mostly occur in connection with extreme events, not the least tropical cyclones. In addition slow and predictable changes in climate are far easier to adapt to. Obviously, slow changes in average conditions and greater variability add up to greater risks of extreme events.

While estimates of changes in both average values and in extreme value statistics may be derived from climate models, the climate system may also display behavior which is not presently possible to reproduce by use of standard climate models. These 'surprises', or 'tipping elements' using the terminology suggested by Lenton et al. [11], of the climate system are associated with its non-linearity, caused by the existence of self-reinforcing mechanisms, or positive feed-backs, which may destabilize the system. A number of possible tipping elements have been identified, and often they involve mechanisms of interaction between the atmosphere, the ocean, sea-/land-ice and the land surface, sometimes associated with changes in bio-geochemical cycles, such as the carbon cycle. When a certain threshold (termed a 'tipping point', see below) is passed, the climate system may 'tip', meaning that it changes its state and behavior substantially within a short period of time, possibly from one 'equilibrium state' to another. Such abrupt climate change with natural causes is known from the paleo-climatic record, documented in ice cores [4]. The current human perturbation of the climate system is believed to make abrupt change far more likely to occur.

Climate change is increasingly being talked about as a 'threat to security' [1, 2]. Such statements require a specification of what is meant by 'security'. Traditional 'security theory' focuses mainly on the national scale and on military threats, yet increasingly the concept is being extended to include other scales, down to the individual and up to the global, and non-military threats such as threats from pandemics, terrorism, loss of access to key resources (e.g. food, water and energy) and environmental and climate change.

Seen from a security perspective, both changes in average values, changes in extreme value statistics and increased risk of 'tipping' of the climate system may be relevant: Slow changes in annual rainfall, in combination with higher temperatures, may have great impacts on rain-fed agriculture as well as on fresh water resources available for irrigated farming, which may have great implications for food security. Gradual sea level rise, as predicted in the IPCC 4AR, may pose threats to food production in coastal areas, deltas and estuaries, as well as threaten the habitability of atolls. Changes in statistics of extreme events will also be of security significance, since they will increase risks of natural disasters, especially in countries exposed to the effects of tropical cyclones. However, we will argue that in the longer term abrupt climate change, associated with so-called 'tipping elements' of the climate system, may well constitute the greatest threat to security: Once a tipping element of the climate system has been triggered, climate change is likely to accelerate and continue to do so for a prolonged period, in some cases bringing the climate system into unknown territory. This may cause failure of available adaptation measures, which function appropriately as responses to slow and gradual climate change. Once started, such self-reinforcing processes are likely to be irreversible, and reactive strategies may therefore be unsuccessful. The prediction of the behavior of the tipping elements of the climate systems is difficult. It may be feared that prediction cannot be made with the certainty required to warrant preventive action, if such preventive action is associated with great costs.

This paper will examine the 'security' implications of the assumed existence of tipping elements of the climate system. The discussion will consider the problem from two sides: After a brief review of the concept of 'tipping elements' and some selected proposed tipping elements, we will discuss in what sense tipping elements constitute a potential threat to security, building on the discussion of the broadening of the concept of security from dealing only with military aspects to including also broader issues, e.g. related to 'human security'. Subsequently we will look at the process of 'securitizing' climate change [1, 3, 5] posing the question of whether it makes sense – and is appropriate and wise – to move the climate change issue from the domain of environmental politics to the security domain.

4.2 Tipping Elements of the Earth System

4.2.1 The Definition of a 'Tipping Element'

While the use of the related concept of a 'tipping point' may be traced back more than 10 years, precise definitions of the concept of 'tipping elements' are relatively recent. Lenton et al. [11] suggest a formal definition of a 'tipping element' which, briefly summarized, includes the following:

1. A 'tipping element' is a component of the Earth System characterized by a rapid increase in the rate of change of the state of the system per unit of change in a 'control variable', e.g. global average temperature, once a certain threshold (the 'tipping point') in the value of the control variable is surpassed, eventually leading to a qualitative change. While the cause of reaching a tipping point may be 'natural' or related to human action (or a combination of both), Lenton et al. [11] consider only tipping elements affected by human actions.
2. Human activities should interfere with the component of the Earth System in question in such a way that a political decision taken within a 'political time horizon' (proposed to be 100 years) could determine whether the tipping point is reached (yet not necessarily within the same time horizon).
3. The maximum time to observe a qualitative change (plus the time to trigger it) should be less than an 'ethical time horizon', suggested to be 1,000 years.
4. A significant number of people should care about the fate of the component in question, which implies that qualitative change should be measured in terms of impact on people.

It is noteworthy that the definition proposed by Lenton et al. [11] not only includes purely bio-geophysical criteria, but explicitly states that mechanisms must be relevant to society in order to qualify. Lenton et al. [11] translates this into requirements on the time perspective within which the mechanism may cause qualitative change in the state of the Earth System.

Several points in this definition may be discussed, and we will return to this in the following brief review of the suggested candidates for tipping elements. Here we will restrict ourselves to tipping elements with significant impacts at time scales less than 100 years, much less than the 'ethical time horizon' of 1,000 years suggested to Lenton et al. [11]. This choice is made because the focus here is on the political and security aspects of climate change, and these aspects, associated with short-term urgency, become extremely speculative beyond a time horizon of 100 years. Political decisions on security issues seldom take time perspectives of more than 100 years into consideration. This implies that certain tipping elements, listed by Lenton et al. [11], become less relevant, e.g. the melting of the Greenland Ice Cap. Restricting ourselves to a 100 year time horizon does not imply that we consider tipping elements with longer time perspectives less important, yet we suggest that these may be better discussed in an ethical, rather than a political, perspective.

4.2.2   Examples of Tipping Elements

Rather than going through Lenton's list of candidates for tipping elements, we will choose a few which illustrate the security implications of tipping elements. These are (1) the destabilization of the West-Antarctic Ice Cap, (2) the acidification of the ocean (and the associated die-back of coral reefs) and (3) the die-back of the Amazon rain forest. Two of these, (1) and (3), are on Lenton's list, while ocean acidification is not. In spite of that it is included here, because we believe it may qualify, and because it has a special political significance, as we shall return to below.

The possibility that a rapid disintegration of the ice cap and ice shelves of West-Antarctica may be triggered by a temperature increase has been discussed for decades [14, 15], but it is still not well understood what is required to trigger such a destabilization. Should it happen it may result in a rapid sea-level rise of 1–5 m (depending on whether the entire or only parts of the ice cap disintegrates) over a relatively short period (possibly in the order of decades or few centuries). Lenton et al. [11] suggest that a collapse within a 300 year time perspective is a 'worst case scenario'. The probability of it happening within 100 years must be considered small, but different from zero, as also indicated by Oppenheimer and Alley [15].

Increased CO_2 concentration in the atmosphere will inevitably lead to a drop in pH of the upper layers of the ocean. A reduction of 0.1 has already been observed, and a further reduction of 0.2 is likely given the expected emissions [6, 7, 9]. This reduction will have a strong negative effect on marine calcifying organisms, including corals. This will have two main consequences: Firstly, it may imply decay of coral reefs at a global scale, especially in combination with the effects of increasing sea temperature and other stressors. The impacts are likely to be large, both in terms of the loss of biodiversity and amenity values, reduction in fisheries, and in terms of reduced coastal protection, all contributing to making low-lying islands less habitable. Secondly, it will reduce the flux of carbon from the atmosphere into the sea water, and further into marine sediments, which is an important CO_2-sink of the Earth System. As mentioned, the process in question is well-underway, and large-scale die-back of calcifying organisms may happen within 50–100 years, yet the probability is difficult to assess, since it involves estimating the capacity of marine organisms and ecosystems to adapt sufficiently quickly. However, the probability may be assumed to be considerably higher than that of disintegration of the West-Antarctic ice cap.

A combination of climate change, and in particular changes in rainfall associated with increased frequency and strength of El Niño – Southern Oscillation – events [10, 12], and human-induced deforestation may cause large parts of the Amazon rainforest to disappear irreversibly, resulting in large emissions of CO_2 to the atmosphere. These emissions will further amplify climate change significantly. The impacts on biodiversity would obviously be great, and other 'ecological services' provided by the rainforest would be lost as well. The probability of irreversible die-back of large parts of the Amazon rainforest is disputed, but it appears likely that a rise in average global temperature of somewhere between 2°C and 4°C (relative to the pre-industrial level), accompanied by continued deforestation, may trigger this change. Temperature increases of this order of magnitude are very likely within the next 100 years, according to the IPCC 4AR, and even greater increases are possible.

It should be noted that while Lenton et al. [11] discuss tipping elements individually, they do interact strongly. Any tipping element, involving a positive feed-back by which higher temperatures will cause increased GHG emissions, will contribute to triggering other tipping elements. One example may be the increased CH_4-emissions from tundra areas, which Lenton et al. [11] do not consider to qualify as a tipping element because it is relatively slow: This process may not lead to great problems in itself at the time scale considered here, but it may certainly have the effect of contributing to triggering other – and faster – tipping elements. In a security perspective such 'chain reactions' must be taken into account.

In the IPCC 4AR the 'tipping element' terminology is not used, yet terms such as 'dangerous' and 'abrupt climate change' and 'large-scale discontinuities', which may be interpreted as corollaries of the 'tipping element' jargon, appear. In the central diagram of the 'Summary for Policymakers' of the IPCC 4AR an indication is given of the intervals of temperature increases that may have detrimental effects. Generally speaking, increased in temperatures greater than 2–4°C above pre-industrial levels (corresponding to approximately 1.3–3.3°C above the present level) are claimed to have strong negative effects on both crop production, water resources, ecosystems, coastal areas and health. This is the background for the present policy goal of limiting the temperature increase to 2°C above the pre-industrial level. The interpretation of the term 'dangerous climate change', which stems from the text of the UNFCCC, is ambiguous, yet points in the direction of the 'securitization' of climate change which we will discuss below.

4.3   Security Aspects of Tipping Elements

The three suggested tipping elements, introduced briefly above, may have quite different 'security implications' in a broad sense of the term:

- The destabilization of the West-Antarctic ice cap is characterized by having a low estimated probability (within the time horizon considered), yet extremely large 'worst case' effects, including flooding of low-lying areas globally. These areas include some of the most fertile and densely populated agricultural lands, e.g. river deltas in Bangla Desh, Burma, Vietnam and China and Egypt. The resulting food shortages and displacements of people can be considered both a direct threat to 'human security' in a broad sense and an indirect threat, through the destabilizing effects of massive food shortages and the associated 'climate refugee' problems, to 'national security' in a narrow sense. While this might be said just to add to the impacts of the general sea level rise, expected without any contribution from West-Antarctica [8, 16], it does involve a significant increase in the speed of the sea level rise, causing problems of adapting fast enough. In addition, it will certainly cause accelerated flooding of most atolls, which include entire nations.
- The acidification (and warming) of the ocean and the reduced function and subsequent death of calcifying organisms have a higher probability, yet in the first place less disastrous impacts in terms of human suffering. Economic losses may be high since incomes generated from coral reefs are considerable, yet the impact on biological diversity may be claimed to be of much greater significance, but very difficult to quantify in economic terms. From a security perspective it should be noted that only a minor fraction of the World's population will be directly affected, at least within the time horizon considered here. On the other hand, it should be noted that whole nations, some of the 'small island states' such as Kiribati and Tuvalu, will be threatened on their existence, which implies that issues of international law and politics become relevant.
- The probability of the Amazon rainforest die-back is debated, both due to lack of precise modeling of the purely bio-physical aspects and due to the importance of local and global human decisions and actions, yet it appears to be the most probable of the three mentioned. With respect to the importance of human decisions and actions, the 'Reduced Emissions from Deforestation and forest Degradation' (REDD) mechanism of the UNFCCC appears to have a considerable potential for influencing whether a tipping point is reached. The potential economic losses from a die-back are likely to be considerable, yet – like in the case of ocean acidification – the losses of biodiversity and ecological services, which are extremely difficult to predict and quantify, may be claimed to be far more significant. In addition, such a die-back would undermine the livelihoods of indigenous people. Seen from a security perspective the Amazon rainforest die-back may serve as a test-case: Will the international political system be able to act with sufficient effectiveness to reduce the human contribution to the threat of irreversible change in the Earth System ? The costs of (possibly) avoiding the die-back through the

REDD mechanism or other means of stopping deforestation are likely to be less than it would be for most other tipping elements, since it involves mostly local action, rather than a transformation of the energy system at global scale. If this cannot be achieved it would demonstrate the inability of the international political and regulatory system to deal with the greater challenges presented by other tipping elements.

4.4 Communicating Risk and Uncertainty Concerning CC

The communication of uncertainty and risk in relation to climate change is controversial. While it is generally acknowledged that we do not and cannot predict future climate change with high precision, among other things because of the 'chaotic' character of the climate system, the ways in which uncertainty is expressed vary widely, as noted above. The study of 'climate change discourses' is presently a field attracting considerable attention [17], and it can only be touched upon briefly here. We will focus on the scientific concept – or metaphor – of 'tipping points/elements' (TP/E) [11], as opposed to more conventional representations of uncertainty and risk (using the terms 'probability', 'risk', 'uncertainty') in the IPCC reports.

What does use of the TP/E terminology entail, above and beyond the IPCC formulations? In the IPCC 4AR WG1 report, dealing with the physical basis of climate change, the uncertainties are represented graphically as intervals of possible outcomes (IPCC 4AR, Summary for Policymakers) and described in terms such as 'likely'/'very likely' which may be translated into quantitative probabilities. As mentioned above, the term 'dangerous climate change' is used along with references to 'abrupt CC' and 'large-scale discontinuities'. This is the closest IPCC comes to the 'tipping element discourse'. It may be argued that the difference between the TP/E-formulation and the various IPCC formulations, and in particular the term 'large-scale discontinuities', is not great, since both imply the existence of non-linearities and positive feed-backs which may cause state-shifts of the climate system. What might make a difference is that the TP/E definition, presented above, implies human interference as well as relatively short time scales.

Three very different questions may be raised in this context:

1. Does our knowledge of the climate system and its possible positive feed-back mechanisms warrant the use of the TP/E terminology /metaphor and is use of this terminology an 'appropriate' way of communicating climate change, since it acknowledges the possible non-linearity of the climate system, as well as the human impact and the short time perspective ?
2. Since the use of the TP/E terminology constitutes a 'speech act', what are the objectives and consequences of this act ?
3. Does the use of the TP/E terminology imply 'securitization' of CC, beyond what the IPCC discourse entails, and if so is this securitization desirable ?

In the following we will focus on the latter question.

4.5 Securitization of CC

As noted above, there has been an increasing tendency in the debate to consider climate change as a 'security issue'. Use of the TP/E metaphor plays a significant role in this context, since it may be interpreted as a deliberate 'act of securitization' [3] aiming at moving the climate change issue from the domain of ordinary politics to the domain of 'security politics', implying that much stronger interventions may be justified. It is obviously difficult to isolate the exact 'securitization effect' of the TP/E terminology in the political discussion of climate change and security, but it is clearly suited as a means of making the point that extraordinary measures are required, outside the traditional realm of international environmental politics.

Securitization of climate change has been ongoing over the last decade as documented by Barnett [1]. It might be useful, however, to distinguish between securitization in the traditional sense, involving the necessity of the use of force ('hard securitization'), and in an extended meaning, not necessarily involving such use ('soft securitization'). In the writings on climate change as a security issue, a number of arguments are given for securitization, the most prominent being the following:

- Climate change will inevitably lead to conflicts over increasingly scarce resources, e.g. of water, food and energy. 'Water-wars' are sometimes mentioned as virtually unavoidable (a case of 'hard securitization'), and large geographical shifts in agricultural potential are by some expected to create international tensions associated with 'food security' (presumably involving 'hard' or 'soft securitization').
- Political tensions will rise between countries particularly seriously affected by climate change, yet not contributing much to it (mostly developing countries), and countries that are not negatively affected to any great extent, but contributing through high GHG emissions or extensive deforestation.

The three examples of candidates for tipping elements of the Earth System have widely different properties as concerns their probabilities, the economic, environmental and political effects and the costs of mitigation. These differences will influence the extent to which they will trigger 'securitization' of climate change, and what sort of securitization might be in question. While the die-back of parts of the Amazon may be a relatively likely event with a great potential to trigger other tipping elements (due to the large CO_2-emissions caused by it), it seems possible that it can be dealt with in the realm of 'normal politics', such as the negotiations within the framework of the UNFCCC. In contrast to this, the much less likely worst-case-scenario of accelerated sea-level rise due to disintegration of the West-Antarctic ice cap, and the consequences this could have in terms flows of 'climate refugees', may be much more difficult to deal with and could cause attempts to securitize climate change. The medium probability event of die-back of coral reefs at global scale has already been labeled a security issue (e.g. by [1]), yet the fact that the immediate threat to human life may be claimed to be limited implies that it may not be able to justify the mobilization of extra-ordinary use of force associated with 'hard securitization'.

4.6 Is Securitization of CC Desirable ?

This question was discussed by Deudney [5], arguing that securitization of climate change and other environmental problems is unlikely to contribute to solving such problems. The only institution presently in a position to react to global 'environmental' threats to security, such as climate change, appears to be the UN Security Council. The types of action taken by the UN Security Council, e.g. international embargos and military intervention, do not immediately lend themselves to the mitigation of climate change, while it may be realistic that the UN could intervene in regional conflicts caused or intensified by climate change. However, the basic causes of climate change are believed to be emissions of GHGs, and these may only be reduced by technological transformation and changes in behavioral and consumption patterns, which the Security Council has little possibility to influence efficiently. This lack of capacity to solve the problem is further emphasized by the fact that the largest GHG-emitters are permanent members of the Security Council with a veto-right. In this situation, 'hard securitization' of climate change may be claimed to be of little consequence – or even counter-productive. On the other hand, the apparent inability of current international political institutions, and more specifically the UNFCCC, to deal with the challenges of climate change points to the necessity to strengthen and reform these institutions, and 'soft securitization' of climate change may (and is probably often intended to) be a means of achieving this.

4.7 Conclusion

Briefly summarized, we conclude that:

- Tipping elements of the climate system are likely to be real and to constitute the greatest 'threats to security', in a broad understanding of the term, in a 100 year perspective – and even more so in a longer perspective.
- The three tipping elements discussed are very different in terms of (1) their probabilities of being triggered within a 'political' time perspective, (2) the types of impacts on society and ecosystems that they are likely to have, (3) who will be the most affected and (4) the implications for security and securitization.
- Both 'hard' and 'soft' securitization of climate change are likely to result from the fear for tipping of the climate system, as already evident from the scientific and public debate, yet its precise character will depend on the expected consequences of the tipping element in question, and not the least on who will be most affected.
- It is unlikely that 'hard securitization' of climate change, involving mobilization of extraordinary (military) means, can further an appropriate global response, yet 'soft securitization' may be required to mobilize the necessary reform and strengthening of international political institutions required to address climate change.

References

1. Barnett J (2003) Security and climate change. Glob Environ Change 13:7–17
2. Brown O, Hammill A, McLeman R (2007) Climate change as the 'new' security threat: implications for Africa. Int Aff 83(6):1141–1154
3. Buzan B, Waever O, de Wilde J (1998) Security: a new framework for analysis. Lynne Rienner, Boulder
4. Dansgaard W, Johnsen SJ, Clausen HB, Dahl-Jensen D, Gundestrup NS, Hammer CU, Hvidberg CS, Steffensen JP, Sveinbjörnsdottir AE, Jouzel J, Bond G (1993) Evidence for general instability of past climate from a 250 kyr ice-core record. Nature 364:218–220
5. Deudney D (1991) Environment and security: muddled thinking. Bull At Sci 47(3):23–28
6. Doney SC, Fabry VJ, Feely RA, Kleypas JA (2009) Ocean acidification: the other CO_2 problem. Ann Rev Mar Sci 1:169–192
7. Fabry VJ, Seibel BA, Feely RA, Orr JC (2008) Impacts of ocean acidification on marine fauna and ecosystem processes. ICES J Mar Sci 65:414–432
8. Füssel H-M (2009) An updated assessment of the risks from climate change based on research published since the IPCC Fourth Assessment Report. Clim Change 97:469–482
9. Hoegh-Guldberg O, Mumby PJ, Hooten AJ, Steneck RS, Greenfield P, Gomez E, Harvell CD, Sale PF, Edwards AJ, Caldeira K, Knowlton N, Eakin CM, Iglesias-Prieto R, Muthiga N, Bradbury RH, Dubi A, Hatziolos ME (2009) Coral reefs under rapid climate change and ocean acidification. Science 318:737–742
10. Huntingford C, Fisher RA, Mercado L, Booth BBB, Sitch S, Harris PP, Cox PM, Jones CD, Betts RA, Malhi Y, Harris GR, Collins M, Moorcroft P (2008) Towards quantifying uncertainty in predictions of Amazon 'dieback'. Philos Trans R Soc B 363:1857–1864
11. Lenton TM, Held H, Kriegler E, Hall JW, Lucht W, Rahmstorf S, Schelnnhuber HJ (2008) Tipping elements in the Earth's climate system. PNAS 105(6):1786–1793
12. Malhi Y, Aragão LEOC, Galbraith D, Huntingford C, Fisher R, Zelazowski P, Sotch S, McSweeney C, Meir P (2009) Exploring the likelihood and mechanism of a climate-change-induced dieback of the Amazon rainforest. PNAS 106(49):20610–20615
13. Solomon S, Qin D, Manning M, Chen Z, Marquis M, Averyt KB, Tignor M, Miller HL (eds) (2007) Climate change 2007: the physical science basis. Contribution of working group I to the fourth assessment report of the Intergovernmental Panel on Climate Change. Cambridge University Press, Cambridge
14. Mercer JH (1978) West Antarctic ice sheet and CO_2 greenhouse effect: a threat of disaster. Nature 271:321–325
15. Oppenheimer M, Alley RB (2004) The West Antarctic ice sheet and long term climate policy. Clim Change 64:1–10
16. Rahmstorf S (2010) A new view on sea level rise. Nat Rep Clim Change 4:46–47
17. Russill C, Nyssa Z (2008) The tipping point trend in climate change communication. Glob Environ Change 19:336–344

Chapter 5
Interactions of Global-Warming and Urban Heat Islands in Different Climate-Zones

Robert Bornstein, Ruri Styrbicki-Imamura, Jorge E. González, and Bereket Lebassi

Abstract IPCC results show that global-warming over the last 35 years has not been spatially uniform over the globe on both the seasonal and diurnal time-scales. Urban areas likewise produce their own climates, e.g., cities cool less rapidly than their rural surroundings at night, thus forming nocturnal urban heat island (UHIs). As UHIs interact with regional global climate-changes, this paper investigated interactions between these two phenomena in different climate zones around the world. It first focused on the spatial distribution of 2-m summer maximum-temperature trends for the period of 1970–2005 in the highly populated Southern California Air Basin, which exhibited a complex pattern of cooling in low-elevation coastal-areas and warming at inland areas. The coastal cooling resulted as global warming of inland areas produced enhanced cool-air sea breeze intrusions, i.e., a "reverse reaction" to global warming. To investigate interactions between global warming and UHI-growth, pairs of sites were identified near cooling-warming boundaries. The faster each urban area grew, the faster its UHI grew. To determine where UHI-growth and global climate-change either are additive or in opposition, requires understanding of the global distribution of climate types and of the diurnal and

R. Bornstein (✉)
Department of Meteorology and Climate Science, San José State University,
San José, CA, USA
e-mail: pblmodel@hotmail.com

R. Styrbicki-Imamura
Consultant, 7035 Jonquil Ln N, Maple Grove, MN, USA

J.E. González
Mechanical Engineering Department, The City College of New York,
New York, NY, USA

B. Lebassi
Department of Environmental and Earth System Science, Stanford University,
Stanford, CA, USA

H.J.S. Fernando et al. (eds.), *National Security and Human Health Implications of Climate Change*, NATO Science for Peace and Security Series C: Environmental Security, DOI 10.1007/978-94-007-2430-3_5, © Springer Science+Business Media B.V. 2012

seasonal patterns of UHI-formation in each climate type. UHI formation is a function of the thermal inertia (TI) of adjacent rural surfaces. Coastal-cooling is most likely in west-coast marine-Mediterranean climates. Urban cool islands counter global-warming and reduces thermal-stress events in dry rural-soil climates, while heat-stress events are most-likely in wet rural-soil climates.

Keywords Global climate change • Urban climate • Temperature trends

5.1 Introduction

Global warming arises from an enhanced greenhouse effect, as incoming short-wave solar energy mostly passes-through the atmosphere without attenuation, but outgoing terrestrial long-wave infrared-red (IR) energy is mostly-absorbed by anthropogenic greenhouse gases (GHGs), the most important of which is CO_2. Observed global temperature changes over the last 100 years include natural- and anthropogenic-induced changes. The latter arises from both greenhouse gas (GHG) induced warming and aerosol induced cooling (due to reflection of incoming solar energy). A complete summary of global warming over the last 100 years is given in IPCC [5] at IPCC_Report.htm.

The Intergovernmental Panel on Climate Change (IPCC) report shows that global-averaged 2-m temperature warming rates have not been uniform over the last 100 years, with:

- warming from 1900 to 1945
- cooling from 1945 to 1970
- warming from 1970 to the present.

The report also shows that these changes have not been uniform over globe (on both the continental and regional scales) on both the seasonal and diurnal time scales. The spatial distribution of climate change is thus a function of latitude, longitude, altitude, and distance from the sea.

Impacts from global warming (that adversely affect local security issues) include:

- higher atmosphere and ocean temperatures, which reduce land and sea ice and thus induce coastal flooding
- increased frequencies of ever more severe-storms
- movement of tropical insect-borne diseases pole-ward and towards higher elevations (an adaptation limited by topographic height)
- higher rates of: human heat stress, summer energy-use, surface ozone levels, and, wildfires
- increased or decreased precipitation, and thus water supply, producing winners and losers (via droughts or floods)
- pole-ward and up-slope movements of crop-growth areas also produces winners and losers.

Positive impacts from global warming include:

- fewer frost-days
- lower-rates of winter energy-use

Urban areas likewise produce their own weather and climate regimes, which form as:

- grass and soil surfaces become concrete and buildings, which alter surface sensible and latent heat fluxes to the atmosphere
- fossil fuel consumption (in high latitude cities) produces atmospheric pollutants and heat
- atmospheric pollutant layers and building walls reduce both incoming solar and outgoing IR radiant fluxes.

A composite effect of some of these impacts is that cities cool less rapidly than their rural surroundings, i.e., they remain warmer at night. Nocturnal urban heat island (UHIs) thus form [1], as a localized global-warming analogy. Urban cool islands (UCIs) also form under certain conditions (e.g., in urban-canyon shadows). Urban areas also either increase or decrease wind speeds and precipitation amounts, dependent on geographic location and on background large-scale weather and climate conditions. Evaluation of global warming trends at urban sites involves subtraction of its "urban bias," i.e., growth-trend of its UHI [6].

As UHIs interact with regional global climate-changes, this paper investigates the interactions (i.e., where they are additive or in opposition) of these two phenomena in different climate zones around the world. As many adverse climate-changes discussed above affect human migration patterns, potentially leading to conflict and war, this interaction is of extreme importance to global -and urban-scale security issues.

5.2 Local Climate-Change Case Study

This part of the current effort focused on two highly populated near sea-level California coastal urban air basins: South Coast Air Basin (SoCAB, Fig. 5.1) and San Francisco Bay Area (SFBA), which includes the northern Central Valley (CenV). National Climate Data Center (NCDC) data for 52 SFBA and 28 SoCAB Cooperative Observational (COOP) sites consisted of 1970–2005 summer (July to August, JJA) 2-m daily maximum air-temperatures (T_{max}), as this period shows the most rapid global warming. The T_{max}-trend at each site in the air basins was plotted at their corresponding station locations, together with an indication of the statistical significance of each trend. Additional analysis-details are found in Lebassi et al. [7].

The spatial distribution of observed SoCAB 1970–2005 JJA T_{max} trend-values (Fig. 5.2) exhibits a complex pattern of cooling in low-elevation coastal-areas open to marine air penetration and warming at both inland and higher-elevation coastal areas; most trend-values have a high statistical significance (1.0-p), i.e., low p-value. Marine air enters at the low-elevation coastal area south of Palos Verdes, and then

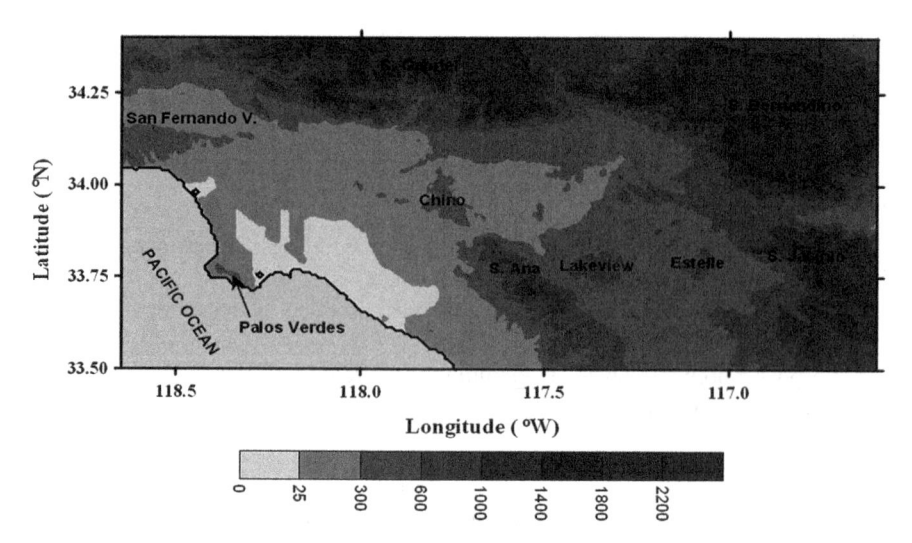

Fig. 5.1 SoCAB geographic areas mentioned in text, where *shading* indicate topographic heights (m) (From Lebassi et al. [7])

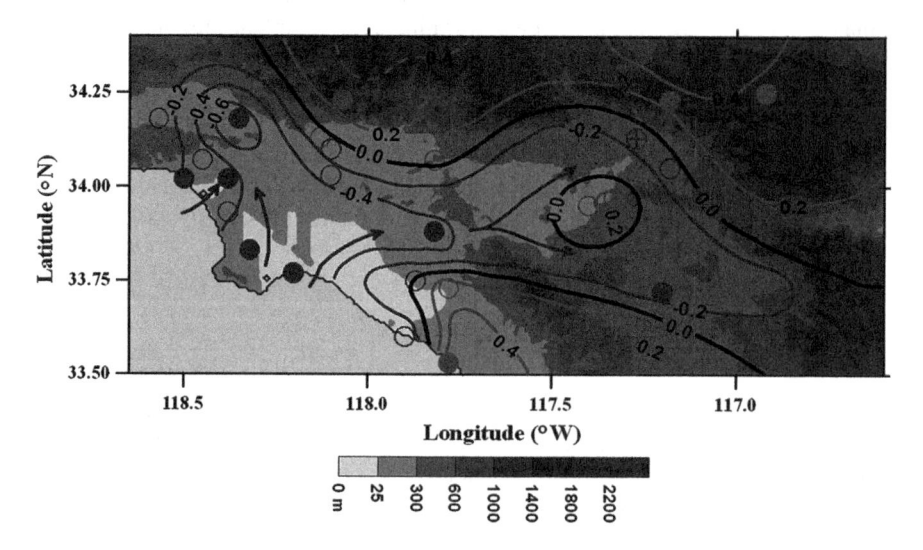

Fig. 5.2 SoCAB 1970–2005 summer 2-m T_{max} warming/cooling trends (°C dec^{-1}), where "?" indicates areas where more data are needed, and where *solid, crossed,* and *open circles* show statistical p-values of <0.01, 0.01–0.05, and >0.05, respectively (From Lebassi et al. [7])

splits northward towards the San Fernando Valley (with a max cooling of −0.99°C dec^{-1}) and eastward towards the Chino hills, where it splits again. One part flows northward, towards the foothills between the San Gabriel and San Bernardino Mts., while its southern branch flows past the Lakeview and Estelle Mts. Additional data would be useful in the three areas with question-marks.

While these regions thus show cooling, higher elevation inland regions (that lack marine-air penetration) show warming, i.e., north of Lakeview (local max of 0.12°C dec^{-1}), over the San Gabriel and San Bernardino Mts. to the north (local max of 0.41°C dec^{-1}), and south of the Santa Ana Mts. (max of 0.64°C dec^{-1}). A similar, but more complex pattern (arising from its more complex topographic pattern) was found in the SFBA. The composite JJA T_{max}-trend for all coastal-cooling sites in both air basins over 1970–2005 was 0.35°C dec^{-1} [7].

The explanation for this coastal cooling is that the expected GHG driven global warming of summer T_{max}-values in the eastern inland CenV and Sierra Nevada Mountains of California produced enhanced coastal-inland pressure and temperature gradients, and hence increased cool-air sea breeze intrusions. These flows thus produced the currently observed cooling of summer T_{max}-values in the low-elevation coastal basins, i.e., it produced a "reverse reaction" to global warming.

To investigate interactions between global warming and UHI-growth, four pairs of sites near the cooling-warming boundaries were identified: two-pairs each in the SoCAB (Fig. 5.2) and the CenV, east of the SFBA (not shown). For each pair, the rural site had shown coastal cooling, while the nearby urban site had shown concurrent warming. It was then assumed that if the urban site had remained rural, it too would have showed the same degree of coastal cooling found at the nearby rural site. This assumption allowed for "true" UHI growth-trends (i.e., warming) at the urban sites to be estimated as the sum of its urban warming trend and the absolute magnitude of the nearby rural coastal-cooling trend. This implies that the observed warming at the urban sites thus resulted only after the nearby rural coastal-cooling trend had been overcome by its UHI-growth.

Results show that the Stockton UHI-growth trend (Fig. 5.3a) is the largest, as it equals its 0.38 K dec^{-1} warming rate plus the 0.17 K dec^{-1} cooling rate at its nearby rural station, for a total of 0.55 K dec^{-1}. Statistical analysis showed that the faster an urban-member of each pair grew, the faster its UHI grew. Both the change in urban aerial-extent (from 21% to 59%) and population (from 40% to 118%) at each of the four sites were thus correlated with its change in UHI-magnitude (0.12–0.55 K dec^{-1}), as seen in Fig. 5.3a–d. It can thus be concluded that the coastal-cooling at SoCAB urban-sites in Fig. 5.1 (as well as those in the SFBA) had thus totally overcome their UHI-growth, and if those areas had remained rural their observed rates of coastal-cooling would have thus been larger, by a rate that can only be determined by mesoscale meteorological model-simulations that sequentially include urbanization and global-warming effects.

5.3 Interactions of Global-Warming and UHIs

To determine the locations where UHI-growth and global climate-change will strengthen or cancel each other, one must utilize both the:

- global distribution of Köppen climate types (the most widely used system)
- diurnal and seasonal patterns of UHI-formation in each of those climate types

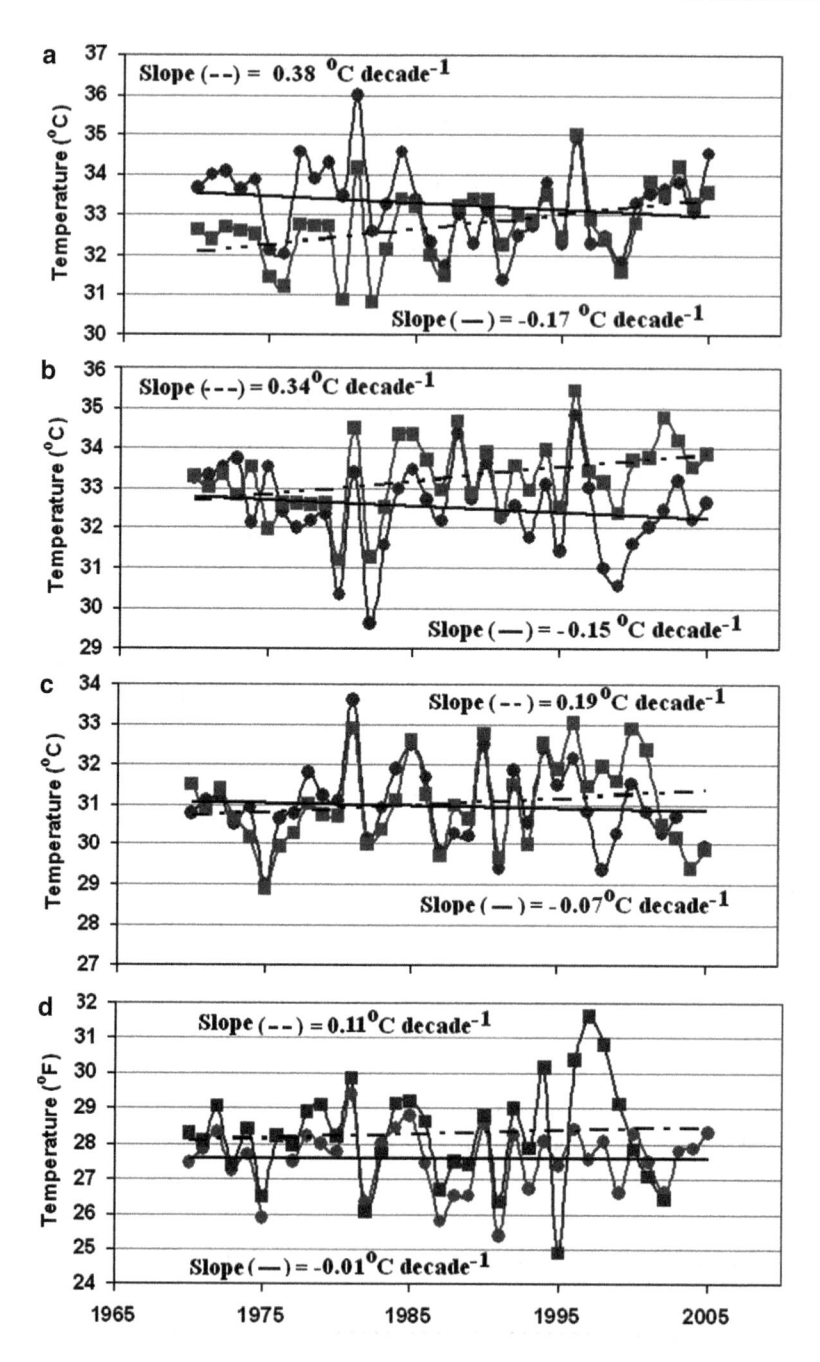

Fig. 5.3 Summer 1970–2005 average 2-m T$_{max}$ warming/cooling trends (°C dec^{-1}) for four-pairs of adjacent urban (*squares, solid*) and rural (*circles, dashed*) sites

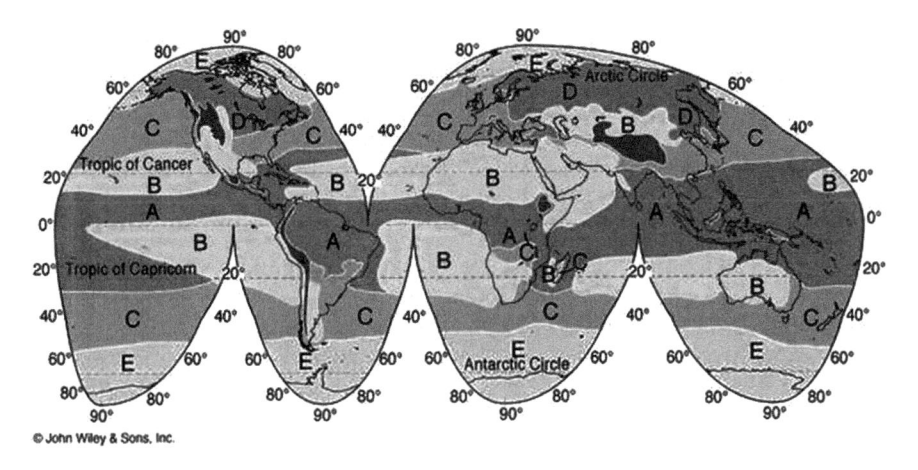

© John Wiley & Sons, Inc.

Fig. 5.4 Köppen global-climate classification system

The original Köppen global climate classification system was based on the requirements for certain plant-types to grow, as estimated by local monthly-average temperature and precipitation values. The modified (with the addition of the H-climate type, as explained below) global distribution of Köppen climate types (Fig. 5.4) shows the following main regions:

- cold high-altitude H-climates (shown on this older map as E, defined below) in Tibet and the Andes
- mid-latitude dry B-climates in the Sahara and on the west-side of the continents
- hot tropical A-climates at the equator
- middle-latitude mild-winter C-climates
- cold snowy-winter high-latitude D-climates in most of Canada and Siberia
- cold polar E-climates in Antarctica, Greenland, and the northern reaches of Canada and Asia.

Sub-divisions of interest to the current study include:

- warm (a) wet (f)summer Cfa Mediterranean C-climates on the east-side of the continents
- cool (b) dry (s) summer Csb Marine Mediterranean C-climates on the west-side of the continents; parts of the current study area on the west coast of California have this climate, and coastal-cooling is most likely in these climates, as their summer on-shore flows move over cold southward-moving coastal ocean-currents.

Imamura [4] related UHI-formation to the thermal inertia (TI) of adjacent urban and rural surfaces. TI is defined as the square root of the product of subsurface heat conductivity and heat capacity. Literature values show that wet rural soils have higher TI values than urban concrete and building materials, which in turn have higher values than dry rural soils. On the diurnal and annual time cycles, wet rural soils thus heat up and cool-down most-slowly, while the converse is true for dry rural soils; urban rates are intermediate.

Use of this UHI formation theory in conjunction with the Köppen climate-types thus makes it possible to generalize when UHIs will be strongest in a given city, based on the seasonal distribution of its regional precipitation, as follows:

- warm or hot (low latitude) cities surrounded by wet rural-soils (e.g., in A-climates and in Cfa-climates, on the east side of the continents) should have

  - daytime and wet-season maximum UHIs
  - nighttime and dry-season UCIs

- warm-cities within dry rural-soils (e.g., in B-climates and in Csb-climates, on the west-side of continents) should show reverse UHI and UCI diurnal patterns
- cold (in high-altitude H-climates and in high-latitude D- and E-climates) cities, whose UHIs form mainly from anthropogenic heat-fluxes should show

  - winter and nighttime maximum-UHIs
  - summer and daytime minimum-UHIs.

These city-scale UCIs are larger-scale (and form by different physical processes) than those behind daytime shaded-buildings.

The survey of UHI-values from around the globe by Imamura [4] generally verifies the above generalizations, as do her observations in the following cities on three continents (whose populations range from a few thousand to a few hundred thousand): Shimozuma, Japan; Sacramento, California; and the four Brazilian cities in Fig. 5.5. Her results show all nighttime and nighttime UHIs proportional to population (in different non-linear rates), and while all nighttime UHI-values fall on the same curve, daytime values in cities in wet rural-soils have larger UHIs than those in dry rural-soils, for the same population.

It is thus possible to conclude that, as it affects human thermal-stress values:

- daytime-UHIs reinforce global-warming in

  - cool cities, a good (in terms of human thermal-stress levels) result
  - warm-cities in wet rural-soil areas in A- and Cfa-climates, a bad result

- nighttime-UHIs reinforce global-warming in

  - cool-cities, a good result
  - warm-cities in dry rural-soil areas in B-climates and in Csb-climates, a bad result

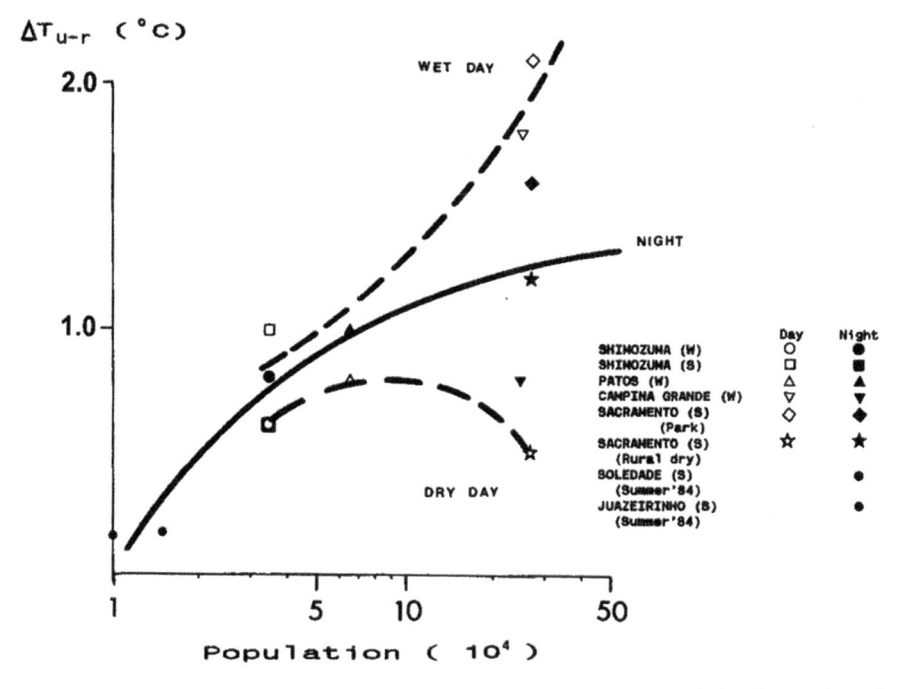

Fig. 5.5 Daytime and nighttime UHI-magnitudes (°C) as a function of population for a variety of cities, as a function of nearby rural soil moisture

- nighttime-UCIs counter global-warming in warm-cities in wet rural-soil areas in A- and Cfa-climates, a good result
- daytime-UCIs counter global-warming in warm-cities in dry rural-soil areas in A-, B- and in Csb-climates, a good result.

The interaction of UHIs and global climate-changes in different climate zones around the world thus effects local-scale security issues, as:

- nighttime and daytime UCIs counter global-warming, and thus reduce thermal-stress events in any climate area
- heat-stress events are most-likely in wet rural-soil areas in A- and Cfa-climates.

5.4   Conclusion

The IPCC showed non-uniform global-warming rates over the last 35 years over the globe, on both the seasonal and diurnal time-scales. The spatial distribution of climate change is thus a function of latitude, longitude, altitude, urbanization, and distance from the sea.

Urban areas also produce their own weather and climate regimes due effects from concrete and buildings, fossil fuel consumption, and atmospheric pollutant layers. One composite effect of some of these impacts is that cities cool less rapidly than their rural surroundings, i.e., they remain warmer at night, and thus nocturnal urban heat island (UHIs) form, as a localized global-warming analogy; urban cool islands (UCIs) also form under certain conditions. The magnitude of UHIs and other urban-induced weather and climate effects is dependent on geographic location, background large-scale weather- and climate-conditions, and rural soil-moisture content. As UHIs interact with regional global climate-changes, this paper investigated the interactions of these two phenomena in different climate zones around the world.

The paper first focused on the highly populated SoCAB California coastal urban air basin. Summer 2-m daily temperatures at 28 sites for 1970–2005 were used to calculate the spatial distribution of T_{max} trend-values. Results showed a complex pattern, with cooling in low-elevation coastal-areas open to marine air penetration and warming at inland areas. The explanation for the coastal cooling was that the expected GHG-driven global warming of summer T_{max}-values in inland California produced enhanced coastal-inland pressure- and temperature-gradients, and hence increased cool-air sea breeze intrusions, producing a "reverse reaction" coastal-cooling to global warming.

To investigate interactions between global warming and UHI-growth, four pairs of sites were identified near cooling-warming boundaries. For each pair, the rural site had shown coastal cooling, while the nearby urban site had shown warming. It was then assumed that if the urban site had remained rural, it too would have showed the same degree of coastal cooling found at the nearby rural site. This assumption allowed for "true" UHI growth-trends (i.e., warming) at the urban sites to be estimated as the sum of its urban warming trend and the absolute magnitude of the nearby rural coastal-cooling trend. This implied that the observed warming at the coastal urban sites resulted only after the nearby rural coastal-cooling trend had been overcome by its UHI-growth.

The faster an urban-member of the pair grew, the faster its UHI increased. Changes in urban aerial-extent and population were both thus correlated with changes in UHI-magnitude. It can also be concluded that the coastal-cooling at urban-sites had thus totally overcome their UHI-growth, and if those areas had remained rural their observed rates of coastal-cooling would have thus been larger, by a rate that can only be determined by mesoscale meteorological model simulations that sequentially include urbanization and global-warming effects. Such simulations were carried out for the SoCAB area by Lebassi et al. [9], who reproduced the general spatial extent of the coastal cooling area, as well as its maximum magnitude of 1.05 K dec^{-1}. They also showed that urban surface-roughness deceleration of SoCAB sea-breeze flows were large than its UHI-induced accelerations.

To determine where UHI-growth and global climate-change will e additive or in opposition, the distribution of climate types around the word and the diurnal and seasonal patterns of UHI-formation in each climate type were studied. UHI-formation is related to the TI of adjacent rural surfaces. Wet rural soils have higher TI values than urban building materials, which in turn have higher values than dry rural soils. On the

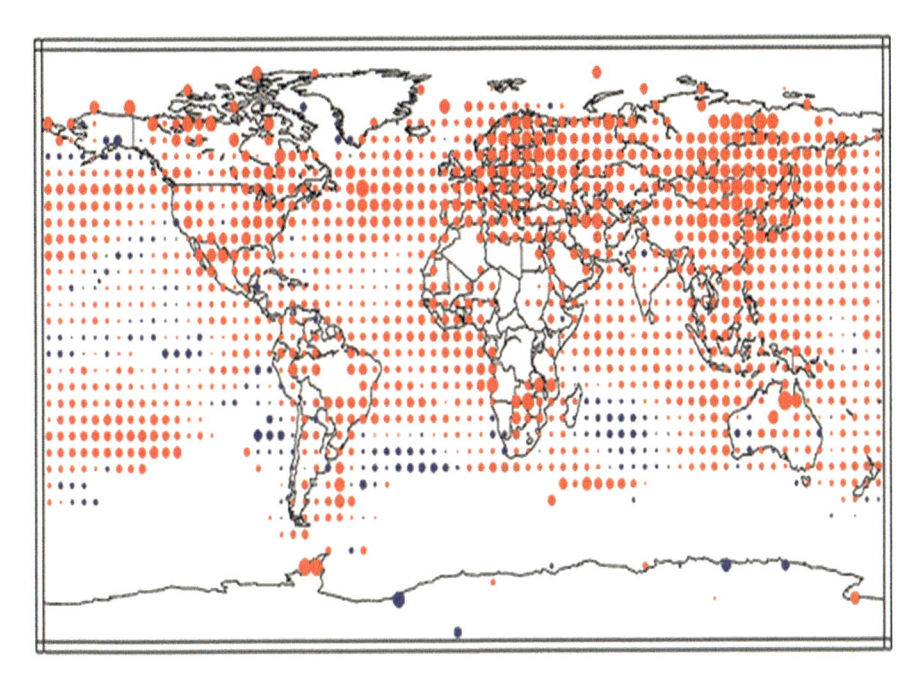

Fig. 5.6 Annual sea surface- and 2-m over-land air- temperature trends from 1970 to 2007, where *red* is warming and *blue* is cooling and where circle-size is proportional to magnitude of change (From IPCC [5])

diurnal and annual time cycles, wet rural soils thus heat up and cool-down most-slowly, while the converse is true for dry rural soils; urban rates are thus intermediate.

It was then possible to generalize when UHIs would be strongest in a given city, based on the seasonal distribution of its regional precipitation, as follows: warm cities surrounded by wet rural-soils have daytime and wet-season maximum UHIs, as well as nighttime and dry-season UCIs, while cities within dry rural-soils have the reverse UHI and UCI diurnal patterns. Cold cities (whose UHIs form mainly from anthropogenic heat-fluxes) have maximum winter- and nighttime-UHIs, as well as minimum summer- and daytime-UHIs.

Coastal-cooling is thus most likely in west-coast marine-Mediterranean climates, with their on-shore flows-over cold southward-moving coastal ocean-currents. In terms of human thermal-stress values, daytime-UHIs reinforce global-warming in cool (high-latitude and high-elevation) cities and in warm-cities in wet rural-soil climates, while nighttime-UHIs reinforce global-warming in cool-cities and in warm-cities in dry rural-soil climates. Nighttime-UCIs counter global-warming in warm-cities in wet rural-soil climates, and daytime-UCIs counter global-warming in warm-cities in dry rural-soil climates. Heat-stress events are most-likely in wet rural-soil climates.

Coastal-cooling reverse-reactions to global-warming are expected in subtropical low-elevation west-coast marine-Mediterranean Csb climate areas (i.e., California, Chile, Australia, South Africa, and Portugal), in which sea breezes strongly influence regional climate. IPCC [5] 1976–2001 annual temperature trends do, in fact, show such cooling (blue dots at west-coast areas in Fig. 5.6) at all these sites, except Portugal. Cordero et al. [2] found similar coastal cooling in their extension the current analysis to all of California, while Falvey et al. [3] observed cooling off the Chilean coast. Oglesby et al. [10] found the phenomenon with 4 km WRF simulations of the west coast of Central America (see its small blue coastal-dot in Fig. 5.6).

Coastal cooling as a regional "reverse-reaction" to global warming produces significant societal impacts, e.g., agricultural production could increase or decrease. Its beneficial effects include decreased summer electricity usage for cooling (as found for the study area by [8]) and reduced maximum O_3 levels due to reduced: fossil-fuel usage for cooling, natural hydro-carbon production, and photochemical photolysis rates. Human thermal-stress rates and mortality would also decrease.

Additional analyses and simulations are needed to evaluate T_{max} cumulative frequency distributions to see if heat-wave frequency might increase, even as average T_{max}-values decrease. The interaction of urban climate and global climate-changes in different climate zones around the world also effect local-scale security issues in other ways not discussed in this paper, e.g., urban effects on precipitation and runoff cause flooding, and thus further observational and modeling efforts need be carried out.

Acknowledgment The authors would like to thank NSF for its funding of this effort.

References

1. Bornstein RD (1968) Observations of the urban heat island effect in New York City. J Appl Meteorol 7:575–582
2. Cordero EC, Kessomkiat W, Abatzoglou J, Mauget S (2011) The identification of distinct patterns in California temperature trends. Climatic Change (in press)
3. Falvey M, Garreaud RD (2009) Regional cooling in a warming world: recent temperature trends in the southeast Pacific and along the west coast of subtropical South America (1979–2006). J Geophys Res 114:D04102. doi:10.1029/2008JD010519
4. Imamura IR (2005) Micrometeorological observations of urban heat islands in different climate zones. Int J Climatol 25:1–31
5. IPCC (2007) Contribution of working group I to the fourth assessment report of the Intergovernmental Panel on Climate Change. In: Solomon S, Qin D, Manning M, Chen Z, Marquis M, Averyt KB, Tignor M, Miller HL (eds). Cambridge University Press, Cambridge/New York
6. Karl TR, Diaz HF, Kukla G (1988) Urbanization: its detection in the United States climate record. J Climate 1:1099–1123
7. Lebassi BH, González JE, Fabris D, Maurer E, Miller NL, Milesi C, Bornstein RD (2009) A global-warming reverse-reaction coastal summer daytime cooling in California. J Climate 22:3558–3573
8. Lebassi BH, González JE, Fabris D, Bornstein RD (2010) Impacts of climate change on degree days and energy demands in coastal California. J Solar Energy Eng 132(3):222. doi:10.1115/1.4001564
9. Lebassi BH, Gonzalez JE, Bornstein R (2011) Modeling of global-warming and urbanization impacts on summer coastal California climate trends. J Geophys Res (in review)
10. Oglesby RJ, Rowe CM, Hays C (2010) Using the WRF regional model to producehigh resolution AR4 simulations of climate change for Mesoamerica. Paper A23F-07, AGU meeting, San Francisco, 14 Dec 2010

Chapter 6
ENSO Forcing of Climate Variability over the North Atlantic/European Region in a Warmer Climate Conditions

Ivana Herceg Bulić

Abstract Changes in winter climate variability in the North Atlantic European (NAE) region associated with El Niño–Southern Oscillation (ENSO) forcing in a warmer climate are investigated. The study is based on two 20-member ensembles of numerical integrations by utilizing an atmospheric general circulation model (AGCM) of intermediate complexity. Current climate experiment is based on simulations forced with observed sea-surface temperatures (SST) for the period 1855–2002. The warmer climate corresponds to the doubled CO_2 concentration with SST forcing represented by the same SST anomalies as in the current climate experiment superimposed on the climatological SST that was obtained from a complex atmosphere–ocean general circulation model forced with the doubled CO_2. A composite analysis of atmospheric response is based on categorization into warm and cold composites according to the strength of SST anomalies in the Niño3.4 region.

In the current climate, ENSO impact on the winter interannual variability in the NAE region is rather weak, but is still discernible and statistically significant. Over the south-western part of Europe warm (cold) ENSO events are mainly associated with warmer (colder) and dryer (wetter) conditions than usual. According to the results of numerical simulations in the climate with doubled CO_2 concentrations, substantial modifications of ENSO influence on the NAE region are found. The spatial pattern of the ENSO impact on the NAE precipitation projects onto the distribution of differences between the warmer climate and the current climate precipitation climatology fields. Therefore, a considerable ENSO impact on temperature and precipitation may be expected in warmer climate conditions implying a possibility of greater importance of tropical-extratropical teleconnections for

I. Herceg Bulić (✉)
Andrija Mohorovičić Geophysical Institute, Department of Geophysics, Faculty of Science, University of Zagreb, Horvatovac 95, Zagreb, Croatia
e-mail: ihercegb@gfz.hr

H.J.S. Fernando et al. (eds.), *National Security and Human Health Implications of Climate Change*, NATO Science for Peace and Security Series C: Environmental Security, DOI 10.1007/978-94-007-2430-3_6, © Springer Science+Business Media B.V. 2012

future climate variability in the NAE region. Since temperature and precipitation are the variables of great interest of policy makers for adaptation and mitigation purposes, their proper representation is essential for climate projections.

Keywords Climate variability • ENSO tropical-extratropical teleconnections • Warmer climate • North Atlantic/European region

6.1 Introduction

Sea-surface temperature (SST) anomalies in tropical oceans are strong generators of climate variability on Earth. It is well known that changes of SST in the tropical Pacific associated with the El Niño-Southern Oscillation (ENSO) considerably affect not only the interannual climate variability in the tropics, but also have an impact on extratropical areas through atmospheric teleconnections (see [25] for a review of mid–latitude teleconnections). Tropical Pacific SST anomalies interact with the overlaying atmosphere and generate anomalies in convection which act as a source of Rossby waves that propagate into higher latitudes. Thus, propagation of Rossby waves that emanate in tropics provides the physical basis of tropical-extratropical teleconnections [15]. The ENSO impact on Pacific-North American (PNA) climate is widely investigated and well documented in scientific literature, but its influence on North Atlantic/European (NAE) region is discussed more controversially. Due to a relative large distance from the tropical Pacific and strong internal variability of the midlatitude atmosphere, the atmospheric response to the ENSO forcing may be masked with some other processes making difficult to isolate ENSO-related signal in remote areas such as NAE region. However, some observational studies indicate a weak but noticeable ENSO signal in European climate anomalies [4, 6]. This also has been confirmed by numerical modelling studies. For example, based on an ensemble of International Centre for Theoretical Physics (ICTP) AGCM simulations, Herceg Bulić and Branković [12] found a detectable atmospheric signal in the NAE region associated with ENSO forcing.

The interest for possible modifications in ENSO and tropical–extratropical teleconnections in a warmer climate has substantially increased in the past few decades because global warming has attracted a considerable scientific attention [16]. It may be expected that the elevated concentrations of greenhouse gases will provoke changes of the mean state in the upper ocean which in turn may affect some ENSO characteristics (i.e. intensity and frequency of occurrence). Some studies indicate that in a warmer climate El Niño events will be stronger than in the present climate [5, 24], while other show weaker events or no changes [19]. The future changes of ENSO will certainly have an impact on tropical-extratropical teleconnections. According to Hannachi and Turner [9], the CO_2 doubling in HadCM3 (Hadley Centre Climate Model) integrations modifies atmospheric preferred structures.

Their model produced weaker El Niño events and the teleconnection with the PNA region was reduced.

Since an earlier study performed by ICTP AGCM has indicated the substantial ENSO impact on climate variability in the NAE region [12], its' possible modulation associated with ENSO forcing in a climate with an increased concentration of CO_2 is analyzed here. Of particular interest for this study is potential changes in precipitation response since precipitation (and associated water availability) is one of the main requests of policy makers for adaptation and mitigation purposes. The structure of the paper is as follows. In Sect. 6.2, the model and the experimental design are described. In Sect. 6.3, composite analysis based on the categorization of ENSO events is presented. Summary discussion and conclusions are given in Sect. 6.4.

6.2   Data and Experimental Design

6.2.1   Model Description

The model used in this study is ICTP AGCM, an model of intermediate complexity, with eight vertical layers and a triangular truncation of horizontal spectral fields at total wave number 30 (T30–L8). An earlier model version with five vertical layers is described in details by Molteni [20] together with its climatology and variability. The model is based on a spectral dynamical core developed at the GFDL[1] [11]. It is a hydrostatic, σ–coordinate, spectral transform model in the vorticity–divergence form described by Bourke [1], with semi–implicit treatment of gravity waves. The parameterised processes include short– and long–wave radiation, large–scale condensation, convection, surface fluxes of momentum, heat and moisture and vertical diffusion. Land and ice temperature anomalies are determined by simple one–layer thermodynamic model. One of the main features of ICTP AGCM is its computational efficiency so it is particularly convenient for creating large ensembles.

Although the ICTP AGCM is a relatively simple model, it was successfully used in a number of studies dealing with various aspects of dynamical climatology (e.g. [2, 10, 12, 13, 17, 18, 20]) as well as climate change [14]. In all these studies ICTP AGCM was able to reproduce climate statistics within the range of observed quantities. Furthermore, it reproduces well the response of the atmospheric circulation to the well–documented SST trends in the tropical oceans [2]. Also, the results reported by Herceg Bulić and Branković [12] are in good agreement with some other modelling as well as observational studies and revealed discernible model ENSO-related response over the NAE region.

---

[1]Geophysical Fluid Dynamics Laboratory, Princeton, New Jersey, U.S.A.

6.2.2   Experimental Design

In this study, the results from the two ICTP AGCM experiments – the control experiment (CTRL) and the warmer climate experiment ($2\times CO2$) – are compared. CTRL represents the current climate simulated with observed SST forcing at the lower boundary and CO_2 at an average concentration for the period 1961–1990. The SST is in a form of monthly averages from a long time series of NOAA_ERSST_V2 data[2] [23] and is blended with the sea–ice monthly climatology from the Hadley Centre [22].

In the $2\times CO2$ experiment, climatological SST and sea-ice fields are taken from the Hadley Centre AOGCM model integrations (HadCM3) forced with doubled CO_2 concentration[3]. HadCM3 model is fully-coupled model composed of atmosphere [8] and ocean [21] components and is extensively used in climate change simulations [16]. Monthly SST anomalies from CTRL are then added to the $2\times CO2$ SST climatology in order to create $2\times CO2$ monthly SSTs. In such a way the SST interannual variability in $2\times CO2$ remains identical to that in CTRL, but climatological fields at the lower boundary (SST and sea ice) are in accordance with the doubling of CO_2.

For both CTRL and $2\times CO2$ experiments, the 20–member ensembles of 149-year long ICTP AGCM integrations were created. The first year of model integrations is discarded so the analysis includes 148 years. A relatively large size of ensembles and the length of integrations give us the confidence that the results are statistically reliable.

The modifications of ENSO impact on climate variability in the NAE region due to a warmer climate is examined by composite analysis considering the sign and strength of the SST anomalies in the Niño3.4 region (5°N–5°S, 170°W–120°W). The composite analysis is a sampling technique based on the condition of a certain event occurring (here El Niño or La Niña) and only those events that satisfy given condition are taken into account. For the purposes of this study, JFM Niño3.4 index (defined as area-averaged anomalies in the Niño3.4 region) of every year is calculated first. Then, categorization of ENSO events into warm (El Niño) and cold (La Niña) is performed in the following way: warm (cold) ENSO years were defined as those years with Niño3.4 greater (smaller) than $1.5 \cdot std$ ($-1.5 \cdot std$), where *std* is standard deviation of SST anomalies. After that, warm (cold) composites are calculated as averages of events that satisfy the condition for warm (cold) ENSO year. Here, the analysis is based on the composites of 18 cold and 9 warm events (see Fig. 6.1). It should be emphasised that the above categorization is valid for the $2\times CO2$ experiment as well. Namely, CTRL and $2\times CO2$ have identical SST anomalies but different climatological means; this difference makes no impact on

---

[2]Provided by the NOAA/OAR/ESRL PSD, Boulder, Colorado, U.S.A. (http://www.cdc.noaa.gov).

[3]Provided through the ENSEMBLES webpage: http://ncas-climate.nerc.ac.uk/research/ensembles-rt4/coord_exp/boundary_conditions.html

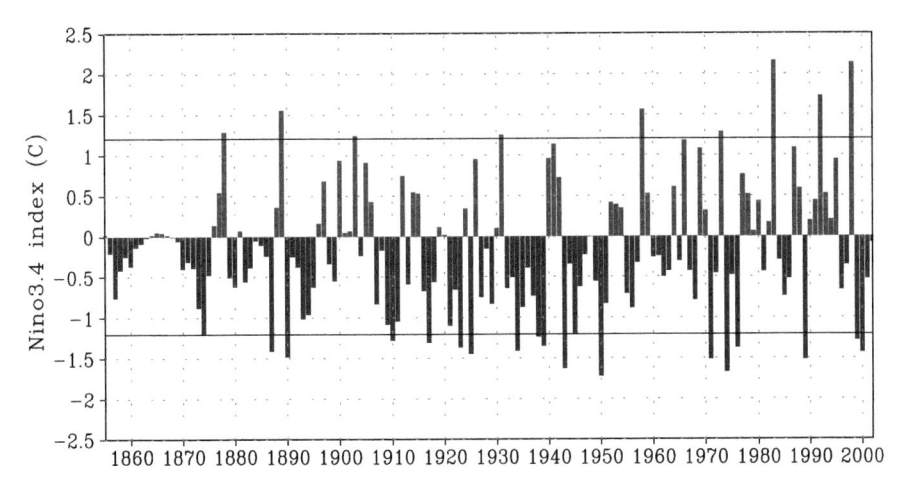

Fig. 6.1 Interannual variability of JFM Niño3.4 index for the period 1855–5002. The *horizontal lines* represents ±1.5·*std*

the computation of sample variance. The model output results were manipulated in the following way. First, for each month ensemble means of various meteorological parameters from 20 independent numerical integrations were calculated. The anomalies of examined atmospheric variables were calculated by subtracting the mean annual cycle. The long-term mean of every calendar month was subtracted from each individual month and winter seasonal anomalies (January, February, March; JFM) were calculated as 3-month averages. Then, seasonal composite analysis according to the above described categorization is applied. The significance of changes between $2 \times CO2$ and CTRL has been evaluated by the two-tailed *t*-test.

6.3 Results

6.3.1 Mean-Sea Level Pressure and Z200 Response

Mean-sea level pressure (*mslp*) responses to ENSO forcing in the current and warmer climate conditions are presented in Fig. 6.2. The spatial distribution of anomalies reveals a monopole structure over the NAE region indicating that warm (cold) ENSO events are associated with decreased (increased) *mslp*. The response to the cold ENSO events (Fig. 6.2b) has the same spatial structure as that for the warm composite (Fig. 6.2a) but with decreased amplitude (the amplitudes of the warm and cold composites are 2.5 and 1 hPa, respectively). The action centres have the same position (*i.e.* there is no phase shift) indicating a linearity of the *mslp* response.

In the warmer climate conditions, the both centres are shifted westward (Fig. 6.2c, d). Amplitude of the warm composite is slightly increased (from 2.5 hPa for the CTRL

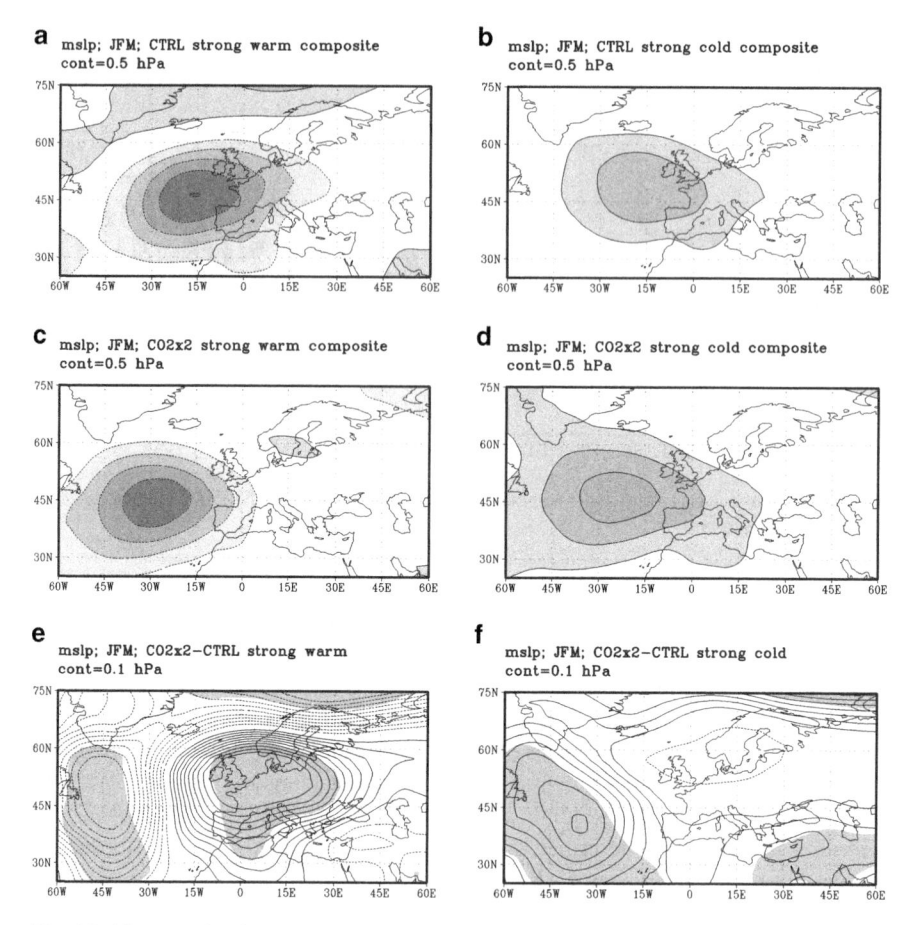

Fig. 6.2 Mean sea-level pressure (mslp) anomalies in JFM for: (**a**) CTRL warm composite, (**b**) CTRL cold composite, (**c**) $2\times CO2$ warm composite, (**d**) $2\times CO2$ cold composite. Mean sea-level pressure differences between $2\times CO2$ and CTRL in JFM for: (**e**) warm composite, and (**f**) cold composite. Contours in (**a–d**) 0.5 hPa, in (**e**) and (**f**) 0.1 hPa. Values exceeding the 95% confidence level of the t statistics are *shaded* in (**e**) and (**f**)

to 2 hPa for the $2\times CO_2$ experiment), while the *mslp* response to the cold ENSO forcing is somewhat strengthen (from 1 hPa for the CTRL to 1.5 hPa for the $2\times CO_2$ experiment). As a result of shifting of the action centre position and/or amplitude change, the modifications of *mslp* response in the warmer climate conditions are statistically significant with stronger response over the eastern subtropical and mid-latitude North Atlantic (Fig. 6.2e, f).

The anomalies of the geopotential height at 200 hPa isobaric level (Z200; Fig. 6.3) coincide with the *mslp* distribution (Fig. 6.2) revealing a barotropic structure of the response over the Atlantic, what is in accordance with the physical

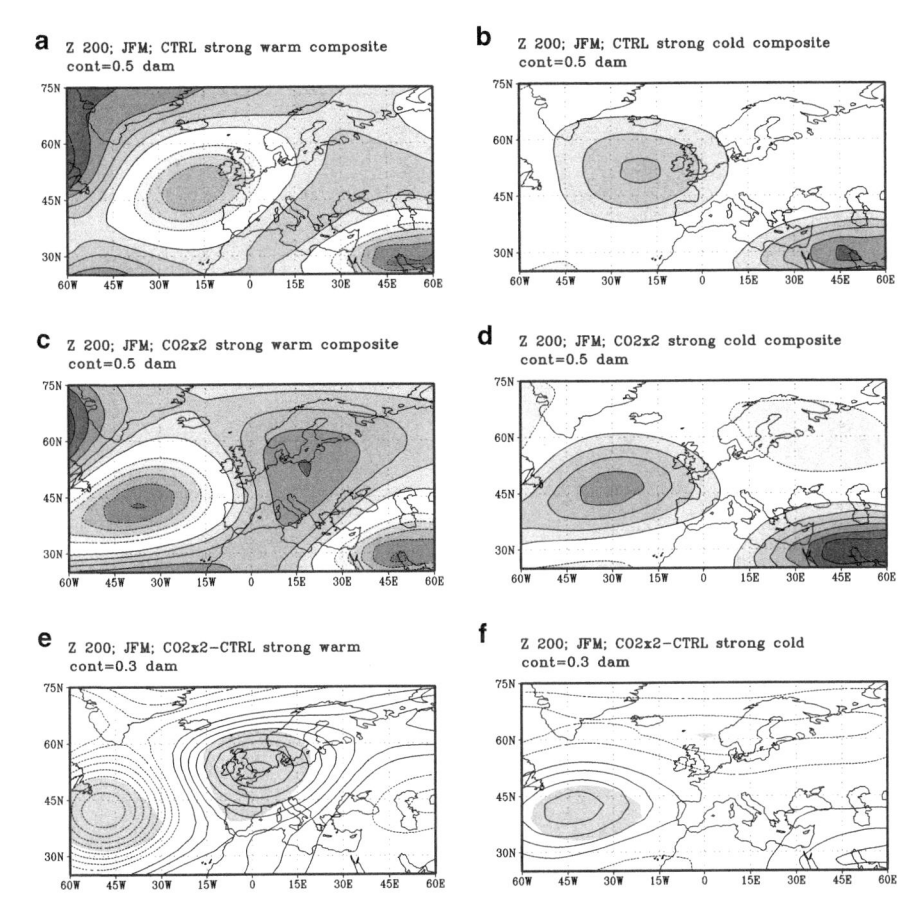

Fig. 6.3 The same as in Fig. 6.2 but for geopotential height at 200 hPa (Z200). Contours in (**a–d**) 0.5 dam, in (**c**) and (**f**) 0.3 dam. Values exceeding the 95% confidence level of the t statistics are *shaded* in (**e**) and (**f**)

mechanism underpinning tropical-extratropical teleconnections (*i.e.* Rossby wave propagation). The amplitudes of the Z200 anomalies achieve the same value for both warm and cold composites (1.5 dam) revealing the linearity of the response regarding both the position and the strength of the response. In addition to the area over the Atlantic, the Z200 response is quite pronounced also over the continental part of the domain forming a bipolar structure which is particularly pronounced for the warm events (Fig. 6.3c, d). According to the difference plots (Fig. 6.3e, f), a statistically significant stronger ENSO impact in the warmer climate is found over the eastern North Atlantic for the warm (Fig. 6.3e) as well as for the cold ENSO events (Fig. 6.3f), while over the western and central Europe the stronger response is found only for the warm composite.

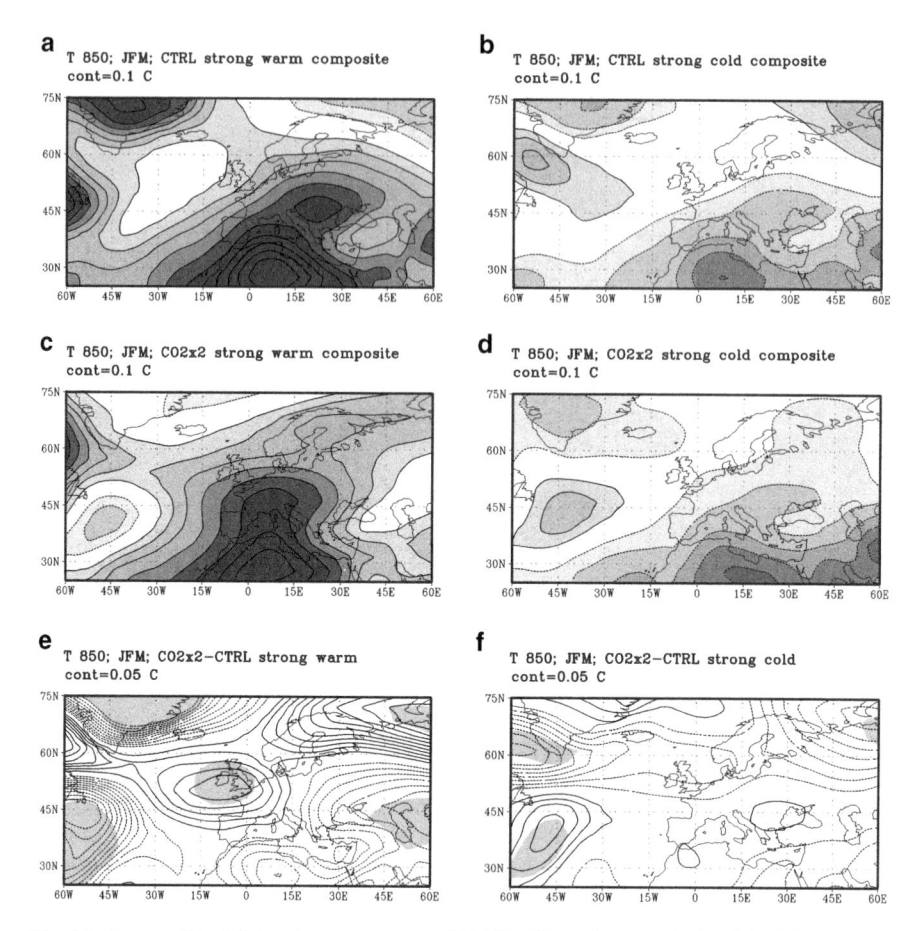

Fig. 6.4 Same as Fig. 6.3, but for temperature at 850 hPa (T850). Contours in (**a–d**) 0.1°C, in (**e**) and (**f**) 0.05°C. Values exceeding the 95% confidence level of the t statistics are *shaded* in (**e**) and (**f**)

6.3.2 Temperature and Precipitation Response

In addition to the temperature, precipitation (and associated water availability) in future (warmer) climate scenarios is one of the main requests of policy makers for adaptation and mitigation purposes. Temperature is of a particular interest for a wide range of climate change-related studies. Thus, potential modifications of ENSO impact on temperature due to warmer climate conditions might be of a great interest for such investigations. Figure 6.4a and b reveal that in the current climate the ENSO impact on the temperature at 850 hPa isobaric level (T850) is mainly linear and El Niño (La Niña) events are associated with warmer (colder) conditions over the continental part of the domain. In the climate with doubled CO_2 levels, some modifications in the temperature pattern are depicted (Fig. 6.4e, f) with

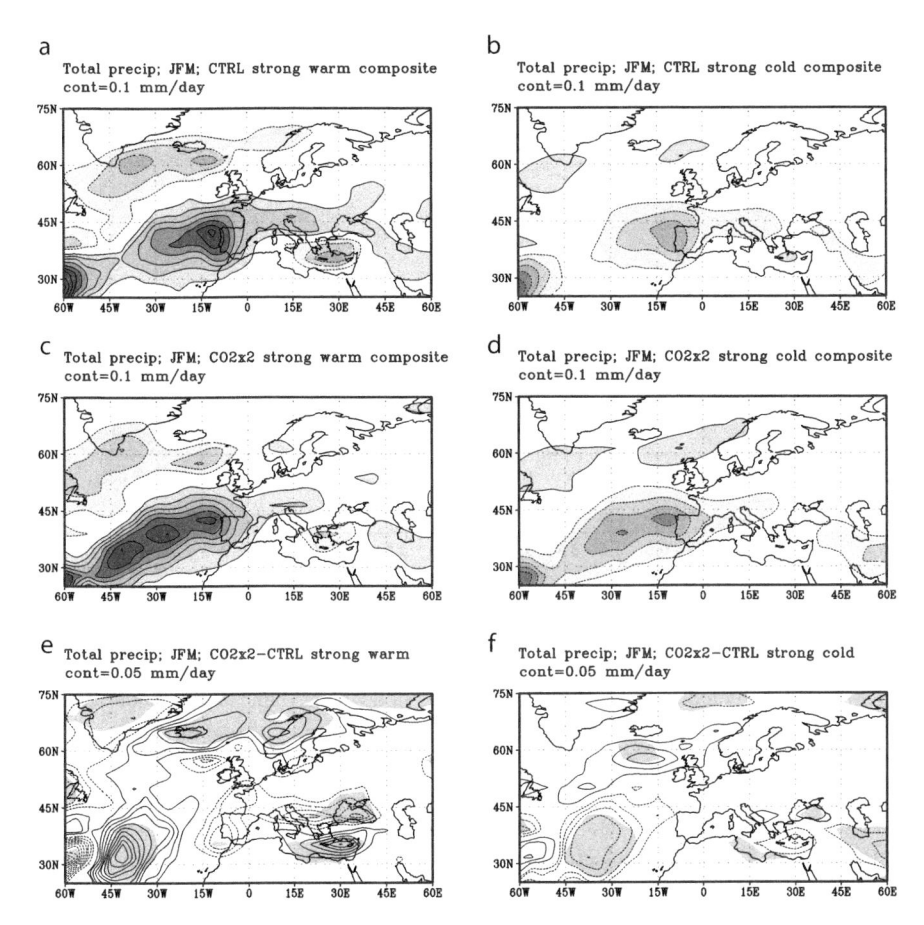

Fig. 6.5 Same as Fig. 6.4, but for total precipitation. Contours in (**a–d**) 0.1 mm/day, in (**e**) and (**f**) 0.05 mm/day. Values exceeding the 95% confidence level of the t statistics are *shaded* in (**e**) and (**f**)

statistically significant differences mainly over the Atlantic. Increased temperature response to the warm ENSO forcing is found over the United Kingdom, west Europe and north Iberian Peninsula (Fig. 6.4e).

Precipitation distribution in the NAE region is also affected by ENSO. In the current climate (CTRL), precipitation response forms two zones of the anomalies: the zone of suppressed (enhanced) precipitation around 60°N and enhanced (suppressed) precipitation around 40°N associated with the warm (cold) ENSO events (Fig. 6.5a, b). The precipitation response to the El Niño events (with the amplitude of 0.7 mm/day) is stronger than the response to the La Niña events (with the amplitude of 0.3 mm/day). In the warmer climate experiment ($2 \times CO_2$), the ENSO signature is stronger, particularly on the southern part of the domain connecting the lower left corner of the domain and Iberian Peninsula (Fig. 6.5c, d). The area that is

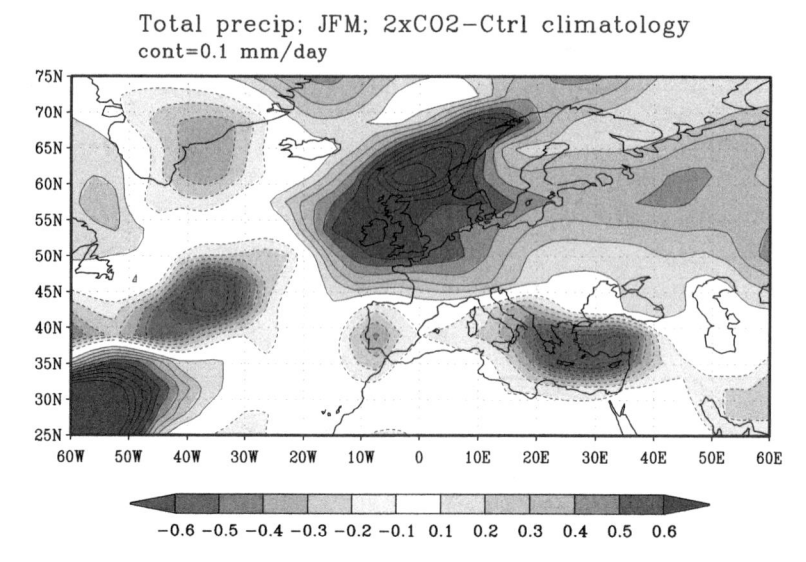

Fig. 6.6 Differences between 2×CO2 and CTRL for JFM climatological precipitation. Contours every 0.1 mm/day

significantly affected with ENSO is also larger in $2 \times CO2$ experiment. Such stronger precipitation response is associated with a strengthening of prevailing westerly winds extending from the Atlantic into the European continent. Namely, as depicted in Figs. 6.2 and 6.3, warm (cold) ENSO events are associated with the trough (ridge) over the Atlantic resulting in a stronger (weaker) Z200 and *mslp* gradient over the Atlantic (around 30°N). Consequently, zonal wind is strengthened (reduced) leading to the increased (decreased) onshore moisture advection and enhanced (suppressed) precipitation as reflected in Fig. 6.5.

The spatial distribution of change in JFM climatological precipitation (Fig. 6.6) reveals a bipolar distribution of anomalies with positive values at the north and negative values at the south indicting a wettening of the continental part of Europe and a drying of the Mediterranean region during the winter as a consequence of doubled CO_2 concentrations. Furthermore, the ENSO influence on winter precipitation (Fig. 6.5a–d) projects onto the spatial distribution of the differences between the $2 \times CO_2$ and CTRL climatological precipitation fields shown in Fig. 6.6. That suggests possibility that in future climate (for which the stronger ENSO impact is found) warm (cold) ENSO events could decrease (increase) the north-south contrast of the precipitation change.

6.4 Summary and Concluding Remarks

In this paper, possible modifications of tropical-extratropical teleconnections with the particular interest on NAE region are examined. The analysis is focused on winter (JFM) season and is based on ensemble averages of ICTP AGCM simulations performed for the current climate conditions (CTRL experiment) and for the warmer climate conditions ($2 \times CO_2$ experiment associated with doubled CO_2 levels). Detectable ENSO impact on the European climate anomalies is found for CTRL experiment. It is generally associated with decreased (increased) *mslp* over the eastern North Atlantic and west Europe. The response is barotropic over the Atlantic. Temperature response reveals warmer (colder) than usual conditions in the continental Europe associated with warm (cold) ENSO events. According to the obtained changes of *mslp*, Z200, T850 and precipitation winter anomalies, ENSO impact on the winter climate variability in the NAE region is considerably modified in the climate conditions associated with doubled CO_2 levels. In the warmer climate, significantly stronger temperature response to the both ENSO phases is found over the Atlantic, but also over the United Kingdom, west Europe and north Iberian Peninsula for the El Niño events. According to the available climate change studies based on different climate scenarios, a substantial temperature increase in a warmer climate is simulated in this part of Europe for winter season [3, 7]. In addition to temperature change caused by increased concentrations of the green-house gases, winters associated with developed El Niño events may be even warmer because of tropical-extratropical teleconnections. Also, a strong La Niña may temporarily suppress (to a certain extent) the effects of global warming.

Precipitation distribution in the NAE region is affected by ENSO as well. Thus, the zone of enhanced (suppressed) precipitation is found in the southern part of the domain and the zone of suppressed (enhanced) precipitation is found in the northern part of the domain as a result of warm (cold) ENSO forcing. Precipitation change (i.e. differences between the CTRL and $2 \times CO_2$ in JFM climatological fields for precipitation) indicates a wettening of the northern Europe and drying of the Mediterranean region. The similar spatial distribution of precipitation change associated with the warmer climate was simulated with the EH5OM climate coupled model under the IPCC A2 scenario as reported by Branković et al. [3]. They found an increase (decrease) in total precipitation north (south) of 45°N with the amplitude about 0.5 mm/day. Generally, they obtained precipitation change that corresponds to that presented here coinciding with the warm and cold composites of the precipitation shown in Fig. 6.5a–d in both the spatial distribution and the amplitude. Therefore, relative strong ENSO impact on NAE precipitation as it is found here for warmer climate conditions may be superimposed onto the change of total precipitation resulting in a weaker (stronger) precipitation modification as a consequence of a warm (cold) ENSO forcing.

At the end, it may be concluded that although the ENSO impact on climate variability in the NAE region is relatively weak, it is still detectable and statistically significant. For the warmer climate conditions associated with doubled CO_2 concentrations, even stronger impact is found over some parts of the domain. Furthermore, the precipitation response to the ENSO forcing project onto the pattern of the precipitation change provided by different climate impact studies. Thus, precipitation in future climate may be considerably affected by ENSO and therefore a proper representation of tropical-extratropical teleconnections may lead to improved climate projections.

Acknowledgements This work has been supported by the Ministry of Science, Educational and Sports of the Republic of Croatia (grants No. 119-1193086-1323).

References

1. Bourke W (1974) A multilevel spectral model. I. Formulation and hemispheric integrations. Mon Weather Rev 102:687–701
2. Bracco A, Kucharski F, Kallummal R, Molteni F (2004) Internal variability, external forcing and climate trends in multidecadal AGCM ensembles. Clim Dyn 23:659–678
3. Branković Č, Srnec L, Patarčić M (2010) An assessment of global and regional climate change based on the EH5OM climate model ensemble. Clim Change 98:21–49. doi:10.1007/s10584-009-9731-y
4. Brönnimann S (2007) Impact of El Niño-Southern oscillation on European climate. Rev Geophys 45:RG3003. doi:10.1029/2006RG000199
5. Collins M (2000) Understanding uncertainties in the response of ENSO to greenhouse warming. Geophys Res Lett 27(21):3509–3512
6. Fraedrich K (1994) An ENSO impact on Europe? A review. Tellus 46A:541–552
7. Giorgi F (2006) Climate change hot-spots. Geophys Res Lett 33:L08707
8. Gordon C, Cooper C, Senior CA, Banks HT, Gregory JM, Johns TC, Mitchell JFB, Wood RA (2000) The simulation of SST, sea ice extents and ocean heat transports in a version of the Hadley Centre coupled model without flux adjustments. Clim Dyn 16:147–168
9. Hannachi A, Turner AG (2008) Preferred structures in large scale circulation and the effect of doubling greenhouse gas concentration in HadCM3. Q J R Meteorol Soc 134:469–480
10. Hazeleger W, Severijns C, Seager R, Molteni F (2005) Tropical Pacific-driven decadel energy transport variability. J Clim 18:2037–2051
11. Held IM, Suarez MJ (1994) A proposal for the intercomparison of dynamical cores of atmospheric general circulation models. Bull Am Meteorol Soc 75:1825–1830
12. Herceg Bulić I, Branković Č (2007) ENSO forcing of the Northern Hemisphere climate in a large ensemble model simulations. Clim Dyn 28:231–254
13. Herceg Bulić I (2010) The sensitivity of climate response to the wintertime Niño3.4 sea surface temperature anomalies of 1855–2002. Int J Climatol, n/a. doi: 10.1002/joc.2255
14. Herceg Bulić I, Branković ttt, Kucharski F (2011) Winter ENSO teleconnections in a warmer climate. Clim Dyn. doi:10.1007/s00382-010-0987-8
15. Hoskins BJ, Karoly DJ (1981) The steady linear response of a spherical atmosphere to thermal and orographic forcing. J Atmos Sci 38:1179–1196
16. IPCC (2007) Climate change 2007: the physical science basis. In: Solomon S, Qin D, Manning M, Chen Z, Marquis M, Averyt KB, Tignor M, Miller HL (eds) Contribution of working group I to the fourth assessment report of the intergovernmental panel on climate change. Cambridge University Press, Cambridge, p 996
17. Kucharski F, Molteni F, Bracco A (2006) Decadal interactions between the western tropical Pacific and the North Atlantic Oscillation. Clim Dyn 26:79–91
18. Kucharski F, Bracco A, Yoo JH, Tompkins AM, Feudale L, Ruti P, Dell'Aquila A (2009) A Gill-Matsuno-type mechanism explains the tropical Atlantic influence on African and Indian monsoon rainfall. Q J R Meteorol Soc 135:569–579. doi:10.1002/qj.406
19. Meehl GA, Teng H, Branstator G (2006) Future changes of El Niño in two global coupled climate model. Clim Dyn 26:549–566. doi:10.1007/s00382-005-0098-0
20. Molteni F (2003) Atmospheric simulations using a GCM with simplified physical parameterizations. I: model climatology and variability in multi-decadal experiments. Clim Dyn 20:175–191

21. Pope V, Gallani ML, Rowntree PR, Stratton RA (2000) The impact of new physical parameterizations in the Hadley Centre climate model: HadAM3. Clim Dyn 16:123–146
22. Rayner NA, Parker DE, Horton EB, Folland CK, Alexander LV, Rowell DP, Kent EC, Kaplan A (2003) Global analyses of SST, sea ice, and night marine air temperature since late nineteenth century. J Geophys Res 108:4407. doi:10.1029/2002JD002670
23. Smith TM, Reynolds RW (2004) Improved extended reconstruction of SST (1855–1997). J Clim 17:2466–2477
24. Timmermann A, Oberhuber J, Bacher A, Esch M, Latif M, Roeckner E (1999) Increased El Niño frequency in a climate model forced by future greenhouse warming. Nature 398:694–696
25. Trenberth KE, Branstator GW, Karoly D, Kumar A, Lau NC, Ropelewski C (1998) Progress during TOGA in understanding and modelling global teleconnections associated with tropical sea surface temperatures. J Geophys Res 103:14291–14324

Chapter 7
Climate Variation or Climate Change? Evidence in Favour in the Northern Adriatic Area, Croatia

Ana Alebić-Juretić

Abstract Global warming is getting ever more concerning. Meteorological parameters such as yearly mean temperature, minimum and maximum temperature, relative humidity, yearly precipitation depth as well as number of days with rain obtained at meteorological station in Rijeka (120 m a.s.l.) were analysed. The time series include data from 1977 to 2010. The data analyses showed increase in yearly average temperature by 0.3–0.6°C, and increase of daily maximum by approx. 10°C over the period studied. In spite of declining trend in precipitation depths until 2007, the number of rainy days increased for approx 30 days at the same time, suggesting switching to warmer and more humid climate. The last 3 years data show the opposing trend thus smoothing these trends. The first evidence of climate variation and/or climate change might be the appearance of Asian tiger mosquitoes (Aedes albopictus) recently, as well as sea level elevation in the Northern Adriatic.

Keywords Climate variation • Global warming • Temperature • Precipitation depth

7.1 Introduction

The dilemma of whether the global warming is due to natural causes (e.g., Milenkovic Cycles) or due to increase of anthropogenic emissions of greenhouse gases, particularly CO_2 since the beginning of industrial revolution in the nineteenth century, is continuing. Extreme events such as strong thunderstorms and heavy draughts suggest that world's climate is subjected to changes that affect human life.

A. Alebić-Juretić (✉)
Teaching Institute of Public Health/School of Medicine, University of Rijeka, Rijeka, Croatia
e-mail: ana.alebic@zzjzpgz.hr; ana.alebic-juretic@medri.hr

H.J.S. Fernando et al. (eds.), *National Security and Human Health Implications of Climate Change*, NATO Science for Peace and Security Series C: Environmental Security, DOI 10.1007/978-94-007-2430-3_7, © Springer Science+Business Media B.V. 2012

Native populations perceive that global warming is taking a toll on their land, and coastal dwellers are subjected to increasing dangers of storm surges, flooding and/or changing river flows that affect agriculture and fisheries. Accelerated melting of ice in polar oceans are causing sea-level rise, floods and coastal erosion, which also introduces of new disease vectors [1]. Effects of global warming are already apparent on ocean organisms, in particular, during the 2005-summer heat wave in the Caribbean that caused elevation of ocean temperature for months. These unusual temperatures disturbed the usual symbiosis between the corals and a type of algae named *Zooxanthellae*. The latter supplies corals the essential nutrients that need to acquire color, since corals are usually colorless. Owing to this elevated sea temperature the algae disappeared, leaving corals to starve that resulted in coral bleaching. The process is reversible, but if the sea temperature continues to increase by additional 2°C by midcentury the coral populations dwindle past the point of no return [2]. Increase in overnight air temperature by 1.2°C during pre-breeding and breeding periods of some amphibian species resulted in altered breeding over a 30 year period in South Carolina [3].

There are three human health-related stresses expected from rising average temperature. Illness and infectious diseases carried by animal hosts and mosquitoes as well as other vectors, heath related illness and death, and health problems due to air (e.g. increased ozone formation) and water pollution (e.g. more frequent heavy downpours; [4]). We already have experience with excess mortality due to heat waves like the one in 2003 in Europe [5, 6] and China. The elevated mortality during these events often connected to air pollution is attributed to cardiovascular and respiratory diseases, especially among the elderly [7]. Evidence of climate–sensitive infectious disease carried by animal hosts or vectors is provided from observations made in China, because of warmer winters *Oncomelania hupensis* (the intermediate host of *Schistosoma japonicum*) may increase its range, thereby spreading schistomiasis to the northern part of China [7]. On the other hand, Hantaviruses are the causative agents of hemorrhagic fever with renal syndrome (HFRS) in Eurasia. Transmission of Hantaviruses from rodents to humans depends on rodent host population that is correlated to climate variability and climate change [8]. Several heat waves experienced in recent years were the principal cause to analyse meteorological data from Rijeka in search of evidence in favor of climate variation or climate change. The period analysed was 1977–2010.

7.2 Methods

Meteorological data for the period 1977–2007 were taken from the Regional statistical yearbooks ([9, 10], 1983–2008), while 2008–2010 data were provided by the State Meteorological and Hydrological Service and State statistical informations ([11], 2010).The data included: annual temperature; temperature maximum and minimum, annual relative humidity annual precipitation depth, and number of rainy days per year. The period analysed was 1977–2010. The given data were compared to few historical datasets on the same parameters that were measured at the meteorological station within Naval Academy (approx. 15 m a.s.l.) in the second half of the 19th [12, 13] and the turn of the twentieth century (Mittheilungen 1899–1904). As the available data are scarce, the obtained results can be only indicative for meteorological trends analysed.

7.3 Results

7.3.1 Temperature

The annual average temperature shows an increase since 1977, but gets somewhat smoother since 2000 (Fig. 7.1a), thus following the global trends. There is almost a 1.5°C rise in annual means within the period studied, ranging from 13.5°C to 15.0°C. The average temperature for a given 34 year period is 14.2°C. The same value is obtained for a 30-year long term climatologic period 1981–2010.

The last two decades appear to be the warmest with average temperature of 14.5°C, while the last decade (2001–2010) has the highest average temperature of 14.7°C, with temperature anomaly of +1.1°C above climate normal 1961–1990 [14], confirming the warmest decade worldwide. Contrary to global data, the year 2010 does not appear to be the warmest [16] in Rijeka, presumably due to highest precipitation depth ever measured. Four of five warmest years are registered in the last decade (2003, 2007 and 2009) and 2000 with average annual temperatures above 15°C, while the fifth was obtained in the early 1872, with absolute maximum of 15.5°C [13]. Unfortunately, missing the original set of data, it is hard to validate the above result. The comparison of annual temperature maxima with a few historical data available shows that they are within the same range (Fig. 7.1b). Similar behaviour of annual mean temperatures with maxima in the second half of nineteenth

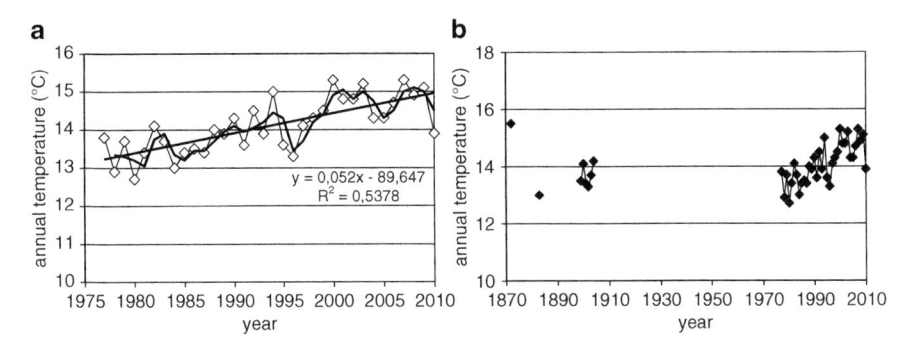

Fig. 7.1 Annual mean temperature: (**a**) for the period 1977–2010 with 2-years average and regression line and (**b**) with historical data included

Table 7.1 Monthly mean temperature profile (°C) for given multi-year periods

Period	1896–1884	1986–1892	1899–1904	1961–1990	1991–2010
January	5.6	5.2	5.4	5.3	5.3
February	6.9	6.2	6.6	6.1	6.4
March	8.6	8.5	8.6	8.5	8.8
April	13.0	12.7	12.9	12.2	12.6
May	16.5	16.9	16.7	16.6	16.4
June	20.6	20.5	20.6	20.1	20.2
July	24.0	23.7	23.9	22.8	23.3
August	23.1	22.9	23.0	22.3	22.6
September	19.2	19.2	19.2	18.9	18.8
October	14.6	14.3	14.5	14.4	14.6
November	9.9	9.6	9.8	9.8	9.2
December	6.9	6.5	6.7	6.5	6.8
Year	14.1	13.9	13.9	13.6[a]	14.5

[a] Reviews, No 10, 2001

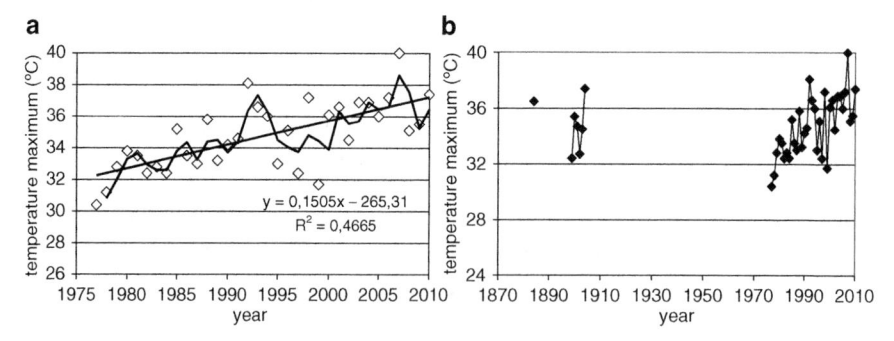

Fig. 7.2 Maximal temperature: (**a**) for the period 1977–2010 with 2-years running average and regression line and (**b**) with historical data included

century and the beginning of twentieth century are found in the complete temperature time series (1861–2009) for Zagreb [16].

The temperature rise is estimated to 0.3–0.6°C comparing the yearly average temperatures profile for period 1977–2010, and/or the warmest two decades (1991–2010) relative to the longest historical data from 1869 to 1892 (Table 7.1), respectively. The monthly profiles are very close for all the multi-year time series available. The comparison of average temperature of summer months (July and August) and climate normal (1961–1990) temperature indicate that Rijeka is the coolest.

The maximum temperatures have also increased from 30.4°C in 1977 to 40°C in 2007 (Fig. 7.2), but maxima above 35°C were recorded at the turn of the twentieth century (36.5°C in 1884; 35.4°C in 1900, 37.4°C in 1904). Minimal temperature felled only ones bellow −10°C (minimum: − 11.0°C in 1985), and also show a slight increasing trend (not reported).

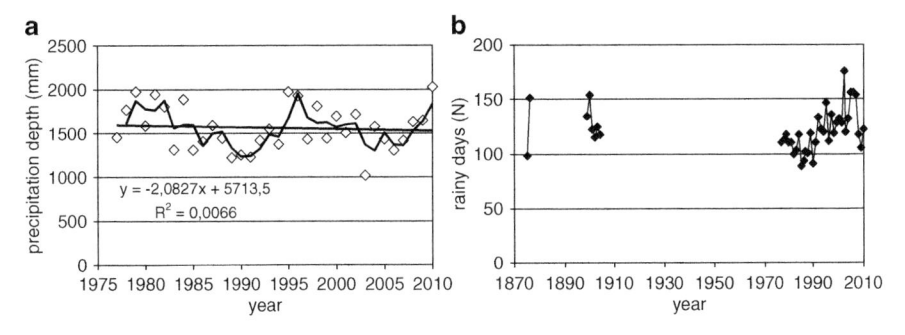

Fig. 7.3 Annual precipitation depth: (**a**) for the period 1977–2010 with 2-years running average and regression line and (**b**) with historical data included

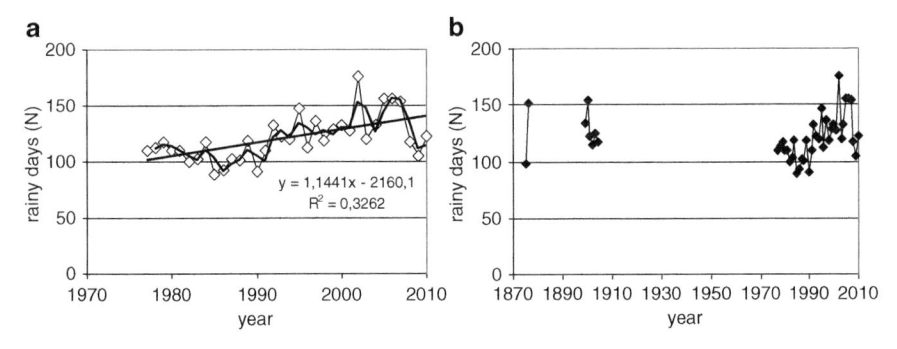

Fig. 7.4 Number of rainy days per year: (**a**) for the period 1977–2010 with 2-years running average and regression line and (**b**) with historical data included

7.3.2 Precipitation

The precipitation depth in Rijeka showed a decline in the period 1977–2010, although this decline is smoothed by the last 3 years (Fig. 7.3). The annual precipitations were in the range 1,021 mm in 2003 and 2,115 mm in 2010. Historical data also are within this range with maximum (2,100 mm) in 1882, practically equal to the 2010 maximum. At the same time, the number of rainy days increased from approx 110 at the end of the 1970s to 156 by '2005–2006, following by a sudden drop to 105–122 in the period 2008–2010 (Fig. 7.4a). The number of rainy days in historical data (Fig. 7.4b) falls in the same range with maxima in 1876 (151 days) and 1900 (154 days). Minimum rainy days are evidenced in 1985 and 1986 (89 and 93, respectively) and 1990 (91 days).

Comparing the monthly precipitation profiles for the multi annual periods with the climate normal 1961–1990 (Fig. 7.5), the change in precipitation is evident for the two oldest time series having drier January and February and wetter April and October. The 6 year time series at the turn of the century shows more precipitation in February and March, and drier November relative to the climate normal

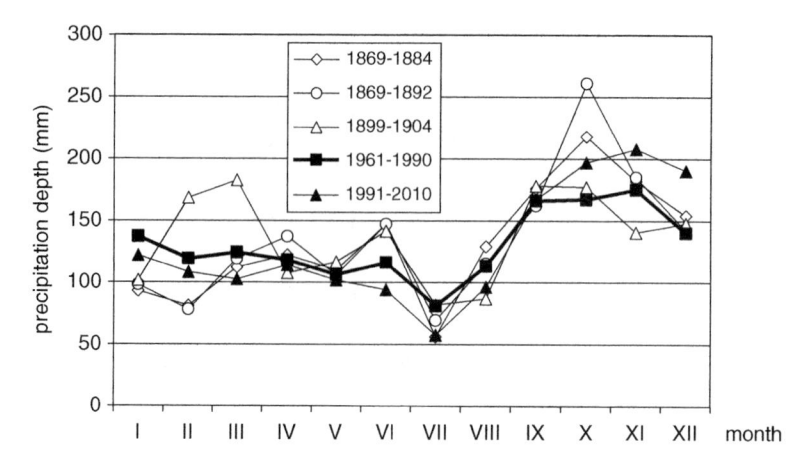

Fig. 7.5 Monthly precipitation depth profiles for the multi year periods analysed

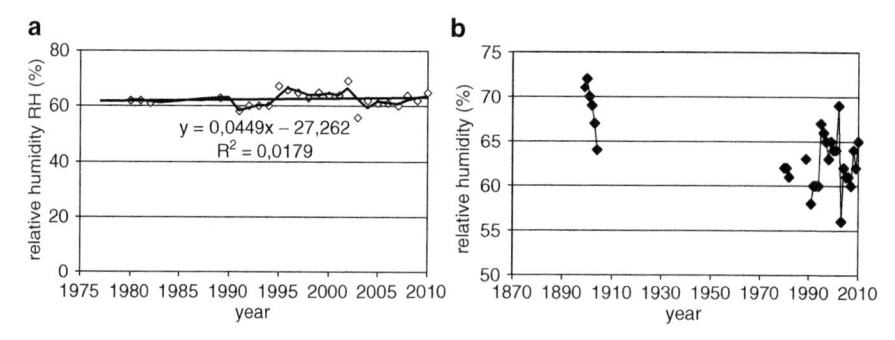

Fig. 7.6 Annual relative humidity (RH %): (**a**) for the period 1977–2010 with 2-years running average and regression line and (**b**) with historical data included

1961–1990. The precipitation profile for 1991–2010 exhibit wetter period from October to December. The general feature for all five time series available is the precipitation minimum in July and maximum in October.

7.3.3 Relative Humidity

The relative humidity (RH %) also shows a slightly increasing trend over the period 1977–2010 (Fig. 7.6a). The RH values are within the range 56–69%. It is interesting to note that two extreme values were observed in two subsequent years: the maximal value was obtained in 2002, a year with maximum rainy days (N = 176), and 2003, a year with second (or third, if historical data considered) highest annual mean temperature (T = 15.2°C), and the minimum precipitation depth. This very same year is known for the extreme hot waves that caused excessive death in Europe.

The historical data exhibit RH above 70% in 1899 and 1900, which looks a bit too high from the nowadays point of view. Since the multi year periods 1869–1884 and 1869–1892 gave the RH means of 68% and 69% respectively [12, 13] the values above 70% seem to be justified (Fig. 7.6b). More details on instrumentation used should be available for validation of these data.

7.3.4 Indication of Warming in the Northern Adriatic Area

The expected consequences of global warming are: draught, changes in precipitation patterns, rise of sea level, spread of insects and rodents as vectors of some infectious diseases. Extension of pollination period due to warmer climate might cause increase in allergic asthma incidence. Furthermore, in combination with air pollution, heat wave episodes may lead to excessive death due to respiratory and cardiovascular disease.

Long term measurements of the sea level at the nearby Bakar (10 km east from Rijeka) showed an increase of 31 mm in the period 1950–1995 i.e. 0.7 mm per year. The increasing trend was lower comparative to the previous period of 1930–1971 (1.56 mm/year), but the variations seem to be caused by local climatic factors. Therefore the author could not find any evidence of global climate change impact [15]. According to some geophysicist, recent measurements indicate again higher (double) sea level rise rate that might be attributed to climate change (www.vjesnik, [16]).

Another evidence of climate changes is the appearance of some insects that were previously absent in the Southern European area. Such a species is an Asian tiger mosquito (Aedes albopictus), known to be a vector for chikungunya virus that causes an infectious diseases. Besides that, its bites can cause strong allergic reaction to sensitive persons. Therefore, Asian tiger mosquitoes represent a public health threat, and its identification is essential for the proper treatment with insecticides. For the first time the Asian Tiger mosquito was identified in Zagreb in 2005, in 2007 in the neighbour Istria peninsula and in 2008 in our County [17].

The excess mortality for the August 2003 heat wave in Rijeka was estimated to 2–3 deaths, but with the particularity that ozone levels were lower and PM_{10} higher relative to values determined in Zagreb [18]. The hospitalization because of asthma (caused by allergy) and gastrointestinal diseases including diarrhea in the last decade (2000–2009) in Rijeka do not support yet the thesis of climate change impact. For such a purpose more precisely statistic records and longer periods are needed.

7.4 Conclusion

With an incomplete meteorological data set it is hard to arrive at conclusions with regards to climate change. There is an increase of annual and maximum temperature during the period 1977–2010, and the same is true for the last decade (2001–2010), which is the warmest period on the globe. Comparing these values with historical data available show that climate normal for the 1961–1990 covered the coolest period. The same is valid for precipitation. The high precipitation during 2008–2010 reduced the decreasing trend of 1977–2007, while the number of rainy days dropped to approximately 110 at the same time but still preserving the increasing trend. Comparing with the historical data, both datasets fall in the same range.

Although sea level elevation rate is approximately double in last two decades comparative to the previous period 1950–1995, it seems to approach the rate of the previous period 1930–1970 explained by local climatic factors.

The spreading of insects like Asian tiger mosquito (Aedes albopictus) originating from warmer climate was already registered in the Northern Adriatic, as a result of global trade. There is no clear evidence yet of harmful effects on human health due to increased air temperature in the last decade, except the estimation for the 2003 heat wave.

All these factors are in favour of climate variation rather than climate change in the Northern Adriatic area.

References

1. Weinhold B (2010) Climate change and health: a native American perspective. Environ Health Perspect 118(2):A64–A65
2. Schmidt CW (2008) In hot water: global warming takes a toll in coral reefs. Environ Health Perspect 116(7):A292–A299
3. Nature (2010) Research highlights; hotter climate, altered breeding 486:1004
4. Coonley CM (2010) Health scenario for a warming world. Environ Health Perspect 118(9):A382
5. Fisher PH, Brunekreef B, Lebert E (2004) Air pollution related deaths during 2003 heat wave in the Netherlands. Atmos Environ 38:1083–1085
6. Stedman JR (2004) The predicted number of air pollution related deaths in the UK during the August 2003 heat wave. Atmos Environ 38:1087–1090
7. Kan H (2011) Climate change and human health in China. Environ Health Perspect 119(2):A60–A61
8. Zhang W-Y, Guo W-D, Fang L-Q, Li C-P, Glass GE, Jiang J-F, Sun S-H, Quian Q, Liu W, Yan L, Yang H, Tong S-L, Cao W-C (2010) Climate variability and hemorrhagic fever with renal syndrome transmission in Northeastern China. Environ Health Perspect 118(7):915–920
9. Statistical yearbook for Istria peninsula, coastal and highlands area, 1989, Rijeka, November 1989; ibid. 1992 (in Croatian)
10. Statistical yearbook of the Littoral-Highlands County (1996) Rijeka, 1996; ibid. 1999, 2002, 2005 and 2008 (in Croatian)
11. Statistical information (2009) Zagreb, 2009; ibid 2010
12. Mittheilungen des Naturwissenschaftlichen Clubs in Fiume (1896), Emidio Mohovich, Fiume: ibid. (1899), Erlau, 1900: ibid. (1900). P. Battara, Fiume; ibid. (1901), P. Battara, 1902; ibid. (1902), P. Battara, Fiume, 1903; ibid. (1903) P. Battara, 1903; ibid. (1904), Eger, 1905
13. Salcher P (1884) Das Klima von Fiume-Abbazia nach meteorologischen Beobachtungen. Emidio Mohovich, Fiume (now Rijeka)
14. Reviews No 10 (2001) Climate monitoring and assessment for 2000, Zagreb, January 2001: ibid. No 21 2011 (in Croatian)

15. Segota T (1996) Sea level of the Adriatic sea indicated by Bakar tide-gauge data, Croatia. Geagrafski glasnik 58:15–32 (in Croatian)
16. Reviews (2011) www: vjesnik.hr/html/2011/02/03 : Adriatic sea level could rise by a meter until 2100 (in Croatian)
17. Susnic V, Vuletic J, Kauzlaric G, Klobučar A (2008) Presence of Aedes albopictus in the littoral – highlands area, professional meeting with international participation "Pest control 2008", Opatija, Nov 6th–Nov 7th, 2008, In: Proceedings of the "Pest control", pp 61–69 (in Croatian)
18. Alebic-Juretic A, Cvitas T, Kezele N, Klasinc L, Pehnec G, Sorgo G (2007) Atmospheric particulate matter and ozone under heat-wave conditions: do they cause an increase of mortality in Croatia? Bull Environ Contam Toxicol 79:469–471

Chapter 8
Atmospheric Aerosol Climatology over the Globe: Emphasis on Dust Storms

Harry D. Kambezidis, Dimitra H. Kambezidou, and Stella-Joanna H. Kampezidou

Abstract Atmospheric aerosols play a major role in climate change science debate. These can influence climate in two ways, directly and indirectly. Since the concentration and composition of atmospheric aerosols are very variable in time and space, their characteristics cannot be studied individually, but in terms of their climatological effects. This study gives some of the main features that characterize atmospheric aerosols. Four locations in the world are selected due to their different environments (weather and atmospheric conditions). The aerosol optical depth and the Ångström-exponent are examined at these sites for the period 2002–2004. Emphasis is given on the atmospheric aerosols in the form of dust. The aerosol optical depth and the aerosol index are presented for a dust storm that occurred over Greece in April 2005.

8.1 Introduction

The cooling effect that aerosols have on the surface of the earth due to the direct reflection of solar radiation is referred to as the *direct effect* or *direct climate forcing*. Aerosol particles also influence the size, abundance and rate of production of cloud droplets. Thus, they influence cloud cover, cloud albedo and cloud lifetime. The effect of aerosols on the radiative properties of earth's cloud cover is referred to as the *indirect effect* of aerosols, or *indirect climate forcing*. Improved aerosol

H.D. Kambezidis (✉)
National Observatory of Athens, Athens, Greece
e-mail: harry@meteo.noa.gr

D.H. Kambezidou
University of the Aegean, Lesbos Island, Greece

S.-J.H. Kampezidou
University of Patras, Patras, Greece

H.J.S. Fernando et al. (eds.), *National Security and Human Health Implications of Climate Change*, NATO Science for Peace and Security Series C: Environmental Security, DOI 10.1007/978-94-007-2430-3_8, © Springer Science+Business Media B.V. 2012

climatologies may enable more accurate estimations of the direct and indirect aerosol forcings [10].

Two parameters are important for the investigation of atmospheric aerosols. (i) The aerosol optical depth (AOD) is an indicator of the aerosol loading in the vertical column of the atmosphere and constitutes the main parameter to assess the aerosol radiative forcing (ARF) and its impact on climate. (ii) The Ångström-exponent α [8, 9] constitutes a good indicator of the aerosol size and its variations [14]. At global scale, four main aerosol types with different characteristics, based on AOD and α estimations can be detected: biomass burning, urban/industrial, maritime and desert-dust aerosols [5].

In the last two decades extensive field measurements and campaigns have been conducted providing significant scientific knowledge about global aerosol properties. Field experiments have provided the most comprehensive data of aerosol properties and their radiative impact on climate. Nevertheless, they are limited in temporal and geographical extent. For this reason, an organized global network of sun-photometers, the Aerosol Robotic Network (AERONET), has been established by NASA aiming at characterizing the aerosol properties and maintaining a continuous database, useful for aerosol climatologies. The AERONET retrieves several aerosol optical properties [2, 4], but it does not give information about the aerosol chemical composition.

Both AOD and Ångström-exponent α vary widely depending on atmospheric turbidity and aerosol size. Thus, high values of AOD are characteristic of turbid atmospheres affected by biomass burning, dust plumes or urban pollution. Values of α near zero correspond to coarse-mode aerosols (sea spray and desert dust) indicating a neutral AOD wavelength dependence, while values of $\alpha > 1.3$ indicate significant presence of fine-mode particles (mainly smoke or urban aerosols). The values of both parameters exhibit strong wavelength dependence [3] and are characteristic of the different aerosol types.

The first part of the present study examines the climatology of aerosols at a rather global scale for a period of almost 3 years over selected locations according to findings by [6]. The data set comes from four locations: (i) Alta Floresta, Brazil, a site directly influenced by biomass burning during the fire season; (ii) Ispra, Italy, an urban/industrial area with significant anthropogenic activities; (iii) Nauru, a remote island in the Pacific Ocean representative of very low aerosol loading; and (iv) Solar Village, a continental remote area in Saudi Arabia with significant contribution of desert particles. A similar study has been published by [2].

The second part of the present study focuses on dust aerosols. Desert-dust aerosols are probably the most abundant and massive type of aerosol particles that are present in the atmosphere worldwide. Therefore, dust, which is a common aerosol type over deserts [13], is considered to be one of the major sources of tropospheric aerosol loading and constitutes an important key parameter in climate aerosol forcing studies [10]. The Sahara desert is the most important dust source in the world and mainly in the Mediterranean region. Exports of dust plumes to North Atlantic and Mediterranean occur every year [12]. The desert-dust aerosol characteristics of a specific dust storm over Greece is examined here following the findings by [8, 9].

8.2 Data Collection and Analysis

The data for the first part of the study have been obtained from the CIMEL sun-photometers of AERONET. The CIMEL sun-photometer takes measurements of the direct-beam radiances in the spectral range 340–1020 nm. Typically, the total uncertainty in spectral AOD is about ±0.01 to ±0.02 and is wavelength dependent with higher errors in UV. The calibration accuracy becomes an obstacle because it causes an error in measuring AOD (±0.01) that is at least of the order of 5–10% of the calculated AOD at 440 nm <0.2 and comparable with the absorption partition in the total optical depth [2]. From the AERONET data, a single Ångström-exponent α is derived in the 440–870-nm band ($\alpha_{440-870}$) from the linear regression of ln(AOD)-ln(λ) data at some wavelengths using the least-squares method; λ is wavelength in mm. The errors in determining α mainly depend on atmospheric turbidity, increasing for low AOD values. For this study, AOD and α data have been used for four selected locations of AERONET, namely Alta Floresta, Ispra, Nauru and Solar Village. Table 8.1 gives the period and number of measurements at each location. The criterion for the selection of the above stations has been the differentiation of the prevailing atmospheric and climatological conditions. Thus, Alta Floresta is a rural site in Brazil affected by biomass-burning aerosols; Ispra is an urban/industrial site in Italy affected by anthropogenic aerosols mainly; Nauru is a remote island in the Pacific affected by maritime aerosols; and Solar Village is a rural area in Saudi Arabia affected by desert-dust aerosols.

The second part of the study examines a severe dust event over Greece occurred in April 2005 (16–17). For this investigation AOD at 550 nm (AOD_{550}) data from MODIS sensor on-board Terra and Aqua spacecrafts have been obtained for April 2005. Also, the Aerosol Index (AI) from TOMS and later OMI instruments on-board Aura spacecraft is used. AI is a measure of the backscattered UV radiation by atmospheric aerosols; AI>0 indicates presence of absorbing aerosols (such as black carbon and dust), while AI<0 presence of scattering ones (such as sulphates). Figure 8.1 shows the "invasion" of Saharan dust over Greece.

Table 8.1 Period of measurements and number of observations [6]

Location	Period	Total	Winter	Spring	Summer	Fall
Alta floresta (9.52° S, 56.06° W, 277 m)	2.1.2002–4.12.2004	561	90	124	181	166
Ispra (45.48° N, 8.37° E, 235 m)	2.1.2002–4.3.2004	653	148	165	217	123
Nauru (0.31° S, 166.54° E, 7 m)	2.3.2002–4.5.2004	363	78	131	79	75
Solar village (24.54° N, 46.24° E, 650 m)	2.1.2002–4.8.2004	792	169	242	209	172

Fig. 8.1 Image from Aqua-MODIS sensor on 17 April 2005 (11.40 UTC). The *yellowish* stream on the *right* is the Saharan dust transported over Greece

8.3 Results

8.3.1 Aerosol Climatology

This section gives the daily as well as the monthly variation of AOD_{500} and $\alpha_{440-870}$. These two parameters establish the differences in the aerosol climatologies that exist at the selected four locations. Figure 8.2 shows the variability of the daily-averaged AOD_{500} at each site.

The daily mean AOD_{500} values at Alta Floresta show little variation except for the biomass-burning periods, where maximum values occur. The highest values are observed in 2002 (0.44) followed by those in 2004 (0.37) and 2003 (0.29). The large variability in AOD_{500} in the burning season is attributed to changes in air-mass trajectories, with advection from the burning regions [15] as well as to the intensity of fires, the fire phase, the prevailing wind pattern, the physicochemical characteristics,

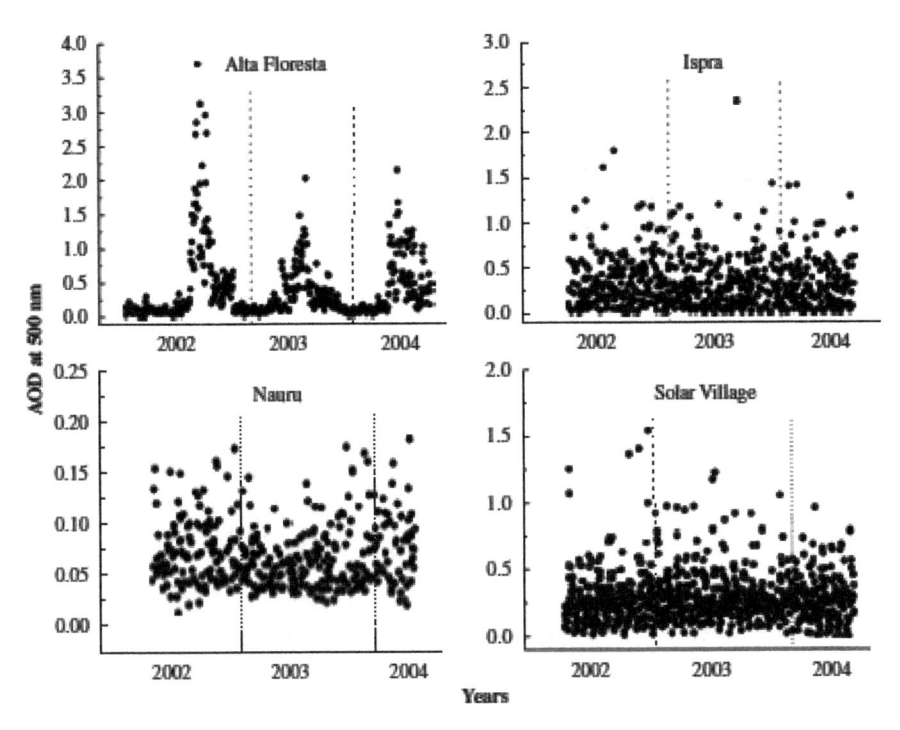

Fig. 8.2 Mean daily values of AOD_{500} for the period of measurements at the selected locations (From [6])

the humidity of the fuel, the duration of the fire and the mixing processes in the atmosphere [14]. In Ispra, the AOD_{500} values show large daily variability attributed to the large variability of the local sources and anthropogenic activities in conjunction with the prevailing meteorological conditions (wind speed/direction, atmospheric stability) and the scavenging processes in the atmosphere such as precipitation. This day-to-day variability is observed in all seasons in all 3 years. In Nauru, the AOD_{500} variability is mainly driven by the prevailing weather conditions affecting the area. Here the AOD_{500} values are relatively low [a mean annual value of 0.07 close to that observed in Pacific maritime environments, [17]]. In Solar Village, the AOD_{500} values do not usually exceed the threshold of 0.5. This occurs because the location is a rural remote site and AOD_{500} is, therefore, influenced by natural factors only.

Figure 8.3 shows the daily variation of $\alpha_{440-870}$ at the same sites. A general observation is the large variation of the parameter at all selected locations.

The large variability in $\alpha_{440-870}$ at the site of Alta Floresta shows a discernible pattern with maximum values in the periods of August-September, indicating presence of fine-mode smoke particles coming from the burning areas. Reid, Eck et al. [14] have found that the Ångström exponent for smoke particles in Brazil ranges from 2.5 for fresh smoke to as low as 0.5 for aged smoke. In contrast to Alta Floresta,

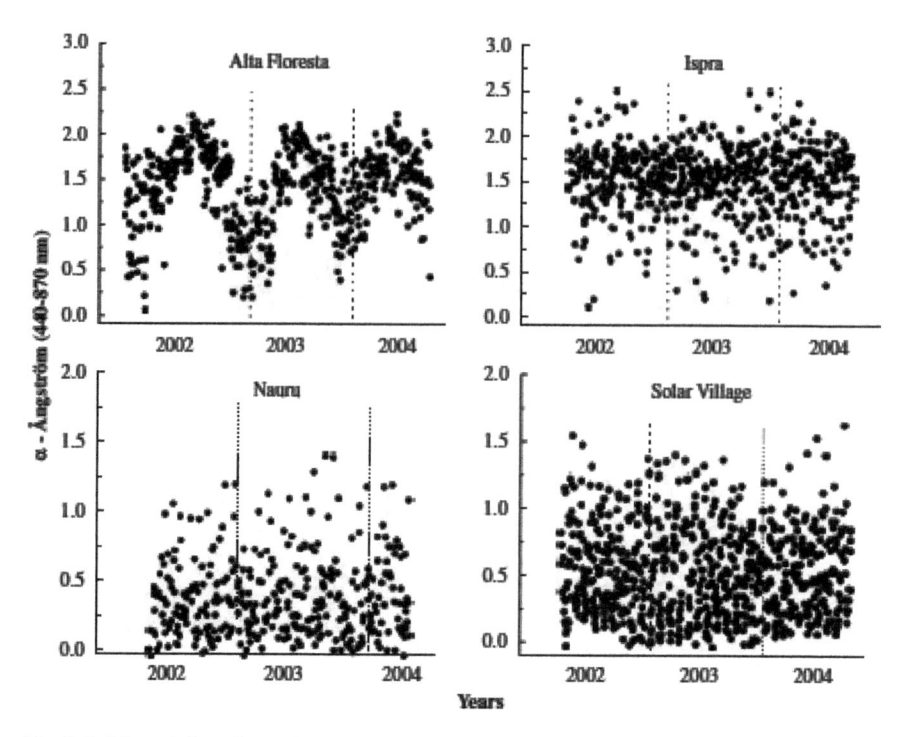

Fig. 8.3 Mean daily values of $\alpha_{440-870}$ for the period of measurements at the selected locations (From [6])

the variation of $\alpha_{440-870}$ at the site of Ispra shows no clear pattern exhibiting large day-to-day variability. The presence of fine-mode aerosols is due to anthropogenic combustion processes, while the coarse-mode ones to the growth of fine particles and their interaction with atmospheric moisture [1]. At Nauru also, there is no clear pattern. Such a great variability, as in Fig. 8.3, has been reported by [17] over oceanic regions. The condition $\alpha_{440-870} > 1$ indicates presence of accumulation- rather than coarse-mode particles. Such particles can be of maritime origin (natural sulphate aerosols), which are smaller in size than sea spray, also exhibiting high values of α [16]. In contrast to the above sites, Solar Village shows extremely high variability in α. This implies a high aerosol-size distribution from pure coarse-mode dust particles ($\alpha < 0.3$) to fine-mode anthropogenic aerosols ($\alpha > 1.5$).

8.3.2   Dust Storm

Figure 8.4 shows the daily values of AOD_{550} (AOD at 550 nm from the MODIS retrievals) for April 2005 as a spatial average covering the area 32–41° N, 20–27° E, which includes continental Greece and the Aegean and Libyan Seas. The Terra-MODIS AOD_{550} values vary between 0.175 and 1.94 and those of the

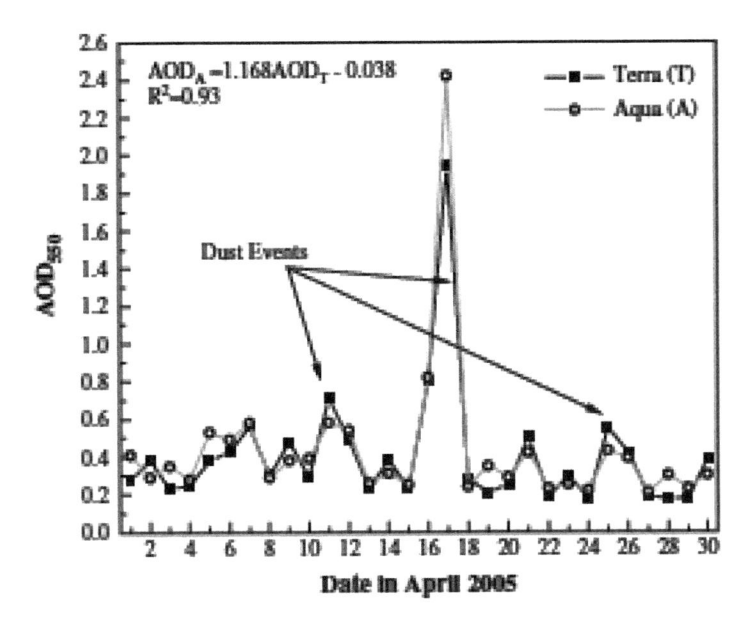

Fig. 8.4 Variation of AOD$_{550}$ from Terra- and Aqua-MODIS sensors over Greece and Aegean/Libyan seas in April 2005 (From [7])

Aqua-MODIS from 0.21 to 2.42. The values of AOD$_{550}$ >0.5 over the area correspond to dust-storm events. The differences in AOD$_{550}$ values between the two sensors are due to their orbiting times and the presence of clouds in the pixels of any of the two satellite images.

Figure 8.5 shows the variation of AI in April 2005 over the same region. These values are spatial averages; they range from 0.01 to 1.58 corresponding to the mixture of absorbing aerosols (dust or soot) that are expected to affect the whole area. The highest values are observed during the dust events.

Specifically for the April 17 intense dust event, the distribution of AOD$_{550}$ and AI over the region is shown in Fig. 8.6. The white gaps in the MODIS figure correspond to lack of data due to cloud contamination of the satellite images. In Fig. 8.6 (left panel) the high AOD$_{550}$ values clearly indicate the dust plume, which covered Mediterranean and Greece and extended over the Balkans. AOD$_{550}$ takes a value of around 2 over Athens on 17 April 2005, causing reduction in visibility and degradation of the air quality [7]. In relation with the MODIS data, the AI values (Fig. 8.6, right panel) show the presence of highly absorbing aerosols in the UV over the region. Such high values correspond to thick dust layers [11, 18].

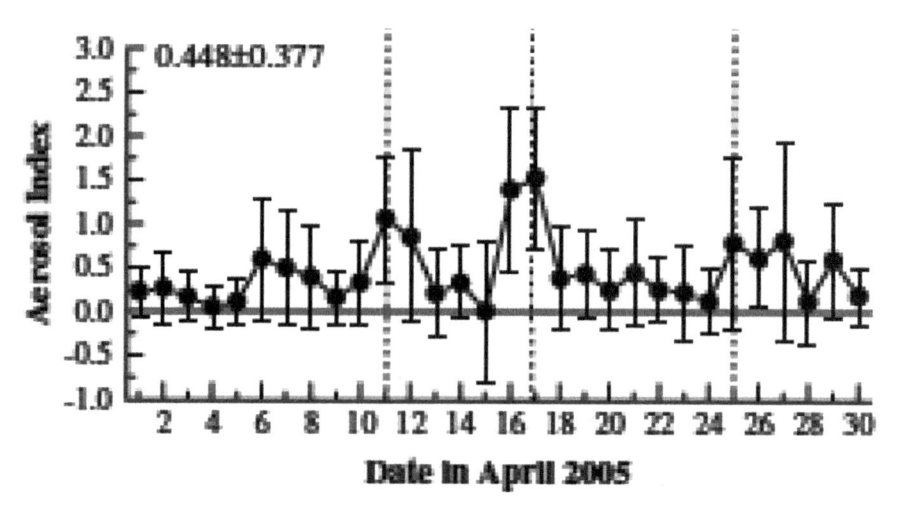

Fig. 8.5 Variation of AI from TOMS sensor over Greece and Aegean/Libyan seas in April 2005 (From [7])

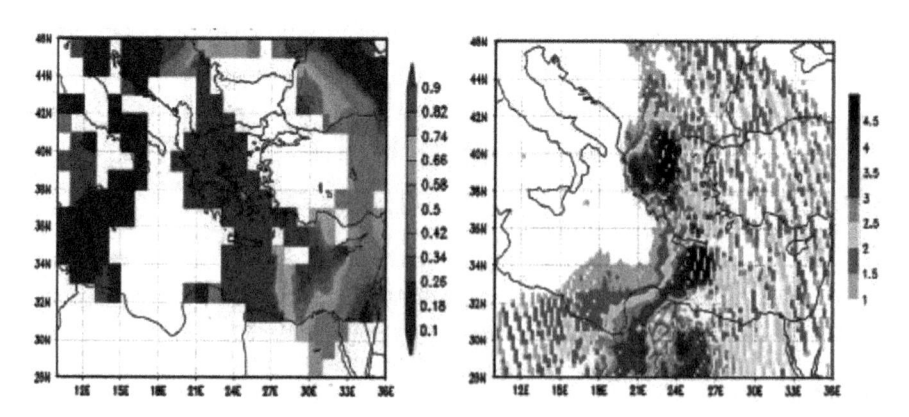

Fig. 8.6 Distribution of AOD$_{550}$ (*left panel*) from Aqua-MODIS sensor and AI (*right panel*) from TOMS instrument over eastern Mediterranean on 17 April 2005 (From [7])

8.4   Conclusions

The present study has shown that different aerosol climatologies exist over the various areas of the world. The reason is the different weather patterns, the differing chemical composition of the suspended particles in the atmosphere as well as the various atmospheric processes (coagulation, scavenging, mixing). All of these result in the prevalence of different types of aerosols over an area. The immediate result of a specific type of aerosols over an area for extended periods of time (e.g., months or seasons) is an influence on the weather known as weather modification; this occurs because of the direct effect (disturbance of the local radiative flux by reflecting or absorbing solar light) and indirect effect (interaction with clouds) of aerosols.

Remote places in the world show little aerosol variation and low values of atmospheric turbidity, expressed in terms of AOD (in this study Alta Floresta, Nauru and Solar Village). In contrast, urban sites have higher atmospheric turbidity and are influenced by anthropogenic type aerosols (in this study Ispra). The AOD500 values at Ispra reached even 2.5 with values of $\alpha > 1.5$.

As for dust storms, these carry dust particles to long distances. Greece suffers from such dust events several times a year. The episode analysed in this study (17 April 2005) was a severe one, causing visibility degradation and problems to human health. The high AI values (around 1.5) indicated the presence of aerosols that highly absorb solar UV radiation.

References

1. Day DE, Malm WC (2001) Aerosol light scattering measurements as a function of relative humidity: a comparison between measurements made at three different sites. Atmos Environ 35(30):5169–5176
2. Dubovik O, Holben B et al (2002) Variability of absorption and optical properties of key aerosol types observed in worldwide locations. J Atmos Sci 59(3):590–608
3. Eck TF, Holben BN et al (1999) Wavelength dependence of the optical depth of biomass burning, urban, and desert dust aerosols. J Geophys Res 104(D24):31333–31349
4. Holben BN, Tanré D et al (2001) An emerging ground-based aerosol climatology: aerosol optical depth from AERONET. Geophys Res 106((D11):12067–12097
5. Ichoku C, Kaufman YJ et al (2004) Global aerosol remote sensing from MODIS. Adv Space Res 34(4):820–827
6. Kambezidis HD, Kaskaoutis DG (2008) Aerosol climatology over four AERONET sites: an overview. Atmos Environ 42(8):1892–1906
7. Kaskaoutis DG, Kambezidis HD et al (2008) Study on an intense dust storm over Greece. Atmos Environ 42(29):6884–6896
8. Kaskaoutis DG, Kambezidis HD (2008) Investigation of the aerosol optical properties under certain conditions in Athens, Greece. Conference SOLARIS. Hong Kong, China, pp 179–186
9. Kaskaoutis DG, Nastos PT et al (2010) Meteorological patterns associated with intense Saharan dust outbreaks over Greece in winter. Conference COMECAP. Patras, Greece, pp 1039–1047, ISBN: 978-960-99254-0-2
10. Kaufman YJ, Tanre D et al (2002) A satellite view of aerosols in the climate system. Nature 419(6903):215–223
11. Koukouli ME, Balis DS et al (2006) Aerosol variability over Thessaloniki using ground based remote sensing observations and the TOMS aerosol index. Atmos Environ 40(28):5367–5378
12. Moulin C, Lambert CE et al (1998) Satellite climatology of African dust transport in the Mediterranean atmosphere. J Geophys Res 103(D11):13137–13144
13. Ogunjobi KO, He Z et al (2008) Spectral aerosol optical properties from AERONET Sun-photometric measurements over West Africa. Atmos Res 88(2):89–107

14. Reid JS, Eck TF et al (1999) Use of the Ångstrom exponent to estimate the variability of optical and physical properties of aging smoke particles in Brazil. J Geophys Res 104(D22)): 27473–27489
15. Remer LA, Kaufman YJ et al (1998) Biomass burning aerosol size distribution and modeled optical properties. J Geophys Res 103(D24)):31879–31891
16. Russell LM, Pandis SN et al (1994) Aerosol production and growth in the marine boundary layer. J Geophys Res 99(D10):20989–21003
17. Smirnov A, Holben BN et al (2002) Optical properties of atmospheric aerosol in maritime environments. J Atmos Sci 59(3):501–523
18. Torres O, Bhartia PK et al (1998) Derivation of aerosol properties from satellite measurements of backscattered ultraviolet radiation: theoretical basis. J Geophys Res 103(D14):17099–17110

Chapter 9
Climate Change and Adoption Strategies – A Report from the Republic of Serbia

Vesela Radovic

Abstract This paper deals with global climate change, which is one of the most pressing environmental, economic, political and social issues of the world. Serbia is no exception, and there is obvious evidence about climate changes in Serbia. The paper presents the way the Serbian Government approaches the Kyoto Protocol, following an approach similar to the EU and the accepted policy of the world. Legislation presented at the beginning of the paper shows the results of these activities. A short overview of climate change impacts on certain business sectors is presented later in the paper. The greatest challenge for Serbia is to find an appropriate response to frequent hydrometeorological events which impact the health and the safety of population and society as a whole. Despite the negative impacts of the global economic crisis, it is expected that the international community will be a confident partner for the Serbian Government in the future as it was many times before.

Keywords Climate change • Legislation • Weather related disasters • Impacts • Emergency situations

9.1 Introduction

In the Republic of Serbia, considerable progress in the context of combating climate change was brought about by the beginning of the process of European Union (UN) accession and the harmonization of national legislation with that of the EU.

V. Radovic (✉)
Faculty for Environmental Protection and Corporative Responsibilities,
Educons University, Sremska Kamenica, Serbia
e-mail: veselaradovic@yahoo.com

H.J.S. Fernando et al. (eds.), *National Security and Human Health Implications of Climate Change*, NATO Science for Peace and Security Series C: Environmental Security, DOI 10.1007/978-94-007-2430-3_9, © Springer Science+Business Media B.V. 2012

The European Parliament in Strasbourg ratified the Stabilization and Association Agreement (SAA) in January 2011. The Serbian population is facing some "new and emerging challenges" in everyday life. Climate change is likely to significantly affect Serbia in the coming decade and there is a need to undertake actions and efforts to contribute to combating climate change. Serbia is engaged in a number of activities in that field. On June 10, 2001, the Republic of Serbia became a member of the United Nation Framework Convention on Climate Change (UNFCC). On January 17, 2008, the Republic of Serbia's membership status in the Kyoto Protocol as a non-Annex party was approved. During 2010, a set of Strategies for Incorporation of the Republic of Serbia into Clean Development Mechanism, jointly with the First National Communication of the Republic of Serbia to the UNFCCC was adopted by the Government. Serbia's First (Initial) National Communication under the UNFCCC adopted by the government on 11 November 2010 represents the first overview in the field of climate change on the national level and envisions national measures to combat climate change.

9.2 What Serbia Has Done in the Past?

During 2009 and 2010, the Serbian Government was actively involved in performing the Designated Authority and other institutions required for the Protocol implementation. Among some key national priorities, Serbia also protects and promotes the environment and in achieving rational use of natural resources. From that point the main activities are to preserve and to enhance the system of environmental protection, to reduce pollution and environmental pressure, and use natural resources in a manner ensuring their availability for future generations.

In the last decade the main strategic documents in the environmental field were developed in accordance with sustainable development principles. During 2010, a number of additional strategic documents were developed [1]. In 2009, a set of 14 Environmental Laws was put in place by the Parliament. Environment-related issues are also covered by other ministries in line with their duties as defined by law: Office of Deputy Prime Minister for EU Integration, Ministry of Agriculture, Forestry and Water Management, Ministry of Economy and Regional Development, Ministry of Energy and Mining, Ministry of Infrastructure, Ministry of Health, etc. Many state organizations focus on various environmental topics, from environmental monitoring to statistical data collection, management – the Statistical Office, Hydrometeorological Institute, Public Health Institute, and so on. Efficient dealing with climate change impacts and adaptation requires involving all relevant sectors to research, measurements and analyses. The Serbian government put in place some mechanisms to address these challenges at local level and at the national level. In many areas international support is needed. *It is not a question of whether to adress climate change but how much to address it* [2].

9.3   The Influence of Climate Changes in Serbia

The influence of climate changes in Serbia is still significantly unknown. The investigative work on climate changes in Serbia is mainly sporadic. The greatest amount of investigative work was performed through state, scientific and other institutions and individuals within scientific-technical programs of the EU in bilateral and multilateral programs. The problem in Serbia is inadequately developed systems of integral monitoring of climate parameter, including the parameters of the environment within forestry, agriculture, energetics, waste management, public health and biodiversity. Research work in the field of climate and climate changes are included as a priority in the period of 2011–2014 [3].

Serbia belongs to the region with known basic influence of climate changes [4]. This influence is seen in temperature and precipitation changes as compared to the average annual values. From statistical data of competent services in the country, the increase in mean annual temperature values is obvious and it reaches up to 0.04% °C per year. For the last 2 years, Serbian agriculture was exposed to frequent periods of drought, which caused great losses. According to the results of investigation on the influence of drought on the yield of crops in the region of Eastern Serbia in the period of 1989–2000, the reduction of crop yield was 40.9% in comparison to average yields in the years without drought. In the future, significant vulerability of agriculture is expected, especially if insufficient capacities for irrigation are taken into account, (only 924 ha is irrigated, it is only 1.7% of total arable land).

The energy sector especially depends upon climate changes both in the field of energy production and in the field of its transportation and distribution to the consumers, respectively. The importance of the infrastructure of electroenergetics for public health is evident after numerous cases all over the world. Disconnections in supply of electrical energy in modern society have manifold consequences [5]. Electric-power companies, being the companies of great importance for the defence of the country (as determined by the decision of the government, upon the proposal of the Ministry of defence) [6], are obliged to have continual cooperation with state institutions in undertaking all necessary measures in the protection of the system. Such measures are also undertaken in the protection from the consequences of climate changes. Therefore, the Sector of Electric-power of Serbia (EPS) for the protection of the environment, prepares a Plan for coordination with climate changes [7].

The experts are of the opinion that the safety of supplying the population in Serbia with energy, drinkable water and basic life products might be significantly endangered in the future. They prove it by many facts and, therefore, it is not permissible to believe that "*lucky circumstances*" would help to avoid possible outcomes.

The health system in Serbia is recognized by the population as the least efficient and the most corrupted. Data show that the population in Serbia is getting older [8]. The experts state that the health status of the population, apart from aging of the population, was also influenced by the social-economic crisis during the last decade of the twentieth century. The health potential of the nation is exhausted, and with all difficulties of the transition, fast restraining of negative health parameters is not likely. Long lasting exposure to stress has caused an increase of mental and coronary illnesses. During long waves of heat for the last few years the increase of heat strokes was recorded. Under such circumstances, even the healthy population had difficulties in tolerance of high temperatures and increased air humidity. The most vulnerable were those in the category of chronic patients and older population [9].

For the last few years there were numerous activities organized in order to build air conditioners in old hospital buildings in order to improve the recovery of the patients and the work of the staff in rooms with temperature over 40°C [10]. None of the hospitals in Serbia have centralized cooling systems because the buildings were built decades ago when such instalations were not needed. It is necessary for Serbia to make clear connections in the future between the way of life and the impact of the environment in a way which is in accordance with the experience of developed countries [11]. Air pollution in Serbia additionally influences the health of the population. The number of respiratory diseases patients in many towns, such as the towns of Bor [12], Pančevo, and Užice [13] represents a huge problem.

There are many activities performed daily within the country in order to improve the health status of the population, but there is still room for improvements. Inclusion of *"health into other politics"* is a great challenge, especially in the field of influence of climate change on health. The government has adopted *The Action Plan of children's health and the environment* and encouraged the work on *Action Plan for heat waves*. The completion of these activities is expected during 2011. The Ministry of Health has continued the work on gathering data and on the analyses of the present state. Serbian health care has increased its rating, but in the years to come, the sustainability of the health system will be impacted by budget restrictions and financial crisis [14].

In the next 10–20 years, it is expected that significant budget funds will have to be given first for the adaptation of the consequences of natural disasters [15]. The strenghtening of sectorial cooperation and inclusion of the issue of climate changes is a key prerequisite for systematic following and monitoring of their consequences. Education, training and strenghtening of the consciousness of the citizens represents a priority in these activities. For the last few years, climate changes have become a subject of the wider interest of the public, particularly because of the media which provided information about climate changes. However, there is still room for action and the executive authorities are aware of this fact [16].

9.4 Emergency Situations in Serbia as Possible Consequences of Global Climate Changes

A clear connection between global climate changes and the frequency of natural disasters has been recorded in many scientific investigations in the world. It is expected that this effect will be significantly increased in the future [17]. Munich Re, the leading world insurance company, is warning: '"The high number of weather-related natural catastrophes and record temperatures both globally and in different regions of the world provide further indications of advancing climate change" [18]. Each meteorological situation that causes considerable damage is a weather-related disaster.

The year 2010 is recorded as the year of natural disasters in the world. The consequences of climate changes on the territory of Serbia are in significant connection with the appearance of hydrometeorological danger. Within the framework of unwanted events the following was recorded: heavy rains leading to floods and lanslides, winter storms, extreme cold and heat, drought, dense fog, phenomena related to storm clouds and freezing.

Over the past years we have witnessed great economic losses in Serbia caused by extreme weather and climate factors. In the country, which is uneven in regional economic development, it is inevitable for the population to have different approaches to basic human needs. The prevailing structure is faced with every day life, huge challenges in providing health services, education, reduction of unemployment, high degree of corruption and with many other problems. One of the basic challenges in fighting against the consequences of climate changes is the reduction of consequences of extreme meteorological climate and hydrological phenomena and the definition of strategies for adaptation. Unlike in most of the European countries, Serbia still does not have clearly defined politics for the adaptation measures for the expected global climate changes.

In Serbia, the Sector for Emergency Management has for the first time been organized within the Ministry of Interior. Laws on emergency situations were passed, but the system still does not have all necessary legal and organizational prerequisites for reaching maximal efficiency. The activities of the Sector for Emergency Management in regional and international cooperation in the field of cooperation and mitigation of consequences of natural disasters are encouraging [19].

It is widely known that emergency situations cause great attention of the media. At the moment they occur there are many who offer help. After some time, however, the burden of sanitation and alleviation of consequences is usually left to the local authorities who are then faced with the lack of skilled and financial capacities. The mayors and the presidents of more than 50 towns in Serbia handed the Government of Serbia a petition requesting the return of full transfers from the budget of the Republic to the local authorities in 2011.

By the law on financing local authorities, which is in effect from 2006, it is anticipated that the transfers from the republican budget to the local authorities is 1.7% of GDP. Because of the economic crisis, these transfers were reduced in twice in 2009 and in 2010. If such practise continues during 2011, the reduction would come to the total of 570 million of euros [20].

The way the local authorities react to the threatening danger is of primary importance for the reduction of consequences. In that light, in Serbia there are many positive examples of strengthening the local capacities of the local authorities in reaction to states of emergency [21]. Certain regions of Serbia record a steady increase of the number of the poor. For them, even human resources represent an important problem because of long lasting migrations to towns. Populations in towns are exposed to stress and to despair, forgotten from all. There is no research work in Serbia which would confirm the influence of floods, landslides and fires and other states of emergency on the mental health of the population of the struck region. The older population in Serbia is especially endangered by such accidents, the poor and the sick, as well as the persons in need. When compared with similar accidents in the world, although different in intensity, it is clear that health help is necessary during and after the event and, in many cases, for years after that.

There are no examples in Serbia of significant psycho-social support after states of emergency. The Red Cross of Serbia has an exceptional role with its universal program for training of volunteers for emergency cases [22]. Climate changes and emergencies caused by them could cause an increase of inter-ethnic tensions, especially in under-developed ethnic regions. Such risk could be recognized in Serbia in the region of Sandzak and in the south of Serbia in the municipalities of Preševo, Medveđa and Bujanovac. Manipulation of national feelings of the poor and endangered sector of the population has already been seen in the immediate past. Similar occurences, therefore, should be prevented, for the sake of everyone.

In autumn 2007, heavy rains caused huge floods in South, South-east and North-east Serbia. The torrents demolished bridges, roads, houses and endangered the health of the population. In a settlement called "Crni Marko" in Vlasotince, there were about 200 Roma inhabitants evacuated into a building of a sport centre. Damages for such a poor municipality were huge, about eight million dollars. Since the unemployment rate in the municipality is almost 50%, twice higher than the average of the Republic, and it had a great number of displaced persons, the actions taken were urgent in order to provide necessary help. USAID also took part in providing help, together with the U.S. Ministry of defense/European command unit, the U.S. Embassy in Serbia with the coordination of the PPES program and with the cooperation of authorized national services and the Red Cross of Serbia.

Other floods occured in 2010 in Bosnia, Croatia, Montenegro and in Serbia. Flooding from the Sava, Drina and Lim rivers devastated great areas and caused enormous economic damages. The floods endangered the populations in Ljubovija, Loznica, Bogatic, Prijepolje and in Priboj. In certain cases it was necessary to evacuate the endangered inhabitants. Having in mind the economic situation in the region, it is obvious that "*capacities of adaptation which depend of social wealth and of the presence of satisfactory health and educational structure of the inhabitants*" are obviously low, which points out the urgent need for their strengthening.

9.5   Conclusion

Challenges of climate changes in Serbia are recognized in a satisfied way. In January 2011, in accordance with international experience, different activities for the reduction of the negative influence of climate changes were taken. The first step in the work on law regulations was realised. Expecting access to the EU, Serbia struggles not to become additional source of problems. Therefore, in the process of decentralization, it is necessary to strenghten the capacities of the local community to enable fast reactions in the protection of critical infrastructure and to reduce the influence of the consequences in cases of natural disasters. Apart from that, it is urgent to start work on the National Action Plan of Adaptation (NAPA) and on the Strategy of reaction in emergency cases. In the present economic situation it is expected that the finances for these needs will be mostly provided by using the European instruments for pre-accession assistance and through bilateral and multirateral cooperation.

References

1. Some of the Serbian Strategies (2011) Available at www.srbija.gov.rs/vesti/dokumenti_sekcija. php?id=45678. Accessed 14 Feb 2011
2. Portney PR, Stavins RN, Scd (eds) (2000) Public policies for environmental protection Washington, DC. In that book authors Shogren and Toman, article: climate change policy, pp 125–168
3. Strategy of scientific and technological development of Serbia for the period 2010–2015 (2010) www.nauka.gov.rs. Accessed 30 Jan 2011
4. Regional perespectives for Central and Eastern Europe (2010) The European environment-state and outlook 2010 (SOER 2010) www.eea.europa.eu/soer. Accessed 30 Jan 2011
5. North American Electric Reliability Council (2004) United States-Canada power system outage task force, interim report: causes of the August 14 blackout in the United States and Canada. http://www.iwar.org.uk/cip/resources/blackout-03. Accessed 14 Feb 2011
6. Law on Defence, article 67, Official Gazette of Republic of Serbia, and no. 116/2007
7. Mr Mihajlo Gavrić (2010) Announced work on a book, so called *White Book*. www.edb.rs/list/pdf/juli_2010.pdf. Accessed 30 Jan 2011
8. World Fact CIA book (estimated 2009) http://www.cia.gov/library/publications/the-world-factbook. Accessed 30 Jan 2011
9. WHO (2008) Global Assessment of National Health Sector Emergency preparedness & response. http://www.who.int/hac/techguidance/preparedness/capacities/en/index.html. Accessed 30 Jan 2011
10. Article in daily news: Heal in hospitals, Serbian (2010) http://www.vesti.rs/Drustvo/PAKAO-U-BOLNICAMA.html. Published at 14 Aug 2010g. Accessed 30 Jan 2011
11. National Public Health Week (NPHW) (2008) Theme was: *Climate change*: Our Health in the Balance. www.nphw.org. Accessed 29 Jan 2011
12. Report on the environmental situation in the Republic of Serbia in 2009 year (2009) http://www.sepa.gov.rs/download/Izvestaj%20o%20stanju%20zivotne%20sredine%20u%20Republici%20Srbiji%20za%202009%20godinu.pdf. Accessed 10 Feb 2011
13. Article: Population in Uzice choking in smoke and sow. Published 11 Jan 2011. www.politika.rs/ilustro/2498/2.htm and www.ue.co.rs. Accessed 30 Jan 2011
14. Statement of minister Milosavljevic (2010) Article: health and happiness at the first place. http://smedia.rs/vesti/detalji.php?id=53560&vest=Milosavljevic:-Zdravlje-I-Sreca-na-prvom-mestu!. Accessed 30 Jan 2011
15. Response to the EU Questionnaire for the Member States on Experiences (2011) Success factors, risks and challenges with regard to objective and themes of UN Conference on Sustainable Development (UNCSD) Attachment C Questionnaire on Addressing new and emerging challenges. www.serbia.gov.rs/?change_lang=en. Accessed 30 Jan 2011

16. Letter of Serbian minister O. Dulić (2010) to Yvo de Boer, Executive Secretary UNFCCC Secretariat, and Date 29 Jan 2010. http://www.ecoplan.gov.rs
17. European Environment Agency (EEA) (2010) Technical report no 13/2010 mapping the impacts of the natural hazards and technological accidents in EU, Publication Office of the EU, Luxemburg
18. Title (2010) Top year for natural disasters published online 01/04/2011: http://www.sustainablebussiness.com/index.cfm/go/news.display/id/21666. Accessed 30 Jan 2011
19. More about international cooperation and many signed agreements at http://prezentacije.mup.gov.rs/sektorzazastituispasavanjel/medjunarodnasaradnja.htm. Accessed 14 Feb 2011
20. Article: Ministers for returning financial resources to local communities (2010) Serbian, published 9 Oct 2010 http://www.b92.net/biz/vesti/srbija.php?yyyy=20108&mm=10&dd=098nar_id=464180. Accessed 31 Jan 2011
21. The Preparedness, Planning and Economic Security Program (PPES) (2011), still ongoing project, a DAI-led project funded by the U.S. Agency for International Development (USAID). http://serbia.usaid.gov/code/navigate.php?Id=657. Accessed 31 Jan 2011
22. Code of Conduct for the International Red Cross and Red Crescent. Movement and Non-Governmental Organizations (NGOs) in Disaster Relief (1994) 31-12-1994 Publication Ref. 1067. www.icrc.org/eng/resources/documents/publication/p1067.htm. Accessed 31 Jan 2011

Chapter 10
Megacities: Urban Environment, Air Pollution, Climate Change and Human Health Interactions

Alexander Baklanov

Abstract Integrated multi-scale modelling concept of urban environment, air pollution, climate change and human health interactions for megacities and overview of integrated modelling frameworks realized in European projects: current MEGAPOLI and previous FUMAPEX, as well as Danish CEEH and European Enviro-RISKS projects, are described in this paper.

Keywords Megacities • Atmospheric pollution • Urban climate • Climate change • Risk

10.1 Introduction

For the past few 100 years, human populations have been clustering in increasingly large settlements. In 2007, for the first time in history, the world's urban population exceeded the rural population. At present, there are about 20 cities worldwide with a population of ten million or greater, and 30 with a population exceeding seven million. Most of them (especially quickly growing cities) are situated in poor and developing countries. These numbers are expected to grow considerably in the near future. From year 2000 to year 2030 the global urban population is expected to rise from 47% (2.9 b) to 60% (5 b). Ninety-five percentage of growth will be in less developed countries. By 2015, 16 of the world's 24 cities with more than ten million people will be located in Asia. The urban transition now underway in Asia involves a volume of population much larger than any other region in the world and is taking place on a scale unprecedented in human

A. Baklanov (✉)
Danish Meteorological Institute, Copenhagen, Denmark
e-mail: alb@dmi.dk

H.J.S. Fernando et al. (eds.), *National Security and Human Health Implications of Climate Change*, NATO Science for Peace and Security Series C: Environmental Security, DOI 10.1007/978-94-007-2430-3_10, © Springer Science+Business Media B.V. 2012

history. Such coherent urban areas with more than five million people are usually called megacities. In Europe there are six major population centres that clearly qualify as megacities: London, Paris, the Rhine-Ruhr region, the Po Valley, Moscow, and Istanbul.

Megacities and heavily urbanized regions produce a large fraction of the national gross domestic product (GDP) (e.g. London, Paris and Mexico City account respectively for 19.9%, 27.9% and 26.7% of the corresponding national GDP. Human activities in megacities lead to serious challenges in municipal management, such as housing, employment, provision of social and health services, the coordination of public and private transport, fluid and solid waste disposal, and local and regional air pollution. This article focuses on the latter, spanning the range from emissions to air quality, effects on regional and global climate, and feedbacks and mitigation potentials. It takes into account the different features and growing trends that characterize cities located in developing countries to highlight their present and future effects on local to global air quality and climate.

Urban respiration (oxygen in/primary and secondary gaseous pollutants and airborne particulates/aerosols/out) represents the direct impacts of urban metabolism on the atmosphere. Emitted urban air pollutants have a significant impact on both regional viability (human health, agricultural/ecosystem productivity, visibility), and global change (climate, ozone depletion, oxidative capacity) issues. Hence, megacities present a major challenge for the global environment. Adaptation by humans to significant climate change in major metropolitan areas is possible, but it is certainly not proved that adaptation to global climate change is the total answer. Well-planned, densely populated settlements can reduce the need for land conversion and provide proximity to infrastructure and services, but sustainable development must also include: (i) appropriate air quality management plans; (ii) adequate access to clean technologies; and (iii) improvement of data collection and assessment. A successful result will be to arrive at integrated control and mitigation strategies that are effectively implemented and embraced by the public.

Methodological aspects to analyse the above mentioned problem of the interactions of the urban environment, air pollution, climate change and human health in and outside megacities, realized in the previous EC FP5 FUMAPEX project 'Integrated Systems for Forecasting Urban Meteorology, Air Pollution and Population Exposure' (see: http://fumapex.dmi.dk) and in the current EC 7FP MEGAPOLI project 'Megacities: Emissions, urban, regional and Global Atmospheric POLlution and climate effects, and Integrated tools for assessment and mitigation' (see: http://megapoli.info) are discussed in the paper. The main MEGAPOLI objectives [13] are (i) to assess impacts of megacities and large air-pollution hot-spots on local, regional and global air quality, (ii) to quantify feedbacks among megacity air quality, local and regional climate, and global climate change, (iii) to develop improved integrated tools for prediction of air pollution in megacities.

10.2 Integrated Modelling Methodology

Processes involving nonlinear interactions and feedbacks between emissions, chemistry and meteorology require coherent and robust approaches using integrated/online methods. This is particularly important where multiple spatial and temporal scales are involved with a complex mixture of pollutants from large sources, as in the case of megacities. The impacts of megacities on the atmospheric environment are tied directly to anthropogenic activities as sources of air pollution.

These impacts act on street, urban, regional and global scales. Previously there were only limited attempts to integrate this wide range of scales for regional and global air quality and climate applications. Indeed, progress on scale and process interactions has been limited because of the tendency to focus mainly on issues arising at specific scales. However the interrelating factors between megacities and their impacts on the environment rely on the whole range of scales and thus should be considered within an integrated framework bringing together the treatment of emissions, chemistry and meteorology in a consistent modelling approach. Numerical weather and air pollution prediction models are now able to approach urban-scale resolution, as detailed input data are becoming more often available. As a result the conventional concepts of down- (and up-) scaling for air pollution prediction need revision along the lines of integration of multi-scale meteorological and chemical transport models. MEGAPOLI aims at developing a comprehensive integrated modelling framework which will be tested and implemented by the research community for a range of megacities within Europe and across the world to increase our understanding of how large urban areas and other hotspots affect air quality and climate on multiple scales [4].

The integration strategy in MEGAPOLI (Fig. 10.1) is not focused on any particular meteorological and/or air pollution modelling system. The approach considers an open integrated framework with flexible architecture and with a possibility of incorporating different meteorological and chemical transport models (see model specifications in [13]).

The following levels of integration and orders of complexity (temporal and spatial scales and ways of integration) are considered:

- Level 1 – Spatial: One way (Global → regional → urban → street); Models: All.
- Level 2 – Spatial: Two way (Global ⇔ regional ⇔ urban); Models: UM-WRF-CMAQ, SILAM, M-SYS, FARM.
- Level 3 – Time integration: Time-scale and direction; Direct and Inverse modelling (Fig. 10.2).
- Order A – off-line coupling, meteorology/emissions → chemistry; Models: All.
- Order B – partly online coupling, meteorology → chemistry & emission; Models: UKCA, DMAT, M-SYS, UM-WRF-Chem, SILAM.
- Order C – fully online integrated with two-way feedbacks, meteorology ⇔ chemistry & emissions; Models: UKCA, WRF-Chem, Enviro-HIRLAM, EMAC (former ECHAM5/MESSy).

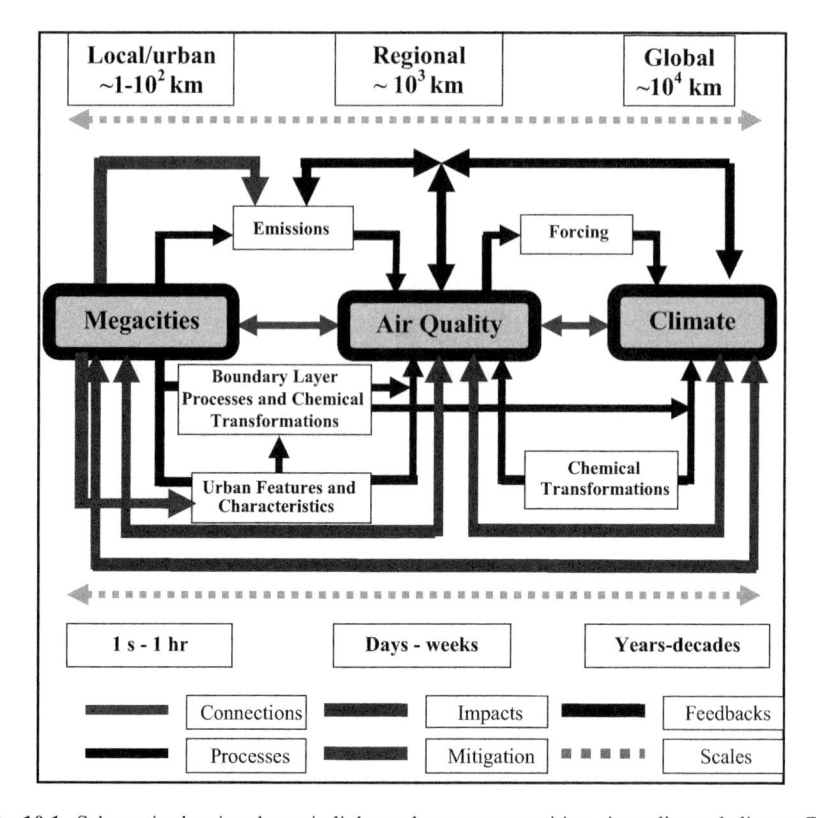

Fig. 10.1 Schematic showing the main linkages between megacities, air quality and climate. The connections and processes are the focus of MEGAPOLI. In addition to the overall connections between megacities, air quality and climate, the figure shows the main feedbacks, ecosystem, health and weather impact pathways, and mitigation routes which will be investigated in MEGAPOLI. The relevant temporal and spatial scales are additionally included [13]

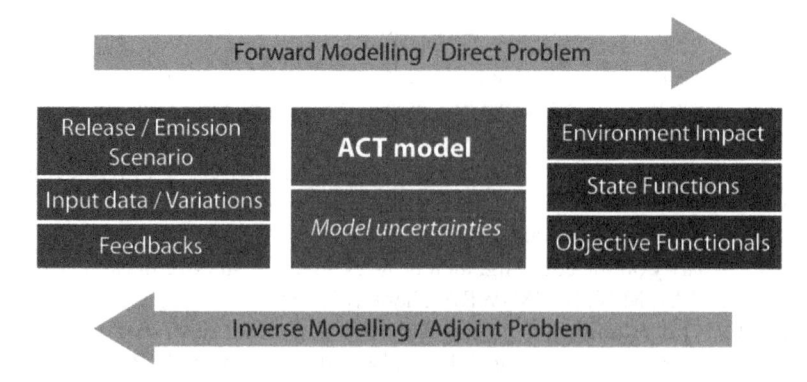

Fig. 10.2 Scheme of environmental risk assessment and mitigation strategy optimization basing on forward/inverse modelling

Multi-scale modelling chain/framework includes nesting of the following characteristics and processes from the global to street scale:

- Land-use characteristics and scenarios,
- Anthropogenic heat fluxes,
- Emission inventories and scenarios,
- Atmospheric processes model down- and up-scaling (two-way nesting, zooming, nudging, parameterizations, urban increment methodology).

Where it is required, new or improved interfaces for coupling (direct links between emissions, chemistry and meteorology at every time step) are developed. Common formats for data exchange (such as GRIB, netCDF formats) are defined to ease the implementation and to help combine the different models via conventional data exchange protocols. The current chemistry schemes (tropospheric, stratospheric and UTLS) are examined for their suitability to simulate the impact of complex emissions from megacities. The coupled model systems are applied to different European megacities during the development phases of the project. The framework will be used and demonstrated for selected models including UKCA (MetO), WRF-Chem (UH-CAIR), Enviro-HIRLAM (DMI), STEM/FARM (ARIANET), M-SYS (UHam) and EMAC (MPIC) on different scales. This part of the work is linked to the requirements and use of simpler tools for assessing air quality impacts within megacities (OSCAR – UH-CAIR, AIRQUIS – NILU, URBIS – TNO, EcoSence – UStutt).

The detailed description of the MEGAPOLI modelling integration framework was done in D7.1 report [4] in a close collaboration with the COST Actions 728 [5, 8] and ES0602 [11]. It is also linked with a new COST Action ES1004: European framework for on-line integrated air quality and meteorology modelling (EuMetCHem: http://eumetchem.info).

10.3 Urbanisation of Models

Urban air pollution (UAP) and atmospheric chemical transport (ACT) models have different requirements in terms of the way in which they represent urbanization (e.g., different importance of low-atmosphere structure details) depending on (i) the scale of the models (e.g., global, regional, city, local, micro) and (ii) the functional type of the model, e.g.:

- Forecasting or assessment models
- Atmospheric pollution models for environmental and air quality applications (mostly for city scale)
- Emergency preparedness models (mostly for city scale or micro-scale)
- Integrated ACT and aerosol models for climate forcing
- Urban-scale research ACT models

Incorporation of the urban effects into urban- and regional-scale models of atmospheric pollution should be carried out, first via improvements in the accuracy of

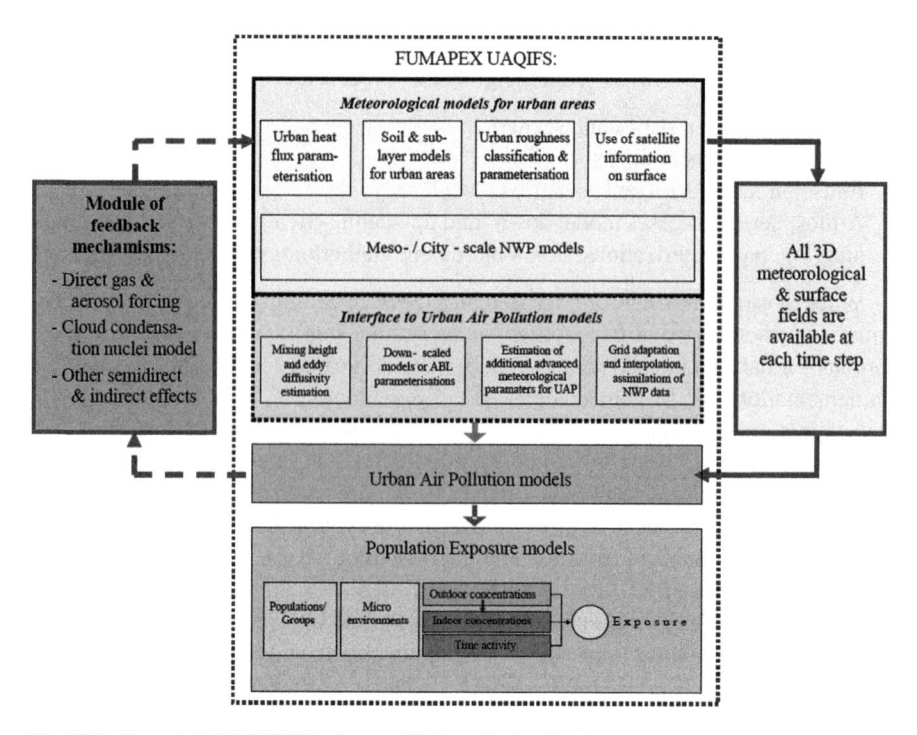

Fig. 10.3 Extended FUMAPEX scheme of Urban Air Quality Information & Forecasting System (UAQIFS) including feedbacks. Improvements of meteorological forecasts (NWP) in urban areas, interfaces and integration with UAP and population exposure models following the off-line or on-line integration (After Baklanov et al. [3])

meteorological parameters (velocity, temperature, turbulence, humidity, cloud water, precipitation) over urban areas. This requires a kind of "urbanization" of meteorological and numerical weather prediction (NWP) models that are used as drivers for urban air quality models or special urban met-preprocessors to improve nonurbanized NWP input data (Fig. 10.3). In MEGAPOLI a hierarchy of urban canopy models/parameterisations for different type and scale models was developed [12].

In comparison with NWP models, the urbanization of UAP models has specific requirements, e.g., better resolution of the urban boundary layer (UBL) vertical structure; by themselves, the correct surface fluxes over the urban canopy are insufficient for UAP runs. Furthermore, for urban air pollution, from traffic emissions and for the modelling of preparedness for emergencies, there is a much greater need for vertical profiles of the main meteorological parameters and the turbulence characteristics within the urban canopy.

Other important characteristics for pollutant turbulent mixing in UAP modelling include the mixing height, which has a strong specificity and heterogeneity over urban areas because of the urban heat island (UHI), internal boundary layers, and blending heights from different urban roughness neighbourhoods [15]. For the modelling of preparedness for emergencies at local scale (e.g., biological, chemical, or nuclear accidental releases or terrorist acts) the statistical description of building structure is suitable only for distances longer than three or four buildings from the release, whereas for the first two to four buildings from the source, more precise obstacle-resolved approaches are needed [2].

Other specific effects of urban features on air pollution in urban areas, which cannot be realized via the urbanization of NWP models, include:

- deposition of pollutants on specific urban surfaces, e.g., on vertical walls, from different building materials and structure, vegetation, etc.
- specific chemical transformations, including increasing the residence times of chemical species (e.g., inside street canyons), the heterogeneity of solar radiation (e.g., street canyon shadows) for photochemical reactions, and specific aerosol dynamics in street canyons (e.g., from the resuspension processes)
- very heterogeneous emission of pollutants at the subgrid scale, especially from traffic emissions, which need to be simulated on detailed urban road structures with taking into account the distribution of transport flows, etc.
- the indoor–outdoor interaction of pollutants (not only via heat fluxes), which requires a more comprehensive description and modelling of emissions

The effects of air pollution on health are the final and most important aim of UAP modelling. It is therefore important to combine the UAP with population exposure modelling, which includes high-resolution databases of urban morphology and population distribution and activity [9]. One of the realisations of such an Urban Air Quality Information and Forecasting Systems (UAQIFS) was done within FUMAPEX (Baklanov 2006). The improved UAQIFS is enhancing the capabilities to successfully describe and predict urban air pollution episodes through improvement and integration of systems for forecasting urban meteorology, air pollution, and population exposure based on modern information technologies (see Fig. 10.3). The UAQIFSs were implemented and demonstrated in seven European cities: Oslo, Norway (urban air quality forecasting mode), Turin, Italy (urban air quality forecasting mode), Helsinki, Finland (urban air quality forecasting mode + public health assessment and exposure prediction mode), Valencia/Castellon, Spain (urban air quality forecasting mode), Bologna, Italy (urban management and planning mode), Copenhagen, Denmark (urban emergency preparedness system), and London, UK (urban air quality forecasting mode).

10.4 Integrated Tools for Mitigation Assessments

The policy questions in MEGAPOLI are also addressed by generating scenarios – descriptions of possible consistent future developments of the megacities and the surrounding regions – and comparing the impacts of the different scenarios [17]. The scenarios are used to generate an emission data set for each scenario, which will then be used as input for the integrated assessment tool. Output of this model will be maps and parameters describing air quality, deposition and impacts on climate change for each scenario. These results will be assessed in the following ways: first the compliance with current existing thresholds will be examined, and using cost-benefit analyses, bundles of measures that fulfil certain aims with least costs can be identified. Secondly, to be able to analyse the importance of air quality and climate changes, the results will be converted into damages using the impact pathway approach developed in the ExternE project series [7] including the currently running projects (e.g. EU FP7 TRANSPHORM, Danish CEEH: ceeh.dk, etc.), and using the integrated tool for environmental impact analysis EcoSence [10]. Health risks are calculated using concentration-response and exposure-response relationships developed and recommended in these projects, different health endpoints can then be aggregated using DALY's (disability adjusted life years). Climate change damage is assessed using results from the FUND model [18] and by analysing studies on climate change impacts, including the IPCC report and DEFRA studies. Damage to ecosystems from acidification and eutrophication is assessed with a method developed in the NEEDS project [16] using 'potentially disappearing fractions' of species as damage indicators.

Figure 10.3 shows one example of the integrated Urban Air Quality Information & Forecasting System (UAQIFS) including chemistry/aerosol feedbacks and heath risk assessments, where improvements of meteorological forecasts in urban areas, interfaces and integration with air quality and population exposure models are implemented following the off-line or on-line integration.

To be able to compare the different damage categories with each other and with costs of measures, the damage indicators should be converted into a common unit; here monetary units are chosen using contingent valuation, which measures the preference of the population, e.g. by surveys about the willingness to pay to avoid/minimise a risk, as means to allocate monetary values to risks and damages. Using the monetized results, cost-benefit analyses can be carried out (for short- and medium-term measures, for long-term measures benefits are calculated). To be able to generate these results efficiently, integrated computer tools / model frameworks are developed. One of the examples of such integrated 'Energy-Environment-Health-Cost' modelling frameworks, realised by the Danish Centre for Energy, Environment and Health (CEEH), is demonstrated in Fig. 10.4 [6].

The first MEGAPOLI analysis asks about the development of impacts from megacities, in case when no additional measures are implemented. For that, the baseline scenario [17], assuming a trend development of activities and emission factors that take into account current legislation and legislation in the pipeline, is used. Available policy options (possibilities for implementing instruments by the policy makers to accomplish their goals), which could be implemented in addition to those of the baseline scenario, will be systematically collected. Assumptions have to be made about how the operators and users of emission sources react to these options, i.e., which abatement and mitigation measures they will implement. Both technical measures (changing emissions factors, e.g., change of fuel or energy types) and non-technical measures (which change the decisions and the behaviour of users of

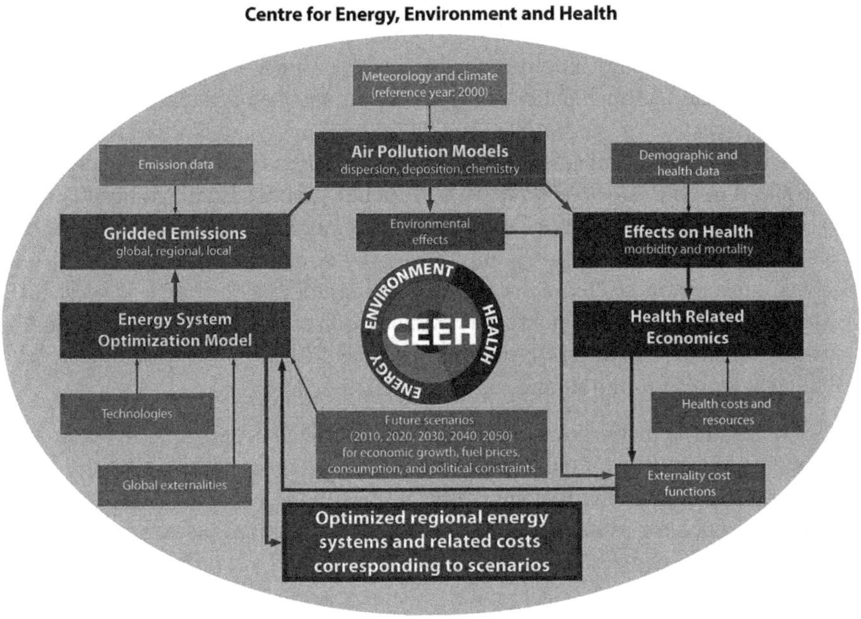

Fig. 10.4 Schematic description of the CEEH integrated 'Energy-Environment-Health-Cost' modelling framework system [6]

emission sources, e.g., by implementing a charge on emissions) will be addressed. Measures to reduce urban drivers of climate change, to be considered in MEGAPOLI, include, e.g.:

- Reducing green-house gases (GHG) and aerosol emissions
- Reducing traffic congestion
- Switch to fuels with less GHG side effects
- Conserving energy and water
- Greater use of passive heating and cooling technology
- More compact city design and greater use of mass transportation
- Intelligent use of trees to shelter or shade
- Increased use of light coloured surfaces in hot climate megacities

10.5 Conclusion

The article overviewed the MEGAPOLI integration concept, including the overall integration strategy and suggested integrated modelling framework.

The following key issues of the integrating modelling were considered:

- Integration and implementation of models with on-line and off-line coupling of meteorological and air quality models with an analysis of the advantages and disadvantages;
- Implementation of feedback mechanisms, direct and indirect effects of aerosols;
- Advanced interfaces between NWP, ACT, population exposure and health effect models with recommendations towards their harmonisation and standardisation;
- Recommended methods for the model up- and down-scaling and nesting from global-, macro-, meso- to micro-scales as well as the scale interaction issues;
- Model ensembles for assessments and evaluation with recommendations for the multi-model ensemble building and testing the ENSEMBLE system for model evaluation;
- Integrated tools for impact and mitigation assessments.

One of the major innovations of MEGAPOLI is that it strives for a full integrated assessment of megacities. Policy options and mitigation measures generally influence the emission of more than one pollutant, thus for assessing such measures all effected impacts have to be taken into account. The relationship between climate change and air pollution is particularly important, but not yet fully analysed. Thus the integration occurs:

- across impacts, especially climate change impacts and air pollution impacts, including health risks and ecosystem damage;
- across pollutants and emission sources, e.g. transport, energy conversion, industry, households, waste, agriculture, natural and biogenic processes; PM10, PM2.5, ozone, acid substances, nutrients, greenhouse gases, and others;
- across scales: local, urban, regional, global; short, medium and long term.

The assessment of policy and mitigation options is based on the simultaneous assessment of all relevant changes in damages and risks caused by the option (and not on the potential, e.g., with regard to the reduction of a single pollutant).

Current results of MEGAPOLI studies [14] are giving the following preliminary answers on the scientific question: "Are the cities to blame for climate change/ global warming?":

- On city- and meso-scales definitely 'Yes' (both via UHI and emissions),
- On regional and continental scale: urban plume extends up to thousands kilometer, so it could effect the climate change,
- On global scale: probably 'No' due to UHI, but 'Yes' due to GHG emissions (anthropogenic CO_2, CFC, CH_4, N_2O and tropospheric ozone),
- Source of aerosols which have both direct and indirect cloud radiative effects (cooling or warming),
- Too early to make final conclusions: new multi-scale studies are necessary.

Acknowledgements This overview article was written based on the EC projects FP7 MEGAPOLI, FP5 FUMAPEX, FP6 Enviro-RISKS, Danish CEEH and COST Actions 728, ES0602, ES1004 and the author acknowledge many scientists involved into these projects for close collaboration, productive discussions and scientific contributions.

References

1. Baklanov A (ed) (2010) Framework for integrating tools. Deliverable D7.1, MEGAPOLI scientific report 10–11, MEGAPOLI-14-REP-2010-03, 68 p, ISBN: 978-87-993898-4-1
2. Baklanov A, Nuterman R (2009) Multi-scale atmospheric environment modelling for urban areas. Adv Sci Res 3:53–57
3. Baklanov A, Hänninen O, Slørdal LH, Kukkonen J, Sørensen JH, Bjergene N, Fay B, Finardi S, Hoe SC, Jantunen M, Karppinen A, Rasmussen A, Skouloudis A, Sokhi RS, Ødegaard V (2006) Integrated systems for forecasting urban meteorology, air pollution and population exposure. Atmos Chem Phys 7:855–874
4. Baklanov A, Lawrence M, Pandis S, Mahura A, Finardi S, Moussiopoulos N, Beekmann M, Laj P, Gomes L, Jaffrezo J-L, Borbon A, Coll I, Gros V, Sciare J, Kukkonen J, Galmarini S, Giorgi F, Grimmond S, Esau I, Stohl A, Denby B, Wagner T, Butler T, Baltensperger U, Builtjes P, van den Hout D, van der Gon HD, Collins B, Schluenzen H, Kulmala M, Zilitinkevich S, Sokhi R, Friedrich R, Theloke J, Kummer U, Jalkinen L, Halenka T, Wiedensholer A, Pyle J, Rossow WB (2010) MEGAPOLI: concept of multi-scale modelling of megacity impact on air quality and climate. Adv Sci Res 4:115–120. doi:10.5194/asr-4-115-2010
5. Baklanov A, Mahura A, Sokhi R (eds) (2011) Integrated systems of meso-meteoro-logical and chemical transport models. Springer, Dordrecht. ISBN 978-3-642-13979-6, 242 p
6. CEEH (2011) Description of the CEEH integrated 'Energy-Environment-Health-Cost' model-ling framework system. In: Baklanov A, Kaas E (eds) Centre for energy, environment and health. CEEH scientific report No 1: www.ceeh.dk\CEEH_Reports\Report_1, ISSN: 1904–7495
7. Extern E (1995) Externalities of energy "EXTERNE" project, vol 2, Methodology. Method for estimation of physical impacts and monetary valuation for priority impact pathways. ETSU Metroeconomica, Harwell Court House, Oxfordshire, 399 p
8. Grell G, Baklanov A (2011) Integrated modeling for forecasting weather and air quality: a call for fully coupled approaches. Atmos Environ. doi:10.1016/j.atmosenv.2011.01.017
9. Hanninen O, Alm S, Katsouyanni K, Kunzli N, Maroni M, Nieuwenhuijsen MJ, Saarela K, Sram RJ, Zmirou D, Jantunen MJ (2004) The EXPOLIS study: implications for exposure research and environmental policy in Europe. J Expo Anal Environ Epidemiol 14:440–456
10. Krewitt W, Trukenmüller P, Mayerhofer R, Freidrich R (1995) EcoSence – an integrated tool for environmental impact analysis. In: Kremers H, Pillman W (eds) Space and time in envi-ronmental information systems, Herausgegeben vom GI-Fachausschuß 4.6 'Informatik im Umweltschutz'. Metropolis, Marburg
11. Kukkonen J, Balk T, Schultz DM, Baklanov A, Klein T, Miranda AI, Monteiro A, Hirtl M, Tarvainen V, Boy M, Peuch V-H, Poupkou A, Kioutsioukis I, Finardi S, Sofiev M, Sokhi R, Lehtinen K, Karatzas K, San José R, Astitha M, Kallos G, Schaap M, Reimer E, Jakobs H, Eben K (2011) Operational, regional-scale, chemical weather forecasting models in Europe. Atmos Chem Phys Discuss 11:5985–6162. doi:10.5194/acpd-11-5985-2011
12. Mahura A, Baklanov A (eds) (2010) Hierarchy of urban canopy parameterisations for different scale models. Deliverable D2.2, MEGAPOLI scientific report 10–04, MEGAPOLI-07-REP-2010-03, 50p, ISBN: 978-87-992924-7-9
13. MEGAPOLI (2008) MEGAPOLI description of work (2008–2011). Baklanov A, Lawrence M, Pandis S (eds). Copenhagen, ISBN: 978-87-992924-0-0, 150 p
14. MEGAPOLI (2010) Second year MEGAPOLI dissemination report. In: Baklanov A, Mahura A (eds) Megacities: emissions, urban, regional and Global Atmospheric POLlution and cli-mate effects, and integrated tools for assessment and mitigation. Sci Rep 10–21, ISBN: 978-87-92731-02-9, 89 p, http://megapoli.dmi.dk/publ/MEGAPOLI_sr10-21.pdf
15. Piringer MS, Joffre AB, Christen A, Deserti M, De Ridder M, Emeis S, Mestayer P, Tombrou M, Middleton D, Baumannstanzer K, Dandou A, Karppinen A, Burzynski J (2007) The sur-face energy balance and the mixing height in urban areas – activities and recommendations of COST Action 715. Bound -Layer Meteor 124:3–24

16. Ricci A (2009) NEEDS: new energy externalities development for sustainability. Annex –
a summary account of the final debate. 22 p. Web-site: http://www.needs-project.org/
17. Theloke J (2011) European and megacity baseline scenarios for 2020, 2030 and 2050
(MEGAPOLI Del 1.3). MEGAPOLI NewsLetter, Mar 2011, #10: 2
18. Tol R (2003) Emission abatement versus development as strategies to reduce vulnerability to
climate change: an application of FUND (working paper). University of Hamburg, Hamburg

Chapter 11
Climate Change Meets Urban Environment

Harindra Joseph S. Fernando, R. Dimitrova, and S. Sentic

Abstract The effects of global warming permeate to local scales in numerous ways, and at times the adverse effects of global change are amplified by urban anthropogenic activities. These local climate influences, however, have not received due attention as current climate discourse mainly focuses on global scales. In this paper, a brief overview is presented on how urban areas bear the brunt of global climate change, in particular, how such climatic signals as sea level rise, desertification, adjustment of hydrological cycle and enhanced cloud cover can have significant repercussions on local climate, thus raising human health and national security concerns. The reduction of diurnal temperature range (DTR) with global warming and its further amplification with urbanization are used as examples to illustrate local impacts of climate change. The possible amplification of urban heat island may even lead to local meteorological regime shifts, which have an important bearing on sustainability of cities. Meteorological variables are related to air pollution, and the relationship between particulate matter and meteorological variables in Phoenix area is used to illustrate possible relationships between human health and climate change.

Keywords Urban-global climate interactions • Heat island • Diurnal temperature range • Human health

H.J.S. Fernando (✉) • R. Dimitrova • S. Sentic
Environmental Fluid Dynamics Laboratories, Departments of Civil Engineering & Geological Sciences and Aerospace & Mechanical Engineering, University of Notre Dame, Notre Dame, IN, USA
e-mail: Harindra.J.Fernando.10@nd.edu

H.J.S. Fernando et al. (eds.), *National Security and Human Health Implications of Climate Change*, NATO Science for Peace and Security Series C: Environmental Security, DOI 10.1007/978-94-007-2430-3_11, © Springer Science+Business Media B.V. 2012

11.1 Introduction

The urbanization of the world continues at an ever increasing rate, and for the first time in human history the urban population exceeded that of rural areas in 2008. Central to urbanization is the emergence of cities, the centers of human activities. The urbanization proffers clear advantages, in that collective human endeavor and coordinated skills may increase the 'quality of human life' in numerous ways. Conversely, urbanization may also bring forth negative consequences. They are at the vanguard in contributing to climate change, being the largest emitters of greenhouse gases (GHG), particularly CO_2, and agents of land-use change [34]. Reciprocally, cities bear the brunt of climate change impacts, such as desertification, outbreak of infectious and vector born deceases, invasive species, and extreme events (e.g. heat waves, intensified hurricanes and storm surges; [20]). For example, intense and narrowly concentrated rainfall may flood, clog and overflow drainage and sewer systems, triggering water-borne diseases, especially in poorly designed cities in developing countries. Mostly at-risk are the poor and disadvantaged, prompting concerns over social and environmental injustice. For example, about 13% low elevation coastal areas (<10 m from the mean sea level) are urban, although they occupy only 2% of the world's coasts. Inhabitants therein are of disproportionately low income, and they encounter sea level rise, coastal erosion, dwindling fish stock and amplified storm surges due to climate change. Cities sprout human conflicts, and climate change is a multiplier of ensuing security threats. Famine, lack of water and dwindling natural resources stoke morbidity, conflict and uprising. In the developing world, the GDP decreases with rising temperature [10], and if the warming trend continues the third world will face huge economic hardships. It is this delicate interaction of urban development and climate that ought to be harnessed for the benefit of human sustainability, capitalizing the fact that urban-based technological and scientific innovation can be used to mitigate climate change drivers [11]. Available scientific evidence points to urban areas as drivers of climate change, which, in turn, predicates urban sustainability [15, 19, 33]. There are claims to the contrary, which also should be taken into serious consideration in scientific work, but often the climate debate is marred by political and special interests from both sides of the issue.

Unfortunately climate change discourse often neglects the urban response, partly because of the perception that climate change is a "big" issue to be fought and won at the highest political levels. The difficulty of modeling urban domains is also a bane, and so is the notion that climate change is not an issue of urgency. Local authorities find that without strong political and financial will as well as unambiguous data it is difficult to launch and sustain climate change mitigation, adaptation and education programs. Simple interpolation of global results to local scales is not feasible within the highly non-linear earth system, requiring sophisticated high fidelity downscaling methods for educing local impacts of global change. When local details are involved, modeling becomes complicated and uncertainties rack up.

Special educational programs may help develop grass-root awareness of possible grave consequences of climate change, but related efforts are stymied by controversy; misrepresentations by some climate scientists have taken a toll on public confidence of scientists [37]. Climate change mitigation and adaptation involve economic impacts, and only a few local governments are willing to make sacrificial steps to a problem, which they consider as 'belonging' to the global community. Establishment of clear cut connections between global and local scales is imperative if progress is to be made on climate change mitigation, adaptation and education.

In this paper, several urban consequences of global climate change are discussed in light of some critical urban phenomena prone to amplification under climate change forcing. The results are mostly qualitative, or confined to data from a limited number of cities, to be verified using modeling efforts and extensive data gathering from a myriad of urban areas.

11.2   Some Urban Consequences of Global Climate Change; Health and Security Implications

The regional and urban responses of climate change are different from those of a global scale predicted on typical ~100 km computational grids. What happens within the grids needs to be educed, for which grid nesting is used wherein meso and smaller scale models are embedded within regional climate models. The upscaling of urban influence is more difficult, and is still in nascence. It is often argued that metropolitan areas are at higher risk for climate impacts, given the possibility of turning natural disasters into human catastrophes due to population concentration and infrastructure failure. Some repercussions include increased frequency of severe storms and hurricanes, flooding, hailstorms, heat waves, droughts, wild fires as well as intensified urban heat island (UHI), storm surges, arable land loss, crop damage, loss of food supplies, biodiversity loss, virus, bacteria and vector born deceases, sea level rise, loss of native species, incursion of foreign species and desertification. Disruption and burdening of infrastructure includes power outages, overflow and damage to drainage systems and increased energy usage. Urban interests are of near term, rather than long term (30-year) averages predicted by global models. Some recent extreme events have been attributed to anthropogenic climate change, and as climate change progresses more extreme events and even local catastrophic regime shifts are possible. For example, IPCC [22] discusses possible increase of tropical cyclone activity and strength, heavy rainfall and droughts, which lead to cascading socioeconomic repercussions. There is no scientific evidence hitherto, however, to connect disaster losses to anthropogenic climate change without equivocation, as demonstrated by Bouwer [6] based on 22 disaster loss studies. Yet, conceptual studies and data point to increased local repercussions of urban-global interaction.

11.2.1   Response of Coastal Urban Areas

Coastal response of global warming includes sea-level rise, salinity intrusions and increased moisture. Since the early 1970s, the global moisture content has increased gradually [42], although urbanization in inland areas decrease moisture because of engineered impervious surfaces. High resolution climate downscaling calculations predict that warming may bring forth a fewer but fierce hurricanes [5, 28], which is supported by trends observed during the past four decades [48]. Enhanced evaporation from warmer oceans intensifies hurricanes, although their total number may not increase. Increased moisture in coastal areas leads to intense rain (sometimes in narrow coastal belts), resulting in flooding and landslides. Eustatic and glacier-melt induced sea-level rise is an on-going phenomenon that limits freshwater supply and enhances the height and inland penetration of storm surges. An example may be the recent flooding over many geographic areas, ranging from California to Pakistan to Sri Lanka. In the latter, eastern coasts received the highest rainfall in two decades as a part of winter monsoons, affecting two million people and destroying 80% of the agriculture. The intense rainfall in Pakistan during summer monsoons in late July 2010 may be a case of extreme events (although Monsoons themselves are weakened by climate change). Whether these events are directly related to global warming or interannual (e.g. la Nina) events is yet to be determined, but they clearly demonstrate the vulnerability of urban areas to extreme events. Impacts on coastal areas transpire heavy economic impacts, for example, on tourism and fisheries assets. Ocean warming has already damaged coral reefs in countries such as Sri Lanka, and sea-level rise is threatening to submerge Maldives Islands in their entirety.

11.2.2   Inequitable Community Impacts; Global Environmental Justice

It is estimated that some 75–85% of the cost of anticipated damages due to global warming will be borne by the developing world [49]. Climate change also disproportionally impacts poor, uneducated, disabled, socially isolated and very young members of communities. For example, on a per capita basis, people in Catonou, the economic capital of Benin, emits 1/80th of CO_2 of that in the USA, but perceived climate change impacts such as sea-level rise, coastal retreat and erosion have already impacted the Benin coast. Similarly, in the Asia/Pacific Rim, cyclone prone areas in Bay of Bengal, South China Sea and the Philippines are fronted by poor communities. The per capita CO_2 emissions in Nepal is about 1/150th of the USA, but flash floods, droughts, glacier retreats and glacier-lake outbursts have plagued rural Nepal. The average temperature rise in Nepal is about 0.6°C/decade, higher than the global average [38], which, in combination with intensified warming in mountains cause Himalayan glaciers to melt alarmingly faster, thus threatening the future water availability and river flow in Himalayan basin. Increase of temperature in cities due to urbanization is concomitant with higher air pollution, and communities closer to city centers, which tend to be economically disadvantaged, are at higher risk [14]. Urbanization may also create segregation, leading to social hierarchy; at the bottom, there lays slum dwellers that are prone to environmental risks but with little access to adequate health and social services.

11.2.3 Warm Spells and Heat Waves

According to IPCC [22], heat waves are very likely over most land areas, thus increasing the water demand and inciting water quality problems (e.g. algae blooms), water-born vectors and wild fire threats. Morbidity and mortality rates may rise among at-risk people, and the heat stresses cause reduction in workplace productivity. Some native plants may disappear and new ones arise above certain temperature thresholds, threatening local agriculture. An example is the August 2003 heat spell in Europe that claimed 35,000 lives. In France, the temperatures rose up to 104°F and remained unusually high for 2 weeks, and London recorded the first ever triple digit (°F) temperatures. It has been predicted that on the average between the years 2075–2099 three such heat waves are possible per year [43]. Urban areas are more prone to extreme heat due to high heat capacity of construction material as well as ensuing UHI. The last century witnessed plaguing of large cities by heat waves, for example, in Los Angeles (1955), New York City (1972) and Chicago (1995). Increase of energy consumption during heat waves cause overstressing of power production, leading to power outages.

11.2.4 Health and Urban Climate Change

Chapters across this volume emphasize potential climatic impacts on human health. Vectors that spread diseases operate in narrow temperature ranges, and warming may enhance their perimeter, exposing communities that have not yet developed immunity to them [39]. Pathogens and hosts also may undergo alterations. For example, rainfall and floods expose communities to water-borne bacteria that were previously confined to sewers. Indirect health effects arise due to reduction in food supplies, leading to malnutrition and increasing the risk of contracting diseases. Bacteria that perform useful ecosystem functions (e.g. nitrate fixation, disintegration of waste) can be affected, altering vital ecosystem functions.

11.2.5 Security Implications

The mix of environment and human security poses new challenges to urban sustainability and human development [35]. Human security is broadly defined along the scales: security of individuals, communities, ecosystems (environment), nations and the entire humanity [4]. It concerns freedom from risk of damage as well as well-being of individuals – the basic tenet being the concerns for human life and dignity [44]. Human security encompasses national security, which implies safeguards against aggression of foreign nations or terrorism; it obviously cascades down to urban scales, since cities are where most people live, national assets are concentrated, and goods and services are created. It is within this complex web of security enterprise that cities and national governments should address climate change mitigation and adaptation strategies.

11.3   Climate Change Influence on Urban Areas

Intense modifications to the earth surface occurs through urban development [7], which affects surface energy budget and hence microclimate of cities [8]. A common urban thermodynamic phenomenon is the Urban Heat Island, UHI [12, 31], where slow cooling of built environment causes the urban core to be warmer at night compared to rural areas. In cities such as Phoenix, the nocturnal temperature difference between urban and rural can be as high as ~10°F. Socioeconomic consequences of UHI are profound, including health impacts, enhanced energy consumption, rising of social inequity and decrease of industrial productivity [17, 33].

As discussed by Fernando [13], the increase of UHI in Phoenix, Arizona, above a threshold can trigger a "regime shift" in urban microclimate. Remarkably, urban response to global climate change makes this regime shift viable, although in ensemble mean climate projections there is no indication of abrupt climate change. In addition, individual model outputs reported in literature do not provide any strong suggestions of robust nonlinearities [22]. The amplification of local changes as a result of global climate change is the driver of local regime shifts.

In general, urban areas occupy only a small portion of global land coverage (e.g., only 2–3% land; [24]). IPCC [22] contends that their global climatic impacts are insignificant compared to drivers such as CO_2 and aerosols. To some others, this may be misleading [25], and urban influence can be as substantial as such major interannual events as ENSO [46]. Thermodynamics of cityscape can drastically affect mesoscale flow, which in turn can influence large-scale circulation [50]. On the scale of the entire continental USA, Kalnay & Cai [26] estimated that the change of land cover in agricultural and urban areas are causing a surface warming of 0.27°C per century, whereas Zhou et al. [51] showed that in Southeastern China a warming of ~0.05°C per decade can be attributed to urban sprawl. According to Houghton et al. [18], global radiation balance determines the global climate, and changes therein may cascade down to regional and local scales while the local influence may propagate to global scales. This was substantiated by Hansen et al. [16], who showed that urban areas have a disproportionate impact on climate, in that microclimate modifications at smaller scales can be amplified to produce "urban warming."

11.3.1   Change of DTR

According to IPCC [22], the variability of daily to interannual temperatures is likely to decrease in the winter and increase in the summer for mid-latitude Northern Hemisphere land areas, and daily high temperature extremes are likely to increase as well. Observations show that since 1950 the average surface diurnal temperature range (DTR) has decreased, with both the minimum nighttime temperature and maximum daytime temperature showing an upward trend, the increase in the former being greater than the latter [27]. This is one of the most significant decreases of DTR observed in climatic records, which is attributed to the global increase of cloud cover and trapping of extra long wave radiation due to the increase of GHG. Field studies show that low clouds and soil evaporation also have a significant influence [9, 41]. Nevertheless, model simulations show that GHG forcing alone cannot account for the observed change [40], pointing to the importance of including new boundary layer physics in modeling.

The amount of warming depends on the ABL regime, being greater in the non-turbulent boundary layer and less in the turbulent boundary layer. A striking fact is that increase of GHG forcing or cloud cover, even if slight, can cause the system to transition from a non-turbulent to the turbulent state and produce significant changes in surface temperature [47]. The daytime increase of diurnal temperature maximum T_{max} is a competition between the reduction of incoming radiation due to clouds and outgoing radiation by the increase of cloud cover as well as GHG and moisture, among other factors. At night, increased cloud cover reflects back more long wave radiation, which is absorbed by GHG and moisture, thus decreasing the stability of the Nocturnal Boundary Layer. Enhanced turbulence in weakly stable boundary layer can entrain warmer fluid from aloft, further contributing to higher T_{min}. The overall effect would be a stronger reduction of DTR.

It is instructive to consider how this global trend of DTR is felt locally. With regard to T_{max}, higher albedo of urban surfaces may have some cooling influence [41], which is countered by the long wave radiation reflected back from cloud cover. The higher heat capacity of building material as well as impervious surfaces cause higher temperatures on built surfaces, which transfer heat to adjoining air via establishing convective currents. Building canyons also trap some of the outgoing radiation. The net result is either an increase or even a decrease of T_{max}, depending on governing parameters.

At night, in contrast to rural areas, built elements cool slower, and cooling is further impeded by cloud cover and radiation trapping in urban canyons. The UHI drives a flow toward the city and generates plumes above it, and this circulation interacts with urban elements to produce turbulent mixing and ground-level warming via air entrainment from above. Urban areas therefore are prone to higher increases of T_{min} compared to increase of T_{max}, and DTR in urban areas is expected to decrease faster with global warming and continued urbanization; these effects are catalogued in the Table 11.1.

The phenomena described above are further studied next with reference to two cities: Phoenix (Arizona) and Chicago (Illinois). Chicago represents a humid, mid-latitude climate whereas Phoenix is arid and subtropical. Although the population in Chicago is more than three times that of Phoenix (three million), the rates of infrastructure development and population growth over time have been different. While population in Chicago increased from three million in 1970 to over nine million in 2000, the population growth rate (11.1%) from 1990 to 2000 has been lower than

Table 11.1 Response of maximum and minimum diurnal temperature to climate change, including urban influence

	Changes due to global warming and GHG	Effect on T_{max} and T_{min}	Additional urban effects	Effects on T_{max} and T_{min} (urban)
T_{max}	Cloud cover reduces radiation (in) and (out); increased GHG and water vapor absorbs (out) radiation	T_{max} Increases	Albedo increases (a cooling effect); radiation traps in urban canyons; impervious surfaces reduce moisture	Opposing effects set T_{max}; higher or lower increases in T_{max} are possible compared to rural
T_{min}	Cloud cover and GHG reduce outgoing radiation; stability of NBL is reduced by warming; entrainment occurs from aloft	T_{min} increases faster; DTR decreases	Turbulence from urban elements and UHI promotes entrainment of warm air aloft; radiation is trapped in the canyons	Significant increase of T_{min} (compared to rural); reduction of DTR is larger

that of Phoenix (45.3%) or the USA (15%). From 2000 to 2010, the respective values are 24.2% and 4% [45]. Therefore, the DTR reduction in Chicago is expected to follow more or less the global (background) trend whereas in Phoenix it also encompasses continuing urbanization. In the following, temperatures from routine urban monitoring stations at Chicago O'Hare and Phoenix Sky Harbor airports as well as Phoenix Coronado are compared with their suburban counterparts Dekalb (Chicago) and Sacaton (Phoenix) (see Figs. 11.1 and 11.2).

As illustrated in Fig. 11.2b, the rate of rise of T_{max} for urban Phoenix area is approximately equal to the rural site, which can be explained by the increase of rural T_{max} due to increase of cloud cover [22] as well as more or less similar rise of T_{max} in the urban area as a result of competing factors discussed before. The curious result of same T_{max} in urban and rural areas points to the lack of UHI during the day. This is possibly contributed by the 'oasis effect' as well as slow warming of urban surfaces due to high heat capacity. In Phoenix, the rise of T_{min} is much more dramatic, with Sky Harbor at 10.3°F over five decades, but at Sacaton the increase is slower, 5.6°F over five decades. This is consistent with the postulation that T_{min} is slated to rise in growing urban areas faster than that in rural areas. The result is a clear reduction of DTR at the Sky Harbor site. The rate of increase of T_{min} in rural sites of Phoenix and Chicago has been much slower, being devoid of urban effects; both Sacaton and Dekalb have similar rates of rise of T_{min}, which can be considered as background values (Fig. 11.2a, b). Also, the rates of increase of (rural) T_{max} in Dekalb and Sacaton do not differ, again indicating background values. The rate of increase of T_{min} in Sky Harbor is significantly higher than that of O'Hare, possibly due to continued urbanization in the former. The rate of reduction of urban DTR is higher in both urban cases, but is clearly amplified in Phoenix. The humidity variation of two urban sites in Chicago and Phoenix are shown in Fig. 11.3.

11.3.2 Effects on UHI

The influence of UHI on global temperature is not well understood. Kalnay and Cai [26] argued that urbanization and landuse change can have a positive influence on regional warming, but there are arguments to the contrary. According to Trenberth [41], the increase of albedo due to urbanization may have a cooling influence rather than a warming effect. Parker [32] argued that significant biases have not been introduced into global warming by UHI.

It appears that higher T_{max} and a decrease of DTR have a notable effect on UHI. While UHI is mainly governed by the surface energy balance and the reduction of incoming radiation is uniform over both rural and urban areas, the increase of air temperature during the day and a reduction of building surface temperatures may reduce circulation within urban canyons, thus impeding heat removal from canyons. This increases the air temperature and raises cooling requirements for buildings, thus increasing anthropogenic heat emissions; this is a positive feedback. The increase of albedo and higher urban surface moisture, however, has an opposite

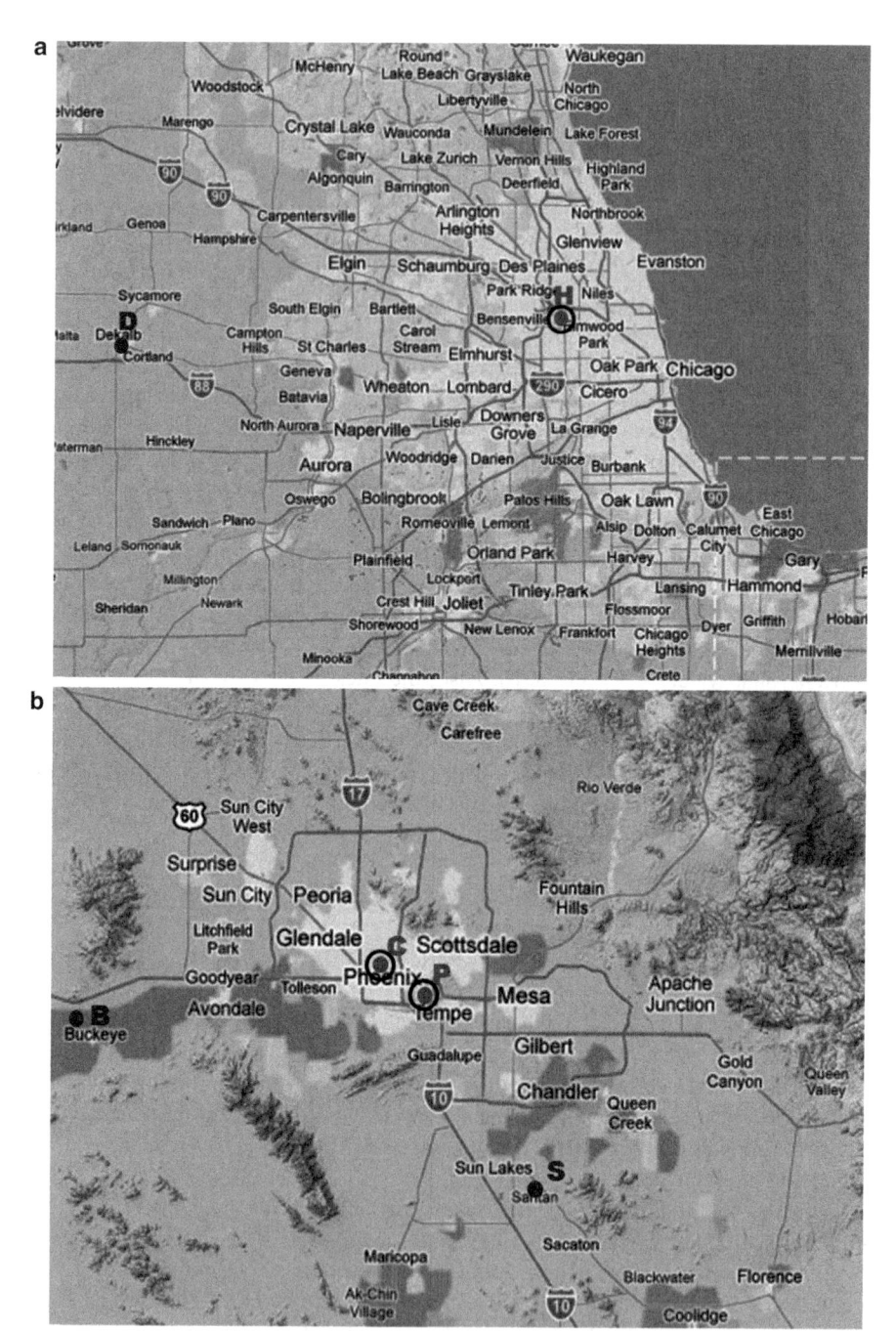

Fig. 11.1 Chicago (**a**) and Phoenix (**b**) metropolitan areas and site locations (*H* O'Hare Airport, *D* Dekalb, *C* Coronado, *P* Sky Harbor Airport, *S* Sacaton, *B* Buckeye); urban monitors are *encircled*

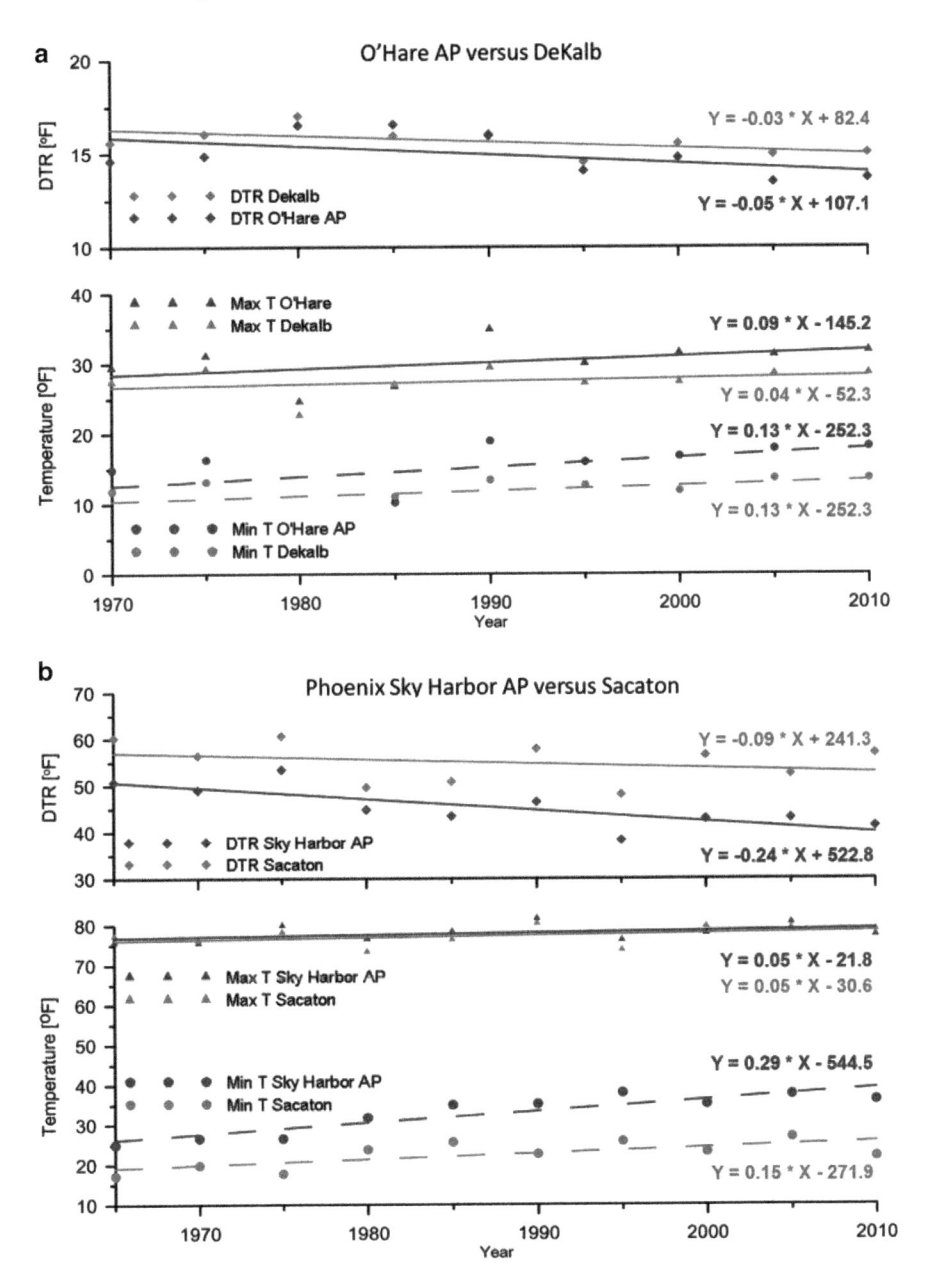

Fig. 11.2 The increase of T_{min} and T_{max} as well as the decrease of DTR for Phoenix (Sky Harbor and Sacaton sites) and Chicago (O'Hare and DeKalb sites), representing 5 year averages

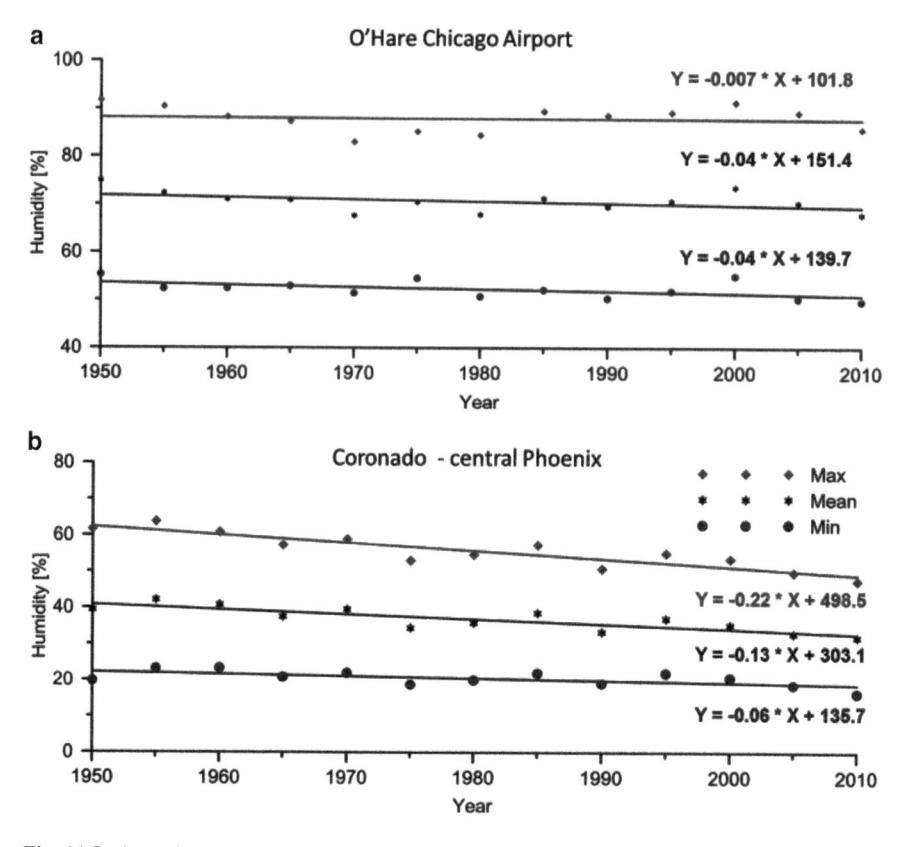

Fig. 11.3 Annual averaged humidity at Chicago – O'Hare International Airport (**a**) and Central Phoenix – Coronado site (**b**)

effect, and as discussed the rise of T_{max} is determined by a competition between different processes. If the ambient temperatures become higher, convective cooling of buildings at night is difficult still, contributing to increase of T_{min}. Therefore, overall, an intensification of UHI can be expected due to global warming, especially at night, unless additional overarching factors determine otherwise.

An increase of UHI has repercussions in micrometeorology and pollution distribution of cities. Figure 11.4 shows how the UHI has changed in Phoenix and Chicago areas, based on the minimum temperature difference (this is a better indication of UHI, given that T_{max} is determined by competing processes). The slow increase of UHI in Chicago and rapid increase in Phoenix are a result of differing local urbanization and global climate change, but it is difficult to separate individual contributions of these two factors without detailed modeling. The slower increase in Chicago is determined by (a) slow urbanization (it is a matured city) and (b) climate change influence that increases T_{min} in the urban core but also in rural areas. In urban Phoenix, continued changes of land use and built environment have more pronounced effect of T_{min}. As far as T_{max} is concerned, Phoenix shows very little UHI due to combined effects of built environment, moisture, radiation trapping and

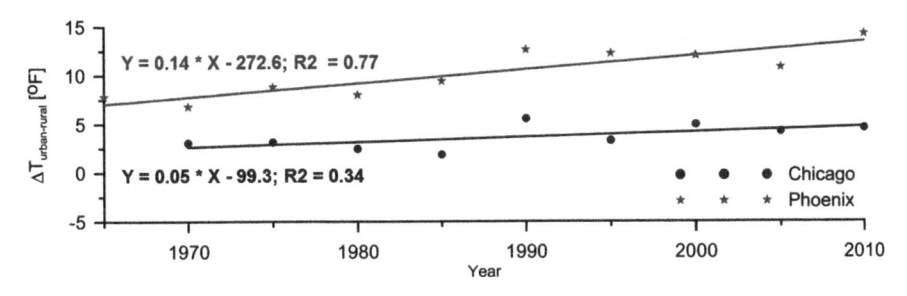

Fig. 11.4 Minimum temperature difference between urban and rural sites (UHI) for Chicago and Phoenix area

climate change, and the oasis effect that can overpower UHI. Conversely, Chicago shows a small (~2°F) UHI that increases gradually with time, that perhaps signifies climate change influence on UHI.

11.3.3 Local Regime Shifts Due to UHI

Based on mesoscale simulations of Lee et al. [29], Fernando [13] pointed out that an increase of UHI by several more degrees may change the characteristics of morning transition in the Phoenix area, which, in turn, can cause a regime shift in air pollutant distribution. The mechanisms responsible for this shift are explained in Fernando [13], wherein the interaction between UHI and background thermal circulation is responsible for an abrupt change. In greater Phoenix, high pollutant emissions during morning are usually carried to the East valley by upslope flow. During transit, ozone is formed and helps create the highest ozone concentration in the east valley. A delay in morning transition from downslope to upslope flow, however, may cause pollutants from the peak emission period to be carried further into the west valley by downslope flow. When downslope to upslope transition occurs at a later time, the pollutant cloud is much closer to and attracted toward the mountainous terrain to the west, rather than to that of the east, thus creating high ozone and PM levels in the west valley. This is undesirable as the west valley is highly populated. If reduction of DTR due to climate change intensifies UHI, in addition to the contribution by urbanization, there is a possibility that climate change may cause a shift in the air pollution distribution regime, thus threatening the quality of life in the area.

11.3.4 Human Health Impacts

Ambient air contains a wide range of pollutants, for example, gases such as ozone, nitrogen oxides and sulfur dioxide as well as particulate matter (PM) of different sizes. Air quality depends on weather, and hence is sensitive to climate change [23].

Air pollutants in urban areas are regulated through the National Ambient Air Quality Standards (NAAQS) that set pollutant concentration limits to protect public health, including the health of "sensitive" populations such as asthmatics, children, and the elderly. Atmospheric particles of aerodynamic diameter less than 10 μm (the so called PM_{10}) are of special concern as they have been directly associated with potential health problems. As such, PM is a principal (criteria) pollutant regulated by the US EPA with a 24 h-averaged mean of 150 μg/m³. Similarly, fine particulates PM2.5, mostly emitted by combustion sources, are also regulated (15 μg/m³ annual and 35 μg/m³ daily). During the past decade, the statewide prevalence of asthma in Arizona has continued to increase, and it is much higher than the national average [30]. Although the causes of childhood asthma are not well understood, air pollution has a clear role in triggering attacks. Environmental exposures may increase the risk of developing asthma, and they may also increase the frequency and severity of asthma incidents. Asthma is a major public health issue of growing concern in Phoenix, especially amongst children of age 0–19, which represent 38% of all asthma cases in Arizona [1]. In 2009, 21% of youth and 15% of adults in Arizona have been diagnosed with asthma [2]. While undisturbed desert is not typically a significant source of dust, human activities of various forms lead to major PM sources, including fugitive dust from paved roads, construction sites and unpaved roads as well as wind-blown dust from agricultural fields and vacant lots. An extensive study of elevated concentrations of PM_{10} and childhood asthma in central Phoenix shows significant correlation. Fernando et al. [14] expressed the risk of adverse health effects as a function of the change from the 25th to 75th percentiles of daily mean PM_{10} (36 μg/m³) and associated it with a 12.6% increase in asthma incidents among children of ages 5–17.

Local climate change of Phoenix, characterized by increasing average air temperature and decreasing humidity, directly affects PM_{10} concentration and thus human health. The PM entrainment by overlying air depends greatly on soil moisture, which is related to local climate. Increasing air temperature and decreasing humidity/precipitation decrease moisture, which facilitates PM_{10} entrainment. In the following, we will attempt to correlate PM_{10} concentration in the Phoenix area with governing meteorological variables, and infer how future climate change and atmospheric moisture would affect asthma incidences.

In our work, relationships between monitored air quality parameters and meteorological variables were determined through multivariate regression analyses using a statistical software module from MatLab. The PM_{10} concentration was considered as the dependent variable, while meteorological parameters [temperature (T), wind speed (W) and relative humidity (H)] were the independent variables (see [3]). It was assumed that the dependent variable follows normal distribution and meteorological parameters are independent of each other. A stepwise multiple regression analysis was performed to predict the regression coefficients of a linear equation between dependent and three independent variables. The coefficients of goodness of fit of a linear model (R^2) were obtained, defined as the proportion of total variability

Table 11.2 Regression equations and corresponding R2

Data type	Model	R^2
Hourly	$PM_{10} = 90.953546 + 0.8859 * W - 0.2418 * T - 0.4844 * H$ (model 1)	0.19
Daily	$PM_{10} = 93.971420 - 0.8807 * H$ (model 2)	0.54
Weekly	$PM_{10} = 116.258946 - 3.9291 * W - 1.0113 * H$ (model 3)	0.75
Monthly	$PM_{10} = 96.336819 - 0.9584 * H$ (model 4)	0.83

Table 11.3 Correlation coefficients (r) between concentration and meteorological variables

Data type	Wind speed	Temperature	Relative humidity
Hourly	0.10	0.06	−0.17
Daily	0.03	0.21	−0.54
Weekly	−0.07	0.27	−0.78[*]
Monthly	−0.05	0.37	−0.83[*]

[*]Correlation is significant at the 0.05 level (two-tailed)

in the dependent variable (PM_{10}) that is accounted by the regression equation [21]. Note that R^2 can be expressed as

$$R^2 = 1 - \frac{\sum (\hat{y}_i - \overline{y})^2}{\sum (y_i - \overline{y})^2},$$

where $\hat{y}_i$ is the value of y predicted by the regression line, y_i is the value of y observed, and $\overline{y}$ is the mean of the y_i. Note that $R^2 = 1$ signifies that the fitted equation accounts for the entire variability of dependent variables.

The hourly PM_{10} data were obtained from the Phoenix Buckeye ambient air quality monitoring station, which were used to calculate the daily, weekly and monthly variations of PM_{10} concentration as a function of appropriately averaged T, H and W. Monitored concentrations (O) were further analyzed vis-à-vis. the predicted values (P) by correlation analysis. The correlation coefficient (r) was determined using

$$r = \left[\sum_{i=1}^{N} (O_i - \overline{O})(P_i - \overline{P}) \right] / \left[\sum_{i=1}^{N} (O_i - \overline{O})^2 \sum_{i=1}^{N} (P_i - \overline{P})^2 \right]^{1/2}.$$

The linear regression models obtained using different independent variables and the corresponding coefficients are listed in Tables 11.2 and 11.3. A significant dependence of PM_{10} concentration on relative humidity was found. When the dependence on all variables was included based on hourly data, R^2 was very low ~ 19%, but a significant increase of R^2 was found when only the relative humidity and wind speed (75%) was included with weekly average data or only humidity (83%) with monthly average data.

Figure 11.5 shows the weekly and monthly averaged PM_{10} together with relevant meteorological parameters for the year 2005, which were used for correlation analysis.

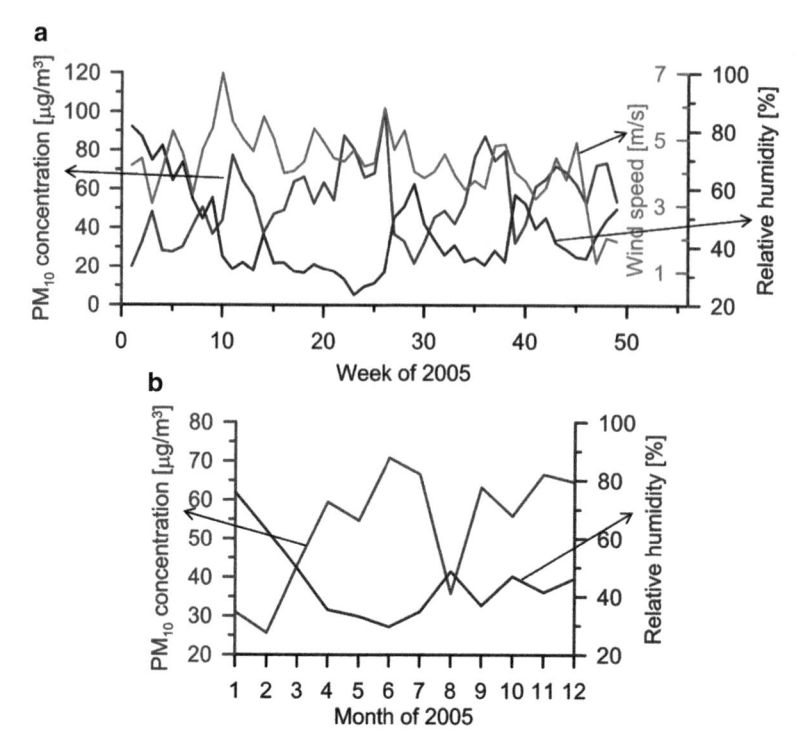

Fig. 11.5 Weekly and monthly variation of the PM$_{10}$ concentration, based on 2005 data (used for establishing regression models)

Figure 11.6 shows the application of models so obtained (Table 11.2) to predict other available data sets (In the adjacent years only 3 months of data were available). For weekly data, model 3 was used whereas model 4, which accounts only for humidity, was applied to predict monthly data. The data trends are well predicted by the regression models. The PM$_{10}$ concentration increases with decreasing relative humidity and according to Fig. 11.3b, Phoenix area is expected to increase its PM$_{10}$ contribution in the future accompanied by an increase of asthma occurrences. The decrease of moisture in this case is a local effect, contributed by land use change. Alternatively, atmospheric warming due to climate change is expected to increase global moisture by about 7.5% per °C of warming [36, 41], and hence may reduce PM$_{10}$.

11.4   Conclusions

It appears that urban response to global climate change is largely an amplified one, which has clear implications for human health and security enterprise. While in the past only limited attention has been given to the linkage of urban and global phenomena, the climate change discourse is rapidly changing to encompass urban

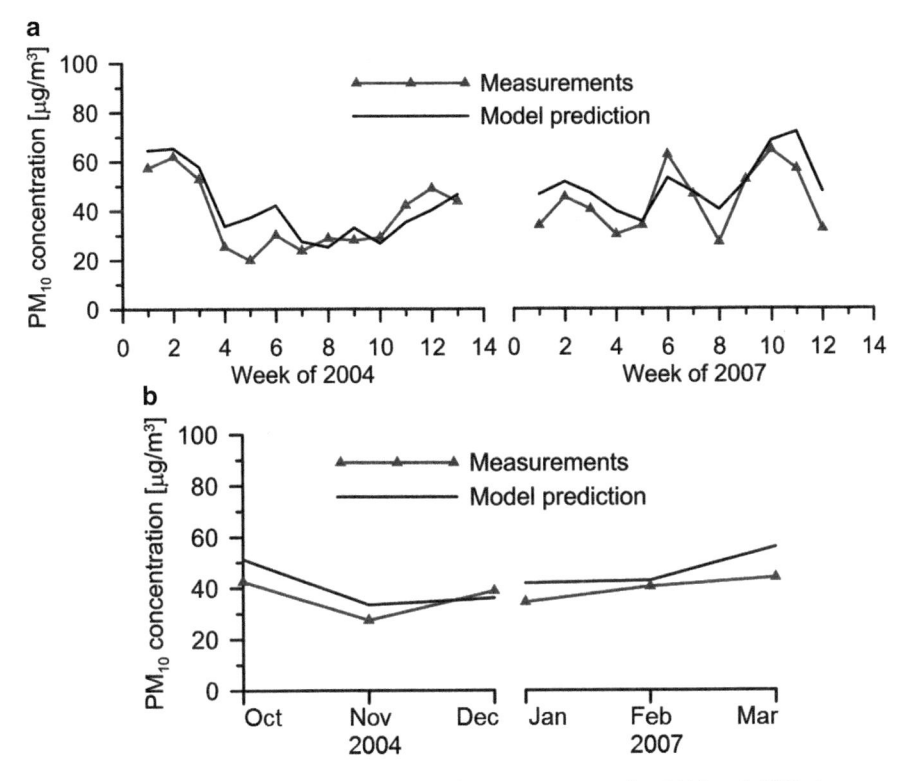

Fig. 11.6 Application of regression models in Table 11.1 to predict 2004 and 2007 data sets (based on weekly and monthly data)

repercussions. In this paper, it was argued that the diurnal temperature range as well as urban heat island may amplify as a result of climate change. Local response also may include increasing health threats and meteorological regime shifts. Modeling effects that connect global and urban scales as well as long term observations are sorely needed to further delineate the urban reaction to global climate change, the later itself being largely contributed by urban actions.

References

1. Arizona Department of Health Services (2009a) Asthma Program. Available from http://www.azdhs.gov/asthma/
2. Arizona Department of Health Services, Division of Public Health (2009b) Characteristics of ER visits and inpatient discharges with the diagnosis of Asthma. Available from http://www.azdhs.gov/plan/hip/for/asthma/index.htm
3. Banerjee T, Singh SB, Srivastava RK (2011) Development and performance of statistical models correlating air pollutants and meteorological variables at Pantnagar, India. Atmos Res 99:505–517

4. Barnett J (2001) The meaning of environmental security. Zed Books, London
5. Bender M, Knutson TR, Tuleya RE, Sirutis JJ, Vecci GA, Garner ST, Held IM (2010) Modeled impact of anthropogenic warming on the frequency of intense Atlantic hurricanes. Science 327(5964):454–458. doi:10.1126/science.1180568
6. Bouwer LM (2010) Have disaster losses increased due to anthropogenic climate change? Bull Am Meteorol Soc 92:39–46
7. Changnon SA Jr (1992) Inadvertent weather modification in urban areas, lessons for global climate change. Bull Am Meteorol Soc 73:619–627
8. Crutzen PJ (2004) New directions: the growing urban heat and pollution "island" effect-impact on chemistry and climate. Atmos Environ 38(21):3539–3540
9. Dai A, Trenberth KE, Karl TR (1999) Effects of clouds, soil moisture, precipitation and water vapor on diurnal temperature range. J Clim 12:2451–2473
10. Dell M, Jones B, Olken B (2009) Temperature and income: reconciling new cross-sectional and panel estimates. AmEcon Rev Pap Proc 99(2):198–204
11. Economist (2010) Facing the consequences, 17 Nov–3 Dec 2010, p 85
12. Emmanuel R, Fernando HJS (2007) Effects of urban form and thermal properties in urban heat island mitigation in hot humid and hot arid climates: the cases of Colombo, Sri Lanka and Phoenix, USA. Clim Res 34:241–251
13. Fernando HJS (2008) Polimetrics: the quantitative study of urban systems (and its applications to atmospheric and hydro environments). J Environ Fluid Mech 8(5–6):397–409. doi:10.1007/s10652-008-9116-1
14. Fernando HJS, Dimitrova R, Ruunger G, Lurponglukana N, Hyde P, Hedquist B, Anderson J (2009) Children's health project: linking Asthma to PM_{10} in Central Phoenix – a report to the Arizona Department of Environmental Quality. Arizona State University – Center for Environmental Fluid Dynamics and Center for Health Information and Research, Feb 2009
15. Grimm NB, Faeth SH, Golubiewski NE, Redman CL, Wu J, Bai X, Briggs JM (2008) Global change and the ecology of cities. Science 319:756–760
16. Hansen J, Ruedy R, Sato M, Imhoff M, Lawrence W, Easterling D, Peterson T, Karl T (2001) A closer look at United States and global surface temperature change. J Geophys Res 106(D20):23,947–23,963
17. Harlan SL, Brazel A, Jenerette GD, Jones NS, Larsen L, Prashad L, Stefanov WL (2008) In the shade of affluence: the inequitable distribution of the urban heat island, equity and the environment. Res Soc Prob Public Policy (Elsevier) 15:173–202
18. Houghton JT, Ding Y, Griggs DJ, Noguer M, Van der Linden PJ, Dai X, Maskell K, Johnson CA (2001) Climate change 2001: the scientific basis. Cambridge University Press, Cambridge, 881
19. Hunt JCR (2004) How can cities mitigate and adapt to climate change? Buil Res Inf 32(1):55–57, Jan–Feb Issue
20. Huq S, Kovtas S, Reid H, Satterwaite D (2007) Editorial: reducing risk to cities from disasters and climate change. Environ Urban 19(1):3–15
21. Ilten N, Selici T (2008) Investigating the impacts of some meteorological parameters on air pollution in Balikesir, Turkey. Environ Monit Assess 140:267–277
22. IPCC (2007) Climate change 2007: the physical bases. In: Solomon S, Qin D, Manning M, Chen Z, Marquis M, Averyt KB (eds) Contribution of working group I to the fourth assessment report of the Intergovernmental Panel on Climate Change. Cambridge University Press, Cambridge, p 996
23. Jacob D, Winner DA (2009) Effect of climate change on air quality. Atmos Environ 43:51–63
24. Jin M, Zhang DL (2002) Changes and interactions between skin temperature and leaf area index in summer 1981–1998. Meteorol Atmos Phys 80:117–129
25. Jin M, Dickinson RE, Zhang DL (2005) The footprint of urban areas on global climate as characterized by MODIS. J Clim 18:1552
26. Kalnay E, Cai M (2003) Impact of urbanization and land-use change on climate. Nature 423:528–531

27. Karl TR, Jones PD, Knight RW, Kukla G, Plummer N, Razvayev V, Gallo KP, Lindseay J, Charlson JRJ, Peterson TC (1993) Asymmetric trends in surface temperature. Bull Am Meteorol Soc 74:1007–1023
28. Kerr R (2010) Models foresee more intense hurricanes in the greenhouse. Science 327:399
29. Lee SM, Fernando HJS, Grossman-Clarke S (2007) Modeling of ozone distribution in the State of Arizona in support of 8-hour non-attainment area boundary designations. Environ Model Pred 12:63–74
30. NCHS National Center for Health Statistics (2003) Current estimates from the National Health Interview survey U.S. Department of Health and Human Services, Public Health Services, Vital and health Statistics, Washington, DC. Available from http://www.cdc.gov/nchs/fastats/asthma.htm
31. Oke TR (1988) Boundary layer climates, 2nd edn. Methuen, New York
32. Parker DE (2004) Large-scale warming is not urban. Nature 432:290
33. Patz JA, Campbell-Lendrum D, Holloway T, Foley JA (2005) Impact of regional climate change on human health. Nature 438:310–317. doi:10.1038/nature04188
34. Romero-Lankao P, Doodman D (2011) Cities in transition: transforming urban centers from hotbeds of GHG emissions and vulnerability to seed beds of sustainability and resilience: introduction and editorial overview. Curr Opin Environ Sustain 3:1–8
35. Sachs W (ed) (1993) Global ecology; a new arena of political conflict. Zed Books, London
36. Santer BD, Mears C, Wentz FJ (2007) Identification of human-induced changes in atmospheric moisture content, vol 104, no. 39. PNAS, 25 Sept 2007, pp 15248–15253
37. Schiermeier S (2010) The real holes in climate science. Nature 463:284–287
38. Shrestha AB, Wake CP, Mayewski PA, Dibb JE (1999) Maximum temperature trends in the Himalaya and its vicinity: an analysis based on temperature records from Nepal for the period 1971–94. J Clim 12:2775–2787
39. Smolinski MS, Hamburg M, Lederberg J (eds) (2003) Microbial threat to health: emergence, detection and response. National Academies Press, Washington, DC
40. Stone DA, Weaver AJ (2003) Factors contributing to diurnal temperature range trends in the twentieth and twenty first century simulations of the CCAma coupled model. Clim Dyn 20:435–445
41. Trenberth KE (2004) Rural land-use change and climate. Nature 427:213
42. Trenberth K, Fausullo J, Smith L (2005) Trends and variability in column-integrated atmospheric water vapor. Clim Dyn 24:741–758
43. Tretkoff E (2010) Urban areas and climate change, EOS. Trans Am Geophys Soc 91(51):503. doi:10.1029/2010EO510004
44. UNDP (United Nations Development Program) (1994) Human development report. Oxford University Press, New York
45. USA Census Bureau (2001) Ranking tables for metropolitan areas: 1990 and 2000. Retrieved on 8 July 2006, 2010. Available from http://www.census.gov/population/cen2000/phc-t3/tab05.txt; http://www.census.gov/population/cen2009
46. Vörösmarty CJ, Green P, Salisbury J, Lammers RB (2000) Global water resources: vulnerability from climate change and population growth. Science 289:284–288
47. Walters JT, McNider RT, Shi X, Norris WB, Christy JR (2007) Positive surface temperature feedback in the stable nocturnal boundary layer. Geophys Res Lett 34:L12709. doi:10.1029/2007GL029505
48. Webster PJ, Holland GJ, Curry JA, Chang H-R (2005) Changes in tropical cyclone number, duration, and intensity in a warming environment. Science 309:1844–1846
49. WDR – World Development Report (2010) Development and climate change. www.worldbank.org
50. Yan H, Anthes RA (1988) The effect of variation in surface moisture on mesoscale circulations. Mon Weather Rev 116:192–208
51. Zhou LM, Dickinson RE, Tian YH, Fang JY, Li QX et al (2004) Evidence for a significant urbanization effect on climate in China. Proc Natl Acad Sci USA 101:9540–9544

Chapter 12
Future Heat Waves over Paris Metropolitan Area

A.L. Beaulant, A. Lemonsu, S. Somot, and V. Masson

Abstract The aim of this study is to analyse urban heat wave events in present climate (1961–1990) and their evolution in a changing future climate (2021–2050, 2071–2100). We used daily observations of temperature from stations in Paris, climate model projections following three SRES scenarios (A2, A1B, B1) and issued from several regional climate models. A heat wave is detected within observed or simulated time-series by a peak, when temperatures exceed the 99.9th percentile. Its duration is determined by all adjacent days to this peak. Events are extracted, then validated within observations and 12 climatic simulations. Over 2071–2100, we count 3.5, 3.8 and 2.1 events per year for A2, A1B and B1 scenario respectively, using one climate model. The ten A1B climate models simulate 1.5 heat waves per year on average. Despite a large variability, HW characteristics show an overall trend to an increase in duration and intensity, which is more pronounced at night-time. Over 2071–2100, extreme events have night-time temperatures of 28°C for A1B (against 20°C in the 2003 heat wave) ; day-time temperatures of 45°C for A1B (against 39°C in 2003) and last up to 1 month.

12.1 Introduction

In the global context of climate change, Intergovernmental Panel on Climate Change (IPCC) experts are agree to announce an increase in temperature during the twenty-first century. These projections suggest that summer heat waves will become more

A.L. Beaulant (✉) • A. Lemonsu • S. Somot • V. Masson
Centre de Recherches Météorologiques, CNRS/Météo France,
42 avenue Gaspard Coriolis, 31057, Toulouse Cedex, France
e-mail: anne-lise.beaulant@meteo.fr

frequent and severe during this century [2]. In this context, cities are particularly vulnerable to heat waves (HWs), in particular because the heat island that characterizes the urban climate [9] exacerbates heat wave effects. The Urban Heat Island (UHI) that develops within and over metropolitan areas is defined by higher air temperatures in cities than in the surrounding rural areas. The difference of temperature can reach up to 10° for large agglomerations ([7] review by [11]).

The higher temperatures in cities can strongly amplify the situation of heat stress, especially at night, during heat waves, and conduct to serious consequences in terms of public health. It has been the case in 2003 in France [10] when a strong heat wave affected West-Europe and caused more than 70,000 casualties. More recently, the heat wave that occurs in Russia during the summer 2010 leads to an excess of nearly more than 56,000 deaths nationwide in comparison to the same period in 2009 (Barriopedro et al. 2011).

The possible evolution of heat wave events in the context of global warming at a local scale such as a city is a crucial subject. This challenging task is the object of our study. It aims at establishing a range of possible future heat waves within climate model projections. The objective is to analyse urban heat waves over the Paris region (France) in present climate and their evolution in a future climate with enhanced greenhouse gases. We use daily observations of temperature from Paris stations, as well as climate model projections following A2, A1B and B1 SRES-scenarios. These projections are issued from several regional climate models forced by several global climate models.

12.2 Heat Wave Definition and Validation

In order to extract HW events within temperature time-series, we must establish a HW definition based on quantifiable variables. In the literature, heat wave definition, detection and duration are numerous [2, 4, 6]. A HW definition is already used in the frame of the French operational warning system called "Plan National Canicule" (PNC). Instead of using daily temperature directly, it is based on minimum and maximum biometeorological indicators (BMIn and BMIx) which are calculated as a moving average of minimum (Tn) and maximum (Tx) daily temperatures, respectively, over three consecutive days (D, D+1, D+2). The approach enables us to account for the heat-stress cumulative effects of HWs. In addition, the thresholds applied to BMIn and BMIx are defined as high percentiles (~99.5%) that are fitted by including statistical information on excess mortality. Based on the literature and on the PNC, a new definition is proposed for our study. A HW event is first detected by a HW peak that is determined when the daily BMI values exceed the 99.9th percentile. This percentile is defined according to the Chartres historical time-series, taken as reference time-series, over the reference time-period (1961–1990). BMI thresholds applied by the PNC for Chartres are18°C and 34°C for BMIn and BMIx, respectively. Then, the HW duration is not imposed but determined by all adjacent days to the peak, for which BMI values are not durably (2 days) smaller

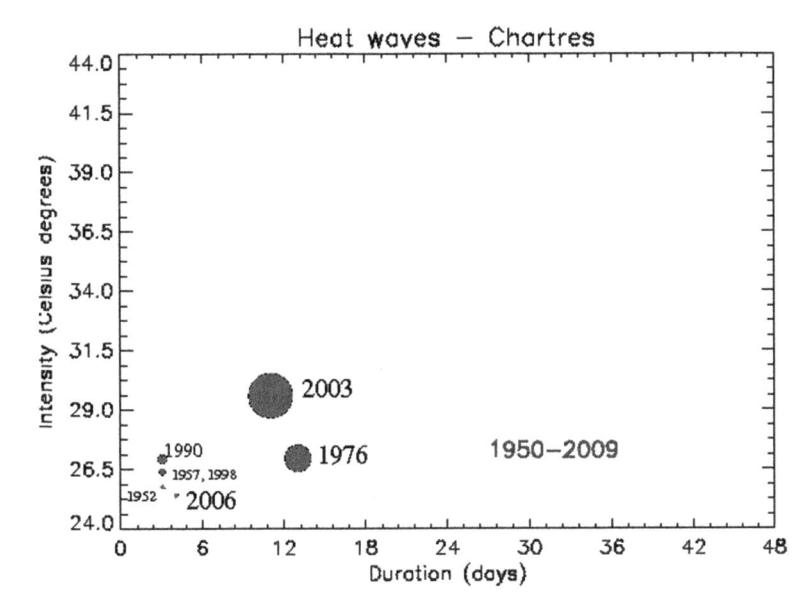

Fig. 12.1 Observed HW events for the 1950–2009 period over the Paris area

than the 99.9th percentile threshold minus 2°C. Finally, only HW events that last at least 3 days are considered. BMI thresholds defined for Paris by the PNC are not chosen here because they account for the UHI and are consequently higher than those applied around the city.

We first ensure that the HW definition enables us to retrieve historical HW events recorded over the Paris area. Meteo France archives count, at the country scale, only one HW event of exceptional intensity (2003), three HW events of strong intensity (1976, 1983 and 2006), and seven of moderate intensity. Following the HW definition, past HW events are extracted from BMI time-series of Chartres over the 1950–2009 time-period. The 99.9th threshold is applied on BMIx and BMIavg=(BMIn+BMIx)/2 (i.e. 34.13°C and 26.26°C, respectively). Thus a HW peak is detected when at least one of these conditions is satisfied. Finally, seven HW events are identified and extracted (see Fig. 12.1).

The events of 2003 and 1976, which are the strongest heat waves affecting the Paris area, are particularly well described in terms of duration. The 1976 event lasts 13 days in our observations against 15 days in the Meteo-France database. For 2003, we identify an event of 11 days instead 13 days in the database. The HW event of 2006 is also identified in our observations, but not the one of 1983. Indeed these two events have particularly affected South-Eastern France and less the Paris region, all the more for the 1983 event. We also extract four HW events (1952, 1957, 1990, 1998) among the seven events listed in the Meteo-France database. We validate the proposed definition of HW event since it enables to retrieve most historical HW events that have affected the Paris area.

12.3 Extraction of Future Heat Waves Within 12 Climatic Projections

12.3.1 Climate Models

We use two main databases of climate simulations that provide daily minimum and maximum temperature time-series over 1950–2100. Climate projections of the first set were conducted with the variable-resolution ARPEGE-Climate-v4 model [3] from Meteo-France that covers Western Europe with a 50-km spatial resolution using a stretched grid approach. These projections were performed using SRES A2, A1B and B1 scenarios [5]. The second set comes from the RT3 regional climate model database of the ENSEMBLES European project, which gathers climate simulations produced by a wide panel of European regional climate models (RCMs) with a 25-km spatial resolution driven by several global climate models (GCMs) following the A1B emission scenario only (see Table 12.1). Details may be found at http://ensemblesrt3.dmi.dk.

12.3.2 Simulated Time-Series

The simulated BMIn and BMIx time-series are calculated from daily minimum and maximum temperature time-series for the closest model grid point to Paris. By combining data from Meteo France and ENSEMBLES databases, we obtain 12 BMIn and BMIx time-series covering 1950–2100. The 1961–1990 control period is used

Table 12.1 Characteristics of climatic simulations

Institute	Scenario	GCM	RCM
Meteo France database (50-km resolution over Europe)			
Meteo France	A2		Arpege Climat VR*
Meteo France	A1B		Arpege Climat VR*
Meteo France	B1		Arpege Climat VR*
ENSEMBLES database (25-km resolution)			
Meteo France	A1B	ARPEGE	RM5.1
Danish Meteorological Institute DMI	A1B	ARPEGE	HIRHAM5
Danish Meteorological Institute DMI	A1B	ECHAM5	HIRHAM5
Royal Netherlands Meteorological Institute KNMI	A1B	ECHAM5	RACMO2
International Center for Theoretical Physics ICTP	A1B	ECHAM5	REGCM3
Suedish Meteorological Hydrological Institute SMHI	A1B	ECHAM5	RCA
Max Planck Institute MPI	A1B	ECHAM5	REMO
ETH Zürich ETHZ	A1B	HadCM3Q0	CLM
MetOffice-Hadley Center METO-HC	A1B	HadCM3Q0	HadRM3Q0

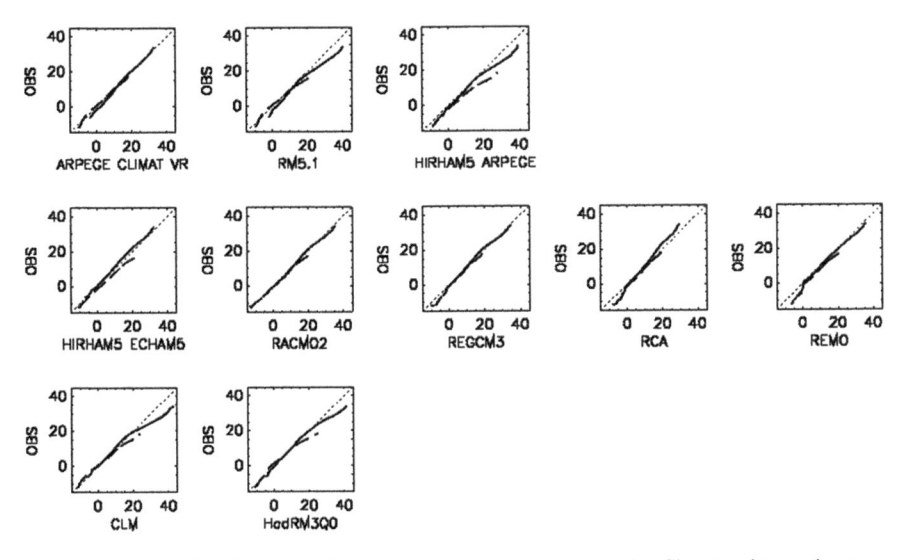

Fig. 12.2 {q-q} plots between all RCMs and observations from the Chartres time-series over 1961–1990, for the minimum biometeorological indicator BMIn in *dashed line* and for the maximum biometeorological indicator BMIx in *solid line*

to evaluate RCMs. Two future time-periods 2021–2050 and 2071–2100 (2099 for a few models) are used to foreseen the evolution of HW events.

Figure 12.2 and Table 12.2 show that all the RCMs overestimate BMIn extreme values, especially the 99.9th percentile. It also shows that most of RCMs (7/10) overestimate BMIx extreme values. We consider here that climate model outputs cannot compare exactly with climatological records. More particularly, RCM simulations may have systematic biases compared to time-series observations that can be adjusted by the quantile-quantile *(q-q)* correction method proposed by [1]. It is assumed that model biases are independent from greenhouse gases emissions [12], and consequently stationary with time. The same correction is thus applied to adjust RCM outputs in future climate.

Here, correction functions are determined for BMIn and BMIx by comparing the observed time-series of Chartres with all modelled time-series over the 1961–1990 control period (Fig. 12.2). We define a correction function in tenth of percentile, that is required for our detection thresholds prescribed to 99.9th. Values of BMI thresholds are listed in Table 12.2.

12.3.3   Evaluation of Past Modelled HWs

The ability of RCMs to simulate HW events is evaluated over the 1961–1990 control period by verifying that modelled heat waves are comparable to past observed heat waves in a statistical point of view. Over this time-period, two HW events are

Table 12.2 99.9th percentile values BMIn, BMIx and BMIavg for the observations and each RCMs

	BMIn	BMIx	BMIavg
Obs Chartres	18.40	34.13	26.26
ARPEGE CLIMAT VR	18.66	32.47	25.56
RM5.1 ARPEGE	23.63	40.28	31.95
HIRHAM5 ARPEGE	27.39	39.08	33.23
HIRHAM5 ECHAM5	23.21	32.20	27.70
RACMO2 ECHAM5	21.43	35.43	28.43
REGCM3 ECHAM5	18.83	34.54	26.68
RCA ECHAM5	19.23	29.54	24.38
REMO ECHAM5	21.88	35.83	28.85
CLM HadCM3Q0	23.90	42.80	33.35
HadRM3Q0 HadCM3Q0	24.77	41.39	33.08

extracted from the historical time-series of Chartres, one event of 13 days and one of 3 days (see Fig. 12.1). In comparison, the ten present climate simulations give up to four HW events with durations ranging from 3 to 10 days. Despite a weak number of events, durations and occurrences of modelled heat waves are realistic in comparison with observations. So, these RCMs are retained to analyse future HWs since we consider that they can simulate past HW events.

12.4   Results: Future HW Occurrence and Characteristics

12.4.1   Occurrence of Future Heat Waves

The 12 climate projections, irrespective of the model and emission scenario used, predict that HW events become more frequent during the twenty-first century, despite a strong variability (Fig. 12.3). Having two or more HW events per year is becoming the norm at the end of the century.

12.4.1.1   Several Climate Models, One Scenario

Over the 1961–1990 control period, the occurrence of heat waves varies from 0.03 to 0.13 $HW \cdot year^{-1}$ (i.e. with return periods between 7.5 and 30 years). In future climate, it varies from 0.16 to 0.97 $HW \cdot year^{-1}$ over 2021–2050, and from 0.30 to 3.77 $HW \cdot year^{-1}$ over 2071–2100.

All climate models taken together, the average number of heat waves per year is 1.62 over the last 30-year time-period. In comparison, models predict 0.096 $HW \cdot year^{-1}$ (i.e. one HW every 10 years) on average in the present time-period. On the mean values, the number of HWs per year is multiplied by a factor of around between present and future time-periods.

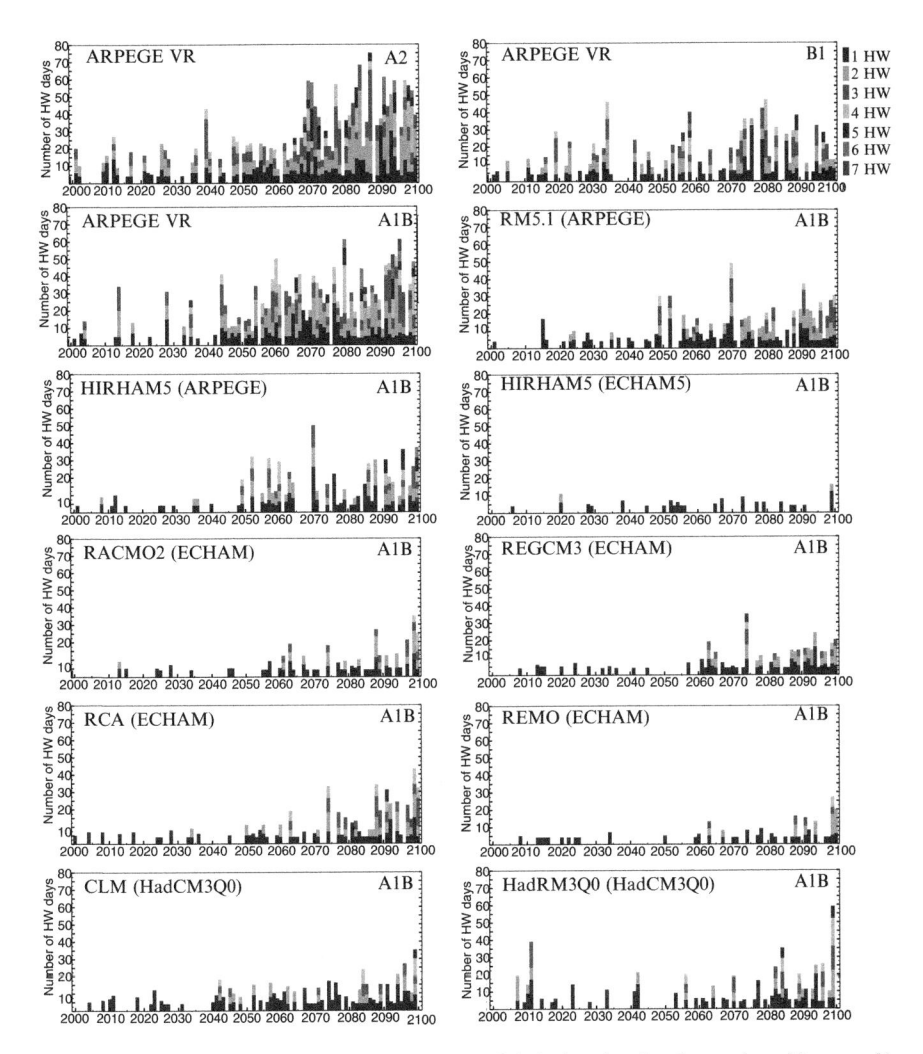

Fig. 12.3 Number of heat wave events per year and their duration for observations (*first graph*), for the Arpege climate model time-series for three scenarios (A1B, A2, B1) and for the nine climate model time-series following the A1B scenario

12.4.1.2   One Climate Model, Three Scenarios

The variability due to the use of different climatic scenarios is also studied through the analysis of the simulations obtained by ARPEGE-CLIMATE-VR for A2, A1B and B1 scenarios (Fig. 12.3). Over 1961–1990, the occurrence of heat waves is the same for the three scenarios since climate projections begin from 2001. It is 0.06

HW·year^{-1} (i.e. a return period of 15 years). Over 2021–2050, occurrences of HWs are 1.46, 0.97 and 1.17 HW·year^{-1} for A2, A1B and B1, respectively. Over 2071–2100, occurrences of HWs are 3.53, 3.77 and 2.1 HW·year^{-1} for A2, A1B and B1, respectively. The evolution of the number of HW events per year from nowadays to the end of the century is quantified by a factor of around 59, 69 and 35 for the A2, A1B and B1 scenario, respectively.

12.4.2 Future HW Characteristics

Future HW characteristics are analysed for the results obtained by several climate models following the A1B scenario, only.

Averaged and maximum values of BMIn and BMIx of future heat waves are presented as boxplots in Fig. 12.4. This figure translates the distribution of average and maximum BMI values of the HW events for each time-period, all model taken together. It shows a progressive increase of both BMIn and BMIx. This increase is more pronounced for BMIn. The median value of average BMIn shows an increase of +0.95°C (from 17.85°C to 18.80°C) from 1961°C to 1990 to 2071–2100, against +0.83°C (from 33.52°C to 34.35°C) for BMIx. This finding tends to reinforce the concern about the health consequences of heat waves. Indeed, considering that night temperatures remain higher in urban than rural areas because of UHI, it is expected that the accumulative effects are more pronounced in cities. Maximum intensities reached in a HW event during the 2071–2100 time-period are particularly high. During the strongest HW event, temperature reaches 29°C at night-time (BMIn) and around 45°C at day-time (BMIx), with median values of 21°C and 37°C, respectively.

In terms of duration, heat waves are also becoming longer. They last on the average 6 days at the end of the century against 4 days in the present climate. Majority of events last between 3 days and a week (25th and 95th percentiles) at the end of the century. However they can last more than 1 month (precisely 32 days) in 2071–2100 period against only 10 days maximum in 1961–1990 period according to climate model simulations (Fig. 12.4).

In order to give a comparison with an historical HW event that affected the Paris metropolitan area, the exceptional 2003 heat wave showed a daily maximum intensity of 20°C at night, 39°C at day and lasted 11 consecutive days. It corresponds approximately to a median heat wave of 2071–2100. It highlights thus the exceptional nature of this event. The combined effects of an increase in both the intensity and duration of most extreme events leads to heat-stress accumulation for population. Since these events are much more frequent at the end of the twenty-first century, the impact on human health may become quite critical.

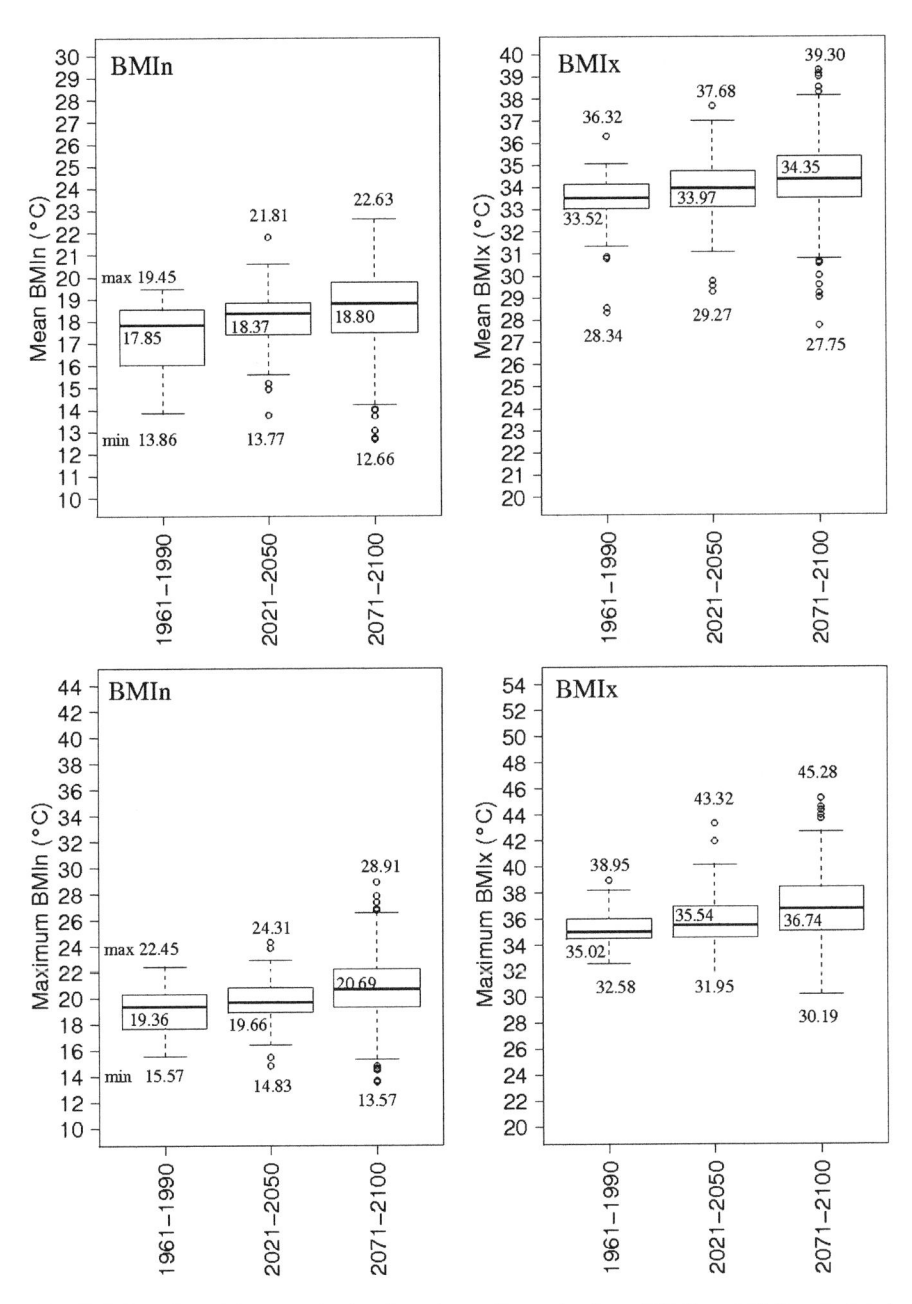

Fig. 12.4 Boxplots of the mean (*top*) and maximum (*bottom*) values of BMIn and BMIx calculated for each HW events detected within the ten climate model time-series following the A1B scenario, and classified by 30-year periods from 1961 to 2100. *Black dash lines* indicates the median. Its value is written within the box

12.5 Conclusion and Further Work

The present work establishes a range of possible future heat waves over Paris by analyzing their evolution within climate model projections. We first establish a HW definition. We use data from Meteo France and ENSEMBLES databases, thus, we obtain a total of 12 BMIn and BMIx simulated time-series covering the whole 1950–2100 period.

Despite the strong variability between models and scenarios, the analysis of the evolution of occurrence shows that they all agree to predict that HW events become more frequent during the twenty-first century. At the end of the century, the frequency is more than 3.5 and 2 HWs per year on average over 30 years, for the A2 and B1 simulations respectively. The set of A1B simulations gives on average 1.5 HWs per year (ranging from 0.3 to 3.77). This evolution must be compared with the present time-period, for which models predict a maximum of one HW event every 7.5 years. The analysis of the evolution of future HW characteristics points out an increase of BMI values, all models taken together, with a more significant increase for the temperature at night. This finding tends to reinforce the concern about the health consequences of heat waves. Indeed, considering that night temperatures remain higher in urban than rural areas because of UHI, it is expected that accumulative effects are more pronounced.

For the strongest future HW event of A1B simulations, night-time temperatures could reach 29°C and day-time temperatures around 45°C. In terms of duration, heat waves are becoming longer and last in average 6 days at the end of the twenty-first century against four at the beginning. However extreme cases are more numerous and the worst can last more than 1 month in 2071–2100 against only 10 days maximum in 1961–1990. For comparison, the 2003 heat wave showed maximum intensities of 20°C at night and 39°C at day, and lasted 11 consecutive days.

Further work will be carried out in order to assess the sensitivity of the Paris urban climate to different future HW events. Synthetic HW events will be built in order to cover the range of future HWs, and will be simulated using a urban-weather model. Then, the impacts in terms of energy consumption and bioclimatic comfort will be analysed and adaptation strategies will be proposed.

References

1. Déqué M (2007) Frequency of precipitation and temperature extremes over France in an anthropogenic scenario: model results and statistical correction according to observed values. Glob Planet Chang 57(1–2):16–26
2. Fischer EM, Schär C (2010) Consistent geographical patterns of changes in high-impact European heat waves. Nat Geosci 3:398–403
3. Gibelin AL, Déqué M (2003) Anthropogenic climate change over the Mediterranean region simulated by a global variable resolution model. Clim Dyn 20:327–339
4. Goodess C (2003) STAtistical and regional dynamical downscaling of extremes for European regions: STARDEX. European geophysical union information newsletter 6

5. IPCC (2007) Climate change 2007, physical science basis, contribution of working group I to the fourth assessment report of the intergovernmental panel on climate change
6. Kuglitsch FG, Toreti A, Xoplaki E, Della-Marta PM, Zerefos CS, T"ukes M, Luterbacher J (2010) Heat wave changes in eastern Mediterranean since 1960. Geophys Res Lett 37:5
7. Lemonsu A, Masson V (2002) Simulation of a summer urban breeze over Paris. Bound Lay Meteorol 104(3):463–490
8. Lenderink G, van den Hurk B, van Meijgaard E, van Ulden A, Cuijpers J (2003) Simulation of present-day climate in RACMO2: first results and model developments, KNMI Technical Report, 252, 24 pp
9. Oke TR (1988) The urban energy balance. Prog Phys Geogr 12:471–508
10. Rousseau D (2005) Analyse the fine des surmortalités pendant la canicule 2003. L'événement météorologique de la nuit du 11 au 12 août 2003 en Ile-de-France. La Météorologie 51:16–22
11. Souch C, Grimmond CSB (2006) Applied climatology: urban climatology. Prog Phys Geogr 30:270–279
12. Wilby RL, Wigley TML, Conway D, Jones PD, Hewitson BC, Main J, Wilks DS (1998) Statistical downscaling of general circulation model output: a comparison of methods. Water Resour Res 34:2995–3008

Chapter 13
The Height of the Atmospheric Planetary Boundary layer: State of the Art and New Development

Sergej S. Zilitinkevich

Abstract The planetary boundary layer (PBL) is defined as the strongly turbulent atmospheric layer immediately affected by dynamic, thermal and other interactions with the Earth's surface. It essentially differs in nature from the weakly turbulent and persistently stably-stratified free atmosphere. To some extent the PBL upper boundary acts as a lid preventing dust, aerosols, gases and any other admixtures released from ground sources to efficiently penetrate upwards, thus blocking them within the PBL. It is conceivable that the air pollution is especially hazardous when associated with shallow PBLs. Likewise, positive or negative perturbations of the heat budget at the Earth's surface immediately impact on the PBL and are almost completely absorbed within the PBL through the very efficient mechanism of turbulent heat transfer. Determination of the PBL height is, therefore, an important aspect of modelling and prediction of air-pollution events and extreme colds or heats dangerous for human health. Because of high sensitivity of shallow PBLs to thermal impacts, variability of the PBL height is an important factor controlling fine features of climate change. Deep convective PBLs strongly impact on the climate system through turbulent entrainment ("ventilation") at the PBL upper boundary, and thus essentially control development of convective clouds. This paper outlines modern knowledge about physical mechanisms and theoretical models of the PBL height and turbulent entrainment, and presents an advanced model of geophysical convective PBL.

S.S. Zilitinkevich (✉)
Finnish Meteorological Institute, PO Box 503, 00101 Helsinki, Finland

Division of Atmospheric Sciences, University of Helsinki, Helsinki, Finland

Nansen Environmental and Remote Sensing Centre, Bergen, Norway

A.M. Obukhov Institute of Atmospheric Physics, Moscow, Russia

Department of Radio Physics, N.I. Lobachevski State University of Nizhniy, Novgorod

Department of Meteorology, Russian State Hydrometeorological University,
St. Petersburg, Russia
e-mail: sergej.zilitinkevich@fmi.fi

H.J.S. Fernando et al. (eds.), *National Security and Human Health Implications of Climate Change*, NATO Science for Peace and Security Series C: Environmental Security, DOI 10.1007/978-94-007-2430-3_13, © Springer Science+Business Media B.V. 2012

Keywords Aerosols • Air pollution • Atmospheric boundary layer • Baroclinic shear • Boundary-layer height • Buoyancy • Colds • Emissions • Entrainment • Diurnal variations • Free atmosphere • Heats • Human health • Turbulence • Stratification • Vertical turbulent fluxes • Wind shear

Abbreviations

CBL	convective boundary layer
CN	conventionally neutral
IGW	internal gravity waves
LES	large-eddy simulation
LS	long-lived stable
NS	nocturnal stable
PBL	planetary boundary layer
SBL	stable boundary layer
TKE	turbulent kinetic energy
TN	truly neutral

13.1 Introduction

The planetary boundary layer (PBL) is defined as an essentially turbulent atmospheric layer immediately affected by dynamic, thermal, aerosol, greenhouse-gas and other impacts from the Earth's surface. In the atmospheric general circulation, it serves as a coupling agent between the land/sea surface and the free atmosphere. The latter is only weakly turbulent (because of its very stable stratification) and experiences the above impacts indirectly, in an aggregated form thorough the PBL integration-and-coupling mechanisms. In particular, the diurnal temperature variations, driven by the diurnal course of the solar irradiation at the surface and controlled by the vertical turbulent mixing, decay together with the intensity of turbulence towards the PBL upper boundary. Similarly, dust, aerosols, gases and any other admixtures released from ground sources are to a large extent blocked within the PBL (Fig. 13.1). The PBL height, varying from a few dozen meters to a few kilometres, is, therefore, the most important parameter controlling extreme colds and heats, heavy air-pollution episodes and local amplification of global warming or consequences of changes in land use (e.g., [9, 28]). Other crucially important PBL parameters are vertical turbulent fluxes of momentum, heat, moisture, aerosols or chemical admixtures through the air-surface interface in their relation to the easily available (measured or calculated in numerical model) mean vertical profiles of the wind, temperature, humidity and concentrations of pollutants, greenhouse gases, etc.

Fig. 13.1 Very shallow long-lived stable PBL visualised by smoke in warm summer day over very cold Lake Teletskoe (Altay, Russia) on 28.08.2010

Because the PBL is the habitat of all of us and contains almost entire terrestrial biosphere, understanding its nature and modelling its basic properties are crucially important from the viewpoint of human health and quality of life. In this context, the most demanding aspects of boundary-layer meteorology are modelling and observation of the PBL height and determination of vertical turbulent fluxes (first of all at the Earth's surface and the PBL upper boundary) through easily available information. This paper focuses on the PBL height and vertical turbulent fluxes due to entrainment at the PBL upper boundary.

13.2   Stable and Neutral Boundary Layers

13.2.1   The Role of Stratification

The PBL height and the intensity of turbulent mixing within the planetary boundary layer (PBL) strongly depend on the stratification of density. In the stable stratification, the buoyancy forces act against production of turbulent kinetic energy (TKE) by the velocity shear, wherefore reduce the intensity of mixing and the height of the stable PBL (SBL). On the contrary, in the unstable stratification the buoyancy forces generate convective motions, contribute to turbulent mixing, and cause deepening of the convective PBL (CBL). Traditionally the sign of the surface buoyancy flux, B_s, was used as the sole indicator of the PBL type: stable for $B_s < 0$, unstable (convective) for $B_s > 0$, and neutral for $B_s = 0$.

Though universally accepted, this classification disregards interactions between the PBL and the free atmosphere. The latter is characterised by the strongly stable stratification (with typical Brunt-Väisälä frequency $N \sim 10^{-2}$ s^{-2}), counteracting vertical extension of the turbulent layer. The decisive effect of N on the growth rate of the CBL has been recognised long ago [30], but for the atmospheric SBL it was generally neglected. The point is that the only type of the SBL considered in boundary-layer meteorology until recently was the *nocturnal stable* (NS) PBL. It develops at mid and low latitudes over land after the Sun set and lives during a few hours, as long as the nocturnal cooling of the land surface sustains $B_s < 0$. In such conditions the NS PBL replaces a comparatively shallow lower part of the residual layer (neutrally or almost neutrally stratified due to strong turbulent mixing during the daytime); so that the rest of the residual layer fully separates the NS PBL from the free atmosphere.

However, such separation does not hold in the absence of pronounced diurnal temperature course. In winter time at high latitudes and throughout the year over oceans, where the sign of the surface buoyancy flux does not change over long periods, PBLs traditionally treated as just neutral or just stable (according to the indication $B_s = 0$ or $B_s < 0$) immediately interact with the free atmosphere. Then the list of the SBL governing parameters, traditionally limited to the friction velocity, u_*, the buoyancy flux at the surface, B_s, and the Coriolis parameter, $f = 2\Omega\sin\varphi$ (where Ω is the angular velocity of the Earth's rotation, and φ is the latitude), should be extended including the free-flow Brunt-Väisälä frequency, N. It has been demonstrated that interactions between the PBL and the free atmosphere cause pronounced stable stratification in the upper portion of the PBL, by this means dramatically reduces the PBL height [1, 23–26, 29], and affects even the surface-layer flux-profile relationships [8, 21, 22, 27].

Accordingly, we now distinguish the four (instead of two) types of neutral and stable PBLs:

- "*truly neutral*" (TN) for $B_s = 0$ and $N = 0$
- "*conventionally neutral*" (CN) for $B_s = 0$ and $N > 0$
- "*nocturnal* stable" (NS) for $B_s < 0$ and $N = 0$
- "*long-lived stable*" (LS) for $B_s < 0$ and $N > 0$

The first three of these types represent simple alternatives: TN – without any stratification, CN – with stable stratification applied from above, and NS – with stable stratification applied from below. They are characterised by the rotational height scale: $h_* = (K_* / |f|)^{1/2}$, where K_* is the eddy-viscosity scale [7]. At given values of the governing parameters, each of these PBLs (characterised by specific expressions of K_*) evolves towards a quasi-steady state characterised by specific equilibrium height, h_E:

$$h_E = \begin{cases} C_R u_* \, | f \, |^{-1} & \text{for TN PBL} & (13.1) \\ C_{CN} u_* \, | fN \, |^{-1/2} & \text{for CN PBL} & (13.2) \\ C_{NS} u_*^2 \, | fB_s \, |^{-1/2} & \text{for NS PBL,} & (13.3) \end{cases}$$

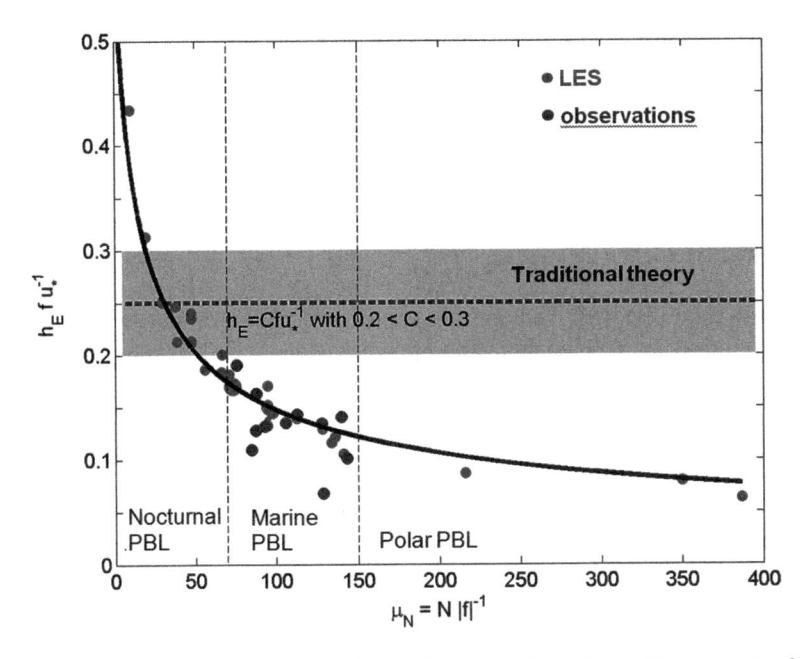

Fig. 13.2 The effect of the free-low Brunt-Väisälä frequency, N, on the equilibrium height, h_E, of the conventionally neutral PBL. The *dashed line* shows Eq. 13.1, the curve shows Eq. 13.2, and *points* show data from LES (*red*) and atmospheric measurements (*blue*)

where $C_R = 0.6$, $C_{CN} = 1.36$ and $C_{NS} = 0.5$ are dimensionless empirical constants determined basically through large-eddy simulation (LES) of a wide range of stable and neutral PBLs. The LS PBL offers properties of both NS and CN PBLs. Its equilibrium height is determined in Sect. 2.3 through appropriate interpolation between Eqs. 13.2 and 13.3.

It is significant that the TN PBL is observed quire rarely. Historically, most of the layers treated as "neutral" just by the indicator $B_s = 0$ were in fact only conditionally neutral. Nevertheless Eq. 13.1 [14] was unconditionally applied to all "neutral" PBLs. It is not surprising that empirical estimates of the coefficient C_R exhibited very wide spread. As follows from Eq. 13.2, the ratio $|f| h_E / u_*$, treated in Eq. 13.1 as universal constant (C_R), is factually strongly variable ($= C_N (f / N)^{1/2}$) and typically about five times smaller than C_R. The suppressing effect of the free-flow stability on the CN PBL height is demonstrated in Fig. 13.2.

Empirical verification of Eq. 13.3 [18] was more successful because most observations in stable stratification represented precisely the mid-latitudinal nocturnal PBL. The above LES-based estimate of the constant $C_{NS} = 0.5$ is generally consistent with its prior estimates based on atmospheric data (e.g., [5, 13]). Equation 2, although convincingly confirmed by LES, needs to be further verified against atmospheric data.

13.2.2   The Effect of Baroclinic Shear

In the above analyses, the friction velocity, u_*, plays the role of the turbulent velocity scale, u_T, for the entire PBL. This is grounded for the barotropic atmosphere characterised by the height-independent horizontal pressure gradient: $\partial p / \partial x$, $\partial p / \partial y = \text{constant}$. In this case, the wind velocity (u, v) tends to the height-constant geostrophic velocity [$u_g \equiv -(\rho f)^{-1} \partial p / \partial y$, $v_g \equiv (\rho f)^{-1} \partial p / \partial x$], and the wind shear diminishes towards the PBL upper boundary. Therefore the TKE is generated dominantly in the surface layer where u_* is the appropriate turbulent velocity scale.

Contrastingly, in the baroclinic atmosphere, the wind shear in the upper part of the PBL approaches the geostrophic shear:

$$\Gamma \equiv \left[\left(\frac{\partial u_g}{\partial z} \right)^2 + \left(\frac{\partial v_g}{\partial z} \right)^2 \right]^{1/2} = \frac{g}{|f| T} \left[\left(\frac{\partial T}{\partial y} \right)^2 + \left(\frac{\partial T}{\partial x} \right)^2 \right]^{1/2}, \tag{13.4}$$

where T is the absolute temperature, and g is the gravitational acceleration [e.g., Sect. 3.4 in Holton [11]]. Clearly, additional TKE generated by baroclinic shear is by no means related to u_* but yet contributes to both turbulent mixing and PBL deepening. Scaling analyses of these mechanisms [26] yields the following "baroclinic corrections" to the turbulent velocity scale:

$$u_T^2 = u_*^2 (1 + C_0 Ri^{-1/2}), \quad Ri = (N / \Gamma)^2, \tag{13.5}$$

and the CN PBL height:

$$h_E = \frac{C_{CN} u_*}{|fN|^{1/2}} \left(1 + C_0 \frac{\Gamma}{N} \right)^{1/2}, \tag{13.6}$$

where Ri is the free-atmosphere Richardson number, and $C_0 \approx 0.7$ is a dimensionless empirical constant (estimated from LES).

13.2.3   Diagnostic and Prognostic Formulations

Generally (including the LS PBL regime with $B_s < 0$ and $N > 0$) the equilibrium SBL height, h_E, is determined accounting for the three stratification mechanisms: rotational (f), "bottom-up" (B_s), and "top-down" (N), through the interpolation:

$$\frac{1}{h_E^2} = \frac{f^2}{(C_R u_*)^2} + \frac{N|f|}{(C_{CN} u_T)^2} + \frac{|fB_s|}{(C_{NS} u_*^2)^2}, \tag{13.7}$$

which automatically gives priority to the strongest suppression mechanism (e.g., [29]).

Knowing h_E, the factual SBL height, h, is expressed through the relaxation equation:

$$\frac{dh}{dt} - w = -\frac{h - h_E}{t_E}. \tag{13.8}$$

Here, $d/dt = \partial/\partial t + u\partial/\partial x + v\partial/\partial y$; u, v and w are components of the wind velocity along horizontal (x, y) and vertical (z) axes taken at the height $z = h(t, x, y)$; t_E is the relaxation time scale:

$$t_E = \frac{h_E}{C_E u_*}; \tag{13.9}$$

and C_E is a dimensionless constant to be determined empirically.

Taking Eqs. 13.8 and 13.9 for the stationary and horizontally homogeneous regime, yields the vertical-motion correction to the equilibrium PBL height:

$$h_{E,w} = h_E \left(1 + \frac{w}{C_E u_*} \right). \tag{13.10}$$

It follows that the PBL heights in high-pressure air-masses (where $w > 0$) are generally higher than in low-pressure air-masses (where $w < 0$), all other factors being the same.

Verification of Eqs. 13.8–13.10 including determination of C_E are subjects of further investigations.

13.3   Convective Boundary Layers

13.3.1   The Concept of Well-Mixed Layer

In unstable stratification the TKE is generated in two ways: "convectively" by the buoyancy forces and "mechanically" by the velocity shear, with the generation rates B and $\tau \cdot \partial \mathbf{u}/\partial z$, respectively, where $B = \overline{b'w'}$ and $\tau = -\overline{\mathbf{u}'w'}$ are vertical turbulent fluxes of buoyancy $b = (g/T)\theta + 0.61gq$ and momentum $\mathbf{u} = (u,v)$; θ is the potential temperature; and q is the specific humidity. The shear-free CBL is characterised by the Deardorff [6] convective velocity scale:

$$W_* = (B_s h)^{1/3}. \tag{13.11}$$

For the sheared CBL, it is customary to determine the squared turbulent velocity scale without proof as a linear combination of W_*^2 and squared friction velocity: $u_*^2 = |\tau|_{z=0}$. This is not necessarily so. Below we demonstrate that more natural counter-scale to W_* is the cubic root of the CBL-overall shear-production of TKE: $(\bar{U}u_*^2)^{1/3}$, where $\bar{U}$ is the CBL-mean wind velocity.

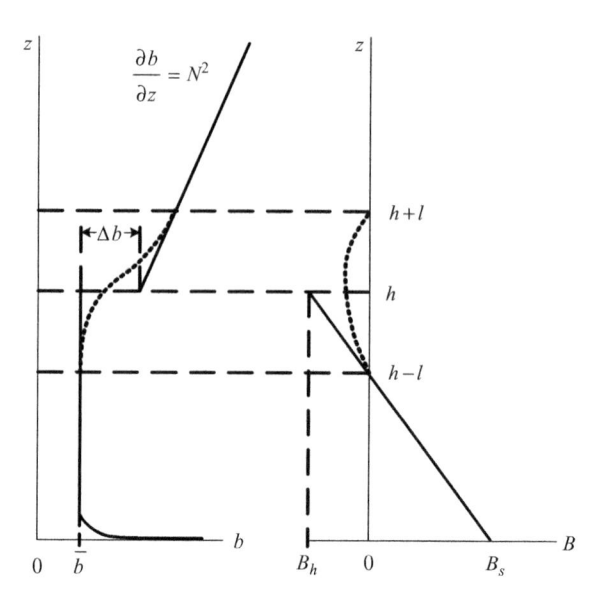

Fig. 13.3 Approximation of vertical profiles of the buoyancy, b, and the buoyancy flux, B, in the well-mixed layer model. More realistic profiles in the entrainment layer are shown by *dotted curves*

With typical atmospheric values of the surface buoyancy flux, $B_s \sim 10^{-3}$ m^{2}s^{-3}, and the CBL height $h \sim 10^3$ m, convective turbulence causes such strong mixing that the buoyancy, b, the wind velocity components, u and v, and concentrations of any admixtures become practically height-independent throughout the CBL, except close vicinity of the surface, $0 < z < \delta$, and thin "turbulent entrainment layer" at the CBL upper boundary (see Fig. 13.3), which in turn results in practically linear vertical profiles of the vertical turbulent fluxes: B and τ.

Furthermore, substituting the CBL eddy-viscosity scale, $K_* \sim W_* h$, in the rotational height scale $h_* = (K_* / |f|)^{1/2}$, yields $h_* \sim B_s^{1/2} / f^{3/2} \sim 30$ km. This estimate is an order of magnitude larger than the heights of deepest observed CBLs, which therefore, should be insensitive to the Earth's rotation. These features enable simplified treatment of the atmospheric CBL as the non-rotational well-mixed layer. As long as the integral $\int_0^h B dz$ remains positive, the CBL height increases in accordance with prognostic equation:

$$\frac{dh}{dt} - w = \dot{h}, \tag{13.12}$$

where $\dot{h}$ is the rate of expansion of the well-mixed CBL into non-turbulent free flow.

13.3.2   Buoyancy-Budget CBL Height Models

The concept of the well-mixed layer assumes the following approximations:

$$b = \begin{cases} \overline{b} & \text{at } \delta < z < h \\ \overline{b} + \Delta b + N^2(z-h) & \text{at } z > h, \end{cases} \tag{13.13}$$

$$B = \begin{cases} B_s(1-z/h) + B_h z/h & \text{at } 0 < z < h \\ 0 & \text{at } z > h, \end{cases} \tag{13.14}$$

where $B_h < 0$ is the buoyancy flux due to entrainment at the upper boundary of expanding CBL in the presence of the buoyancy increment $\Delta b = b_{h+0} - b_{h-0}$:

$$B_h = -\Delta b \dot{h}. \tag{13.15}$$

The upper portion of such layer, $h - l < z < h$, characterised by the negative buoyancy flux, is called the *turbulent entrainment layer*. As shown in Fig. 13.3, its depth scale, l, is expressed as

$$\frac{l}{h-l} = -\frac{B_h}{B_s} \equiv A, \tag{13.16}$$

where the ratio A is called the *entrainment coefficient* (e.g., [16]).

The well-mixed layer model allows roughly estimating h from the mean buoyancy equation. The latter is composed of the equations for potential temperature, θ, and specific humidity, q. We consider its simplest version:

$$\frac{db}{dt} = -\frac{\partial B}{\partial z}, \tag{13.17}$$

neglecting the vertical advection (the term $w db/dt$ on the left hand side), the radiation flux divergence, and condensation/evaporation. Then substituting Eq. 13.13 for b and Eq. 13.14 for B into Eq. 13.17 and then integrating over z from 0 to h, yields the CBL bulk buoyancy-budget equation:

$$\frac{d}{dt}\left(\frac{1}{2}N^2h^2 - h\Delta b\right) = B_s. \tag{13.18}$$

For the horizontally homogeneous CBL developing against the stable stratified free atmosphere without entrainment (at $\Delta b = 0$, $B_h = 0$), Eq. 13.18 immediately yields the *encroachment model* [30]:

$$\frac{dh}{dt} = \frac{B_s}{N^2h}. \tag{13.19}$$

Alternatively, assuming $A = \text{constant} > 0$, Eqs. 13.15, 13.16 and 13.18 yield the simplest *entrainment model* [3, 4, 17]:

$$\frac{dh}{dt} = (1 + 2A)\frac{B_s}{N^2 h}. \tag{13.20}$$

Gryning and Batchvarova [10] and Batchvarova and Gryning [2], generalised Eq. 13.20 accounting for the mean vertical motions (w) and incorporating u_* as additional turbulent velocity scale, employed typical atmospheric value of $A = 0.2$, and validated the resulting equation against atmospheric data. Their model is successfully used in practical applications (e.g., [15]).

13.3.3 Turbulence-Energetics Model of the Laboratory CBL

Generally, atmospheric, laboratory and LES data demonstrate that the entrainment coefficient, A, is essentially variable: $0 < A < 1$ [e.g., Table 3 in Zilitinkevich [19]]. The buoyancy-budget approach presented in Sect. 3.2 is obviously insufficient to explain this variability. Zilitinkevich [19, 20] developed a theoretical model of turbulent entrainment based on the well-mixed layer approximation of the TKE-budget equation:

$$\frac{dE_K}{dt} = \tau \cdot \frac{\partial \mathbf{u}}{\partial z} + B - \frac{\partial F}{\partial z} - \varepsilon, \tag{13.21}$$

where E_K is the TKE, ε is the dissipation rate of TKE, and F is the vertical flux of TKE. The above papers limited to the shear-free CBL ($\tau \cdot \partial \mathbf{u}/\partial z = 0$). Then, in going from Eq. 13.21 to the CBL bulk TKE-budget equation, the integral of B is taken using Eq. 13.14; the integrals of dE_K/dt and ε are taken using the Deardorff [6] similarity theory formulations: $E_K = W_*^2 \Phi_{EK}(\zeta)$, $\varepsilon = (W_*^3/h)\Phi_\varepsilon(\zeta)$, where $\Phi_{EK}(\zeta)$ and $\Phi_\varepsilon(\zeta)$ are universal functions of the dimensionless height $\zeta = z/h$; and the integral of $\partial F/\partial z$ reduces to the energy flux at the CBL upper boundary, F_{h+0}. The latter is generally not negligible because of excitement of internal gravity waves (IGW) in the stably stratified free atmosphere by large-scale convective plumes approaching the CBL upper boundary. According to the linear wave theory, the energy flux carried out by IGW is expressed as $F_{h+0} \propto \lambda^2 \Lambda N^3$, where λ in the wave amplitude, and Λ is the wave length.

To have at hand high quality experimental data, Zilitinkevich considered the shear-free CBL in a laboratory tank, where side walls restrict the lengths of the convection-excited IGW. Accordingly, he estimated both λ and Λ as proportional to the entrainment-layer depth scale, Eq. 13.16:

$$\Lambda \sim \lambda \sim l = \frac{A}{1+A}h. \tag{13.22}$$

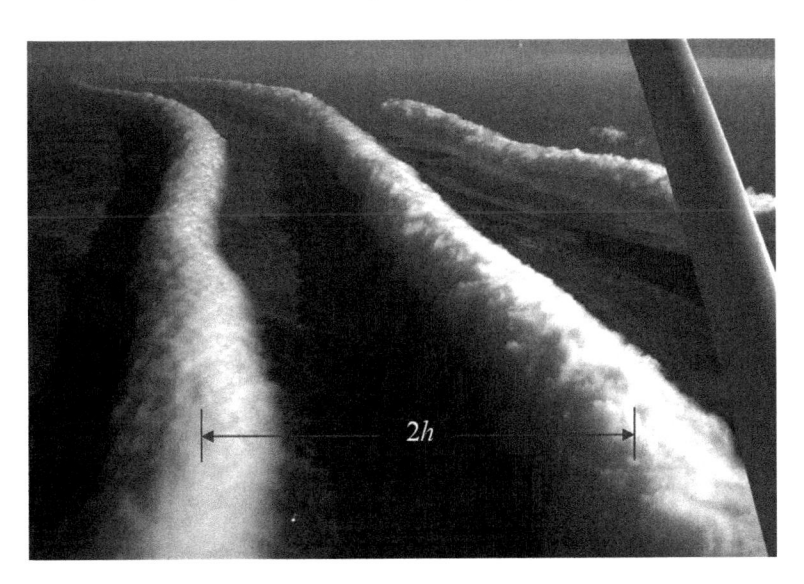

Fig. 13.4 "Cloud streets" visualising updraughts in typical convective rolls: photo by Mick Petroff taken from a plane near Burketown in Queensland (North Coast), Australia; Wikimedia Commons. The updraughts excite internal gravity waves in the stably stratified free atmosphere

This model resulted in the turbulent entrainment equation:

$$\left(C_2 + Ri_1\right)E + C_3 Ri_2^{3/2}\left(\frac{Ri_1 E}{1 + Ri_1 E}\right)^3 = C_1, \qquad (13.23)$$

where $E = \dot{h} / W_*$ is the dimensionless CBL expansion rate; $Ri_1 = h\Delta b / W_*^2$ is the Richardson number based on the buoyancy increment at the CBL upper boundary; $Ri_2 = \frac{1}{7}(Nh / W_*)^2$ is the Richardson number based on the static stability in the free flow; C_1, C_2 and C_3 are empirical dimensionless constants appeared from the two integrals: $\int_0^1 \Phi_{EK}(\zeta)d\zeta$, $\int_0^1 \Phi_\varepsilon(\zeta)d\zeta$ and the omitted proportionality coefficient in Eq. 13.22. Validation of the model against laboratory has given quite accurate estimates of the empirical constants: $C_1 = 0.2$, $C_2 = 0.8$ and $C_3 = 0.1$.

13.3.4   New Turbulence-Energetics Model of the Atmospheric CBL

To account for the difference between atmospheric and laboratory CBLs, we consider the TKE-budget Eq. 13.21 accounting for the shear-generation term $\tau \cdot \partial \mathbf{u} / \partial z > 0$. Because convective structures in the sheared barotropic CBL have the form of large-scale rolls separated by the typical distance ~$2h$ (see Fig. 13.4), the wave length of the IGW excited by these structures is taken proportional to the

CBL height: $\Lambda \sim h$. The estimate of the wave-amplitude, λ, as proportional to the depth of the entrainment layer, l, has no reasonable alternative and is used as before. Then Eq. 13.22 is replaced by

$$\Lambda \sim h, \quad \lambda \sim l = \frac{A}{1+A}h \qquad (13.24)$$

Furthermore, in the sheared CBL the Deardorff velocity scale $W_* = (B_s h)^{1/3}$ (based on the two governing parameters: B_s and h) is no longer unique. To put us on the right track, we recall that W_* is nothing but the velocity scale characterising the doubled global rate of the energy production by the buoyancy forces: $W_*^3 = hB_s \approx 2\int_0^h Bdz$. By analogy, we employ the doubled global rate of the energy production by the velocity shear, namely $2\int_0^h (\tau \cdot \partial \mathbf{u} / \partial z)dz \approx 2\bar{U}u_*^2$, to determine the mechanical velocity scale:

$$V_* = (2\bar{U}u_*^2)^{1/3}, \qquad (13.25)$$

where $\bar{U}$ is the CBL-averaged mean wind velocity. Then generalising the Deardorff similarity theory, we express the TKE, E_K, and its dissipation rate, ε, through the pairs of dimensionless functions [analogous to the functions $\Phi_{EK}(\zeta)$ and $\Phi_\varepsilon(\zeta)$ in the shear-free CBL]:

$$E_K = W_*^2 \Phi_{EK}^{(c)}(\zeta) + V_*^2 \Phi_{EK}^{(m)}(\zeta), \quad \varepsilon = (W_*^3 / h)\Phi_\varepsilon^{(c)}(\zeta) + (V_*^3 / h)\Phi_\varepsilon^{(c)}(\zeta). \qquad (13.26)$$

This model yields the turbulent entrainment equation:

$$\left(C_2 + Ri_1\right)E + C_4 Ri_2^{3/2}\left(\frac{Ri_1 E}{1 + Ri_1 E}\right)^2 = C_1 + C_5 \frac{\bar{U}u_*^2}{w_*^3}. \qquad (13.27)$$

Luckily, empirical constants $C_1 = 0.2$ and $C_2 = 0.8$ are already determined from laboratory experiments. The new constants, C_4 and C_5, are to be determined from atmospheric and LES data for shear-free CBLs (for C_4) and essentially sheared CBLs (for C_5).

Two particular cases of Eq. 13.27 deserve attention. In the early morning, as long as the CBL develops against the weakly stratified residual layer (where $Ri_2 \ll 1$), Eq. 13.27 reduces to

$$E = \frac{C_1 + C_5(V_* / W_*)^3}{C_2 + Ri_1} \qquad (13.28)$$

that corresponds to very fast deepening of the layer. Alternatively, when the CBL approaches the free atmosphere (where $Ri_2 \gg 1$ and $A = Ri_1 E \ll 1$), Eq. 13.27 reduces to

$$E = \frac{[C_1 + C_5(V_* / W_*)^3]^{1/2}}{C_4^{1/2} Ri_1 Ri_2^{3/4}} \qquad (13.29)$$

that corresponds to immensely decreased rate of deepening.

The pair of Eqs. 13.18 and 13.27 determine the CBL expansion rate h, the buoyancy increment at the CBL upper boundary, Δb, and by this means the buoyancy flux due to entrainment at the CBL upper boundary, $B_h = -\Delta b \dot{h}$, through the CBL governing parameters N, h, W_* and V_* (or N, h, B_s, u_* and $\bar{U}$) available in atmospheric models. Then, given $\dot{h}$, the CBL height can be calculated integrating the prognostic Eq. 13.12.

13.4 Concluding Remarks

Generally atmospheric (or oceanic) CBL heights are estimated by the Zubov [30] encroachment model, Eq. 13.19, with reasonable accuracy. The, *predetermined-entrainment* model [3, 4, 17] and its meteorological generalisation [2, 10] based on the assumption $A = B_h / B_s = 0.2$ show quite reasonable accuracy [15]. In the applications that require h, but do not require the fluxes due to entrainment, in particular, the fluxes of temperature, $F_{\theta h}$, and moisture, F_{qh}, comprising $B_h = (g/T)F_{\theta h} + 0.61 \, gF_{qh}$, as well as the fluxes of atmospheric admixtures, the Gryning-Batchvarova model seems to be a plausible approximation.

However, this is not the case in climate modelling. Knight et al. [12] have demonstrated a strong effect of the entrainment coefficient A (currently treated as given empirical parameter) on the simulated cumulus clouds, which causes strong climate sensitivity to the determination of A. Our Eq. 13.27 in combination with the generalised Eq. 13.18 (immediately followed from the full-size equations for temperature and moisture) provide physically grounded tools for the prognostic determination of A (together with the CBL height, h) through parameters available in atmospheric models.

The following problems remain to be solved before Eqs. 13.7–13.9 for the SBL height and Eqs. 13.18 and 13.27 for the CBL height and the entrainment-fluxes, $F_{\theta h}$ and F_{qh}, could be recommended for use in operational models:

- validation of the prognostic SBL height algorithm based on Eqs. 13.7–13.9; and determination of the empirical constant C_E using available data from meteorological observations and LES,
- validation of Eq. 13.27 in terms of dimensionless dependencies of the expansion rate, E, or entrainment coefficient, $A = ERi_1$, on Ri_1, Ri_2 and V_* / W_*; and determination of new empirical constants C_4 and C_5 through properly designed LES,
- validation of the prognostic algorithm for the CBL height, h, and the buoyancy flux due to entrainment, B_h, using data from meteorological observations and LES.

Upon addressing these issues, the physical background for calculation of the PBL height and the fluxes due to entrainment at the CBL upper boundary algorithm will be completed. Prospectively, the PBL height and the turbulent entrainment algorithms are to be implemented into climate, weather-prediction and air-quality models, in particular, for better modelling heavy air-pollution events and extreme colds or heats dangerous for human health.

Acknowledgements This work has been supported by EC FP7 project ERC PBL-PMES (No. 227915) and Federal Targeted Programme "Research and Educational Human Resources of Innovation Russia 2009–2013" (Contract No. 02.740.11.5225); and the Russian Federation Government Grant No. 11.G34.31.0048.

References

1. Baklanov A (2002) Parameterisation of SBL height in atmospheric pollution models. In: Borrego C, Schayes G (eds) Air pollution modelling and its application XV. Kluwer Academic/ Plenum Publishers, New York, pp 415–424
2. Batchvarova E, Gryning S-E (1991) Applied model for the growth of the daytime mixed layer. Bound-Layer Meteorol 56:261–274
3. Betts AK (1973) Non-precipitating cumulus convection and its parameterization. Q J R Meteorol Soc 99:178–196
4. Carson DJ (1973) The development of a dry inversion-capped convectively unstable boundary layer. Q J R Meteorol Soc 99:450–467
5. Caughey SJ, Wyngaard JC, Kaimal JC (1979) Turbulence in the evolving stable boundary layer. J Atmos Sci 6:1041–1052
6. Deardorff JW (1972) Parameterization of planetary boundary layer for use in general circulation models. Mon Weather Rev 2:93–106
7. Ekman VW (1905) On the influence of the earth's rotation on ocean-currents. Arkiv for Matematik, Astronomi och Fysik 2(11):1–52
8. Esau I (2004) An improved parameterization of turbulent exchange coefficients accounting for the non-local effect of large eddies. Ann Geophys 22:3353–3362
9. Esau I, Zilitinkevich S (2010) On the role of the planetary boundary layer depth in climate system. Adv Sci Res 4:63–69
10. Gryning S-E, Batchvarova E (1990) Analytical model for the growth of the coastal internal boundary layer during onshore flow. Q J R Meteorol Soc 116:187–203
11. Holton JR (1972) An introduction to dynamic meteorology. Academic, New York/San Francisco/London, 319 pp
12. Knight CG, Knight SHE, Massey N, Aina T, Christensen C, Frame DJ, Kettleborough JA, Martin A, Pascoe S, Sanderson B, Stainforth DA, Allen MR (2007) Association of parameter, software and hardware variation with large scale behavior across 57,000 climate models. Proc Natl Acad Sci USA 104(30):12259–12264
13. Nieuwstadt FTM, Van Dop H (1982) Atmospheric turbulence and air pollution modelling. Reidel, Dordrecht, pp 107–158
14. Rossby CG, Montgomery RB (1935) The layer of frictional influence in wind and ocean currents. Pap Phys Oceanogr Meteorol MIT Woods Hole Oceanogr Inst 3(3):1–101
15. Seibert P, Beyrich F, Gryning S-E, Joffre S, Rasmussen A, Tercier Ph (2000) Review and intercomparison of operational methods for the determination of the mixing height. Atmos Environ 34:1001–1027
16. Stull R (1976) Mixed-layer depth model based on turbulent energetics. J Atmos Sci 33:1268–1278
17. Tennekes H (1973) A model of the dynamics of the inversion above a convective boundary layer. J Atmos Sci 30:558–567
18. Zilitinkevich SS (1972) On the determination of the height of the Ekman boundary layer. Bound-Layer Meteorol 3:141–145
19. Zilitinkevich SS (1991) Turbulent penetrative convection. Avebury Technical, Aldershot, 179 pp
20. Zilitinkevich SS (1987) Theoretical model of turbulent penetrative convection. Izvestija AN SSSR, FAO 23(6):593–610

21. Zilitinkevich S, Calanca P (2000) An extended similarity-theory for the stably stratified atmospheric surface layer. Q J R Meteorol Soc 126:1913–1923
22. Zilitinkevich S (2002) Third-order transport due to internal waves and non-local turbulence in the stably stratified surface layer. Q J R Meteorol Soc 128:913–925
23. Zilitinkevich S, Baklanov A, Rost J, Smedman A-S, Lykosov V, Calanca P (2002) Diagnostic and prognostic equations for the depth of the stably stratified Ekman boundary layer. Q J R Met Soc 128:25–46
24. Zilitinkevich SS, Baklanov A (2002) Calculation of the height of stable boundary layers in practical applications. Bound-Layer Meteorol 105:389–409
25. Zilitinkevich SS, Esau IN (2002) On integral measures of the neutral, barotropic planetary boundary layers. Bound-Layer Meteorol 104:371–379
26. Zilitinkevich SS, Esau IN (2003) The effect of baroclinicity on the depth of neutral and stable planetary boundary layers. Q J R Meteorol Soc 129:3339–3356
27. Zilitinkevich SS, Esau I (2005) Resistance and heat transfer laws for stable and neutral planetary boundary layers: old theory, advanced and re-evaluated. Q J R Meteorol Soc 131:1863–1892
28. Zilitinkevich SS, Esau IN (2009) Planetary boundary layer feedbacks in climate system and triggering global warming in the night, in winter and at high latitudes. Geogr Environ Sustain 1(2):20–34
29. Zilitinkevich S, Esau I, Baklanov A (2007) Further comments on the equilibrium height of neutral and stable planetary boundary layers. Q J R Meterol Soc 133:265–271
30. Zubov NN (1945) Arctic ice. Glavsevmorput Press, Moscow, 360 pp

Chapter 14
The Influence of Meteorological Conditions on Fine Particle (PM1.0) Levels in the Urban Atmosphere

Zvjezdana Bencetić Klaić

Abstract It is well known that airborne particulate matter (PM) can damage human health and affect climate. Fine particles with an aerodynamic diameter less than 1 μm (PM1.0) are generally more harmful to humans compared to coarser particles. This study investigates the relationships between 1-min average PM1.0 mass concentrations and atmospheric conditions at the same time-scale. Concentrations were measured by the DUSTTRAK™ Aerosol Monitor, which was located in Zagreb's residential quarter, far from major pollution sources. The monitor was placed at a height of 15.8 m above the ground. While the influences of temperature and global radiation remained unclear, it was shown that PM1.0 levels depended on horizontal and vertical wind speed, air pressure and relative humidity. Thus, climate change may at least locally modify PM pollution levels and accordingly affect human health. Finally, results suggested that a nearby road, at a distance of approximately 100 m, with weak to moderate traffic did not affect PM1.0 levels. Instead, recorded concentrations mainly originated from other urban sources that were several kilometers away.

Keywords Ambient concentration • DUSTTRAK™ aerosol monitor • Mass concentration • 1-min average • Particulate matter • Residential

14.1 Introduction

Airborne particulate matter (PM) is a complex and varying mixture of particles suspended in the atmosphere. It is comprised of dust, dirt, soot, smoke, and liquid droplets that are produced by a wide variety of natural processes and human activities. Thus,

Z.B. Klaić (✉)
Andrija Mohorovičić Geophysical Institute, Department of Geophysics,
Faculty of Science, University of Zagreb, Zagreb, Croatia
e-mail: zklaic@gfz.hr

H.J.S. Fernando et al. (eds.), *National Security and Human Health Implications of Climate Change*, NATO Science for Peace and Security Series C: Environmental Security, DOI 10.1007/978-94-007-2430-3_14, © Springer Science+Business Media B.V. 2012

particle sizes and their physical and chemical characteristics depend on formation processes and subsequent reactions in the atmosphere. Although the composition of PM varies, its major components are metals, organic compounds, materials of biologic origin, ions, reactive gases, and the particles' carbon core [1–4].

In the atmosphere, particles have aerodynamic diameters ranging in size from a few nanometers (the size of molecular clusters) to about 100 μm (small enough to be suspended in the air for an appreciable time). According to an early simple trimodal model [5], PM is divided into three distinct size modes: nuclei [1] or ultrafine [4] (size between 0.005 and 0.1 μm), accumulation (0.1–2 μm) and coarse (size above 2 μm).

The transient nuclei (ultrafine) mode contains the highest number of particles and has the smallest mass concentration. It is formed by photochemical reactions on gases in the atmosphere and by combustion. It mainly consists of sulfates, nitrates and organic compounds [6]. Due to its transient nature, this mode is significant only in the immediate vicinity of its sources, because the mode diameter rapidly increases over time [1]. However, the nuclei mode is of great significance to the formation and growth of larger particles [6].

The accumulation mode is mainly anthropogenic. It is produced by complex reactions between gasses emitted into the atmosphere during fossil fuel combustion. This mode is further divided into the condensation mode (mean aerodynamic diameter of 0.2 μm) and the droplet mode (mean aerodynamic diameter of 0.7 μm) [1]. The condensation mode forms and grows by the condensation of gases that either directly or indirectly coagulate with nuclei mode particles; the rate of growth decreases with increasing condensation mode particle sizes. Particles in droplet mode are formed in aqueous phase reactions involving sulphur, and where reaction rates decrease with increasing particle sizes.

Both the nuclei and accumulation modes together (i.e., particles with diameters smaller than 2 μm) were labeled as fine particles in the past [5]. More recent distinctions between the fine and the coarse particles use thresholds of 2.5 μm [1] and 1 μm [4, 7, 8]. In the urban atmosphere, fine particles mainly originate from vehicular exhaust [9, 10], where the influence of traffic increases with decreasing particle sizes [10], and from industrial plants [11]. Generally, they are comprised of sulfates, organic ammonium, nitrates, carbon, lead, and some trace constituents [1].

A coarse fraction is produced by mechanical processes, and it contains crustal material (i.e., silicon compounds, iron, aluminum), sea salt and plant particles [1]. It originates from human activities, wind-blown dust, sea spray, plants and volcanoes.

The impact of PM on human health generally depends on the sizes and surfaces of particles and their number and chemical compositions. While particles larger than 10 μm are eliminated from the respiratory system through coughing, sneezing or swallowing, smaller ones enter the system. Those between 1 and 10 μm deposit mainly in the upper respiratory tract, while fine, submicron and ultrafine particles (smaller than 0.1 μm) may reach the lung alveoli and eventually, deliver the harmful chemicals to the blood system. Numerous studies report associations between fine particles and mortality [12, 13], toxicity, allergies, and cardiovascular and respiratory problems [4, 14]. Thus, for example, patients with asthma may be more sensitive to PM pollution, compared to healthy persons, due to enhanced diffusional deposition of ultrafine particles in their distal airways and alveoli [15]. Finally, chronic exposure to traffic-related PM may be involved in the development of mild cognitive impairment, that is, in the pathogenesis of Alzheimer's disease [16].

Depending on their sizes and chemical compositions, PM also influences global climate through both warming and cooling effects. The direct effect of particles on the planetary energy balance is related to the scattering and absorption of solar radiation [6]. Due to scattering, some portion of solar energy returns to space, resulting in net cooling of Earth. Conversely, absorption of radiation by particles results in a warming similar to that produced by greenhouse gases. The indirect effect of particles on climate is associated with the fog, haze and cloud formation [17] because hygroscopic particles can act as condensation nuclei. Due to their exceptionally complex chemical composition, size distribution and physical characteristics (e.g., index of refraction), the exact effects of PM on the global climate are still insufficiently known. However, it is known that sulfate particles, which are extremely abundant, produce a net cooling. In contrast, soot, minerals and some organics lead to warming due to their absorption of solar radiation [6, 7].

Because smaller particles are generally more harmful to human health then larger ones [4], we will focus here on urban fine particles, namely particles with aerodynamic diameters less than 1 μm [4, 7, 8] (PM1.0). The aim of the study is to investigate the possible influence of ambient meteorology on PM1.0 levels in the urban residential quarter, where there are no major sources of pollution.

14.2   Measuring Site and Measurements

The measuring site is located in the northern, residential part of Zagreb, about 1.5 km north-northeast of the city center and about 8–9 km northwest of the industrial zone. The site is in a moderately hilly terrain on a south-facing slope of Medvednica, an approximately 1 km high mountain. The mountain is oriented in the SW-NE direction. The measuring site belongs to the Department of Geophysics (DG) (Faculty of Science). Apart from several nearby 3–4-story campus buildings that are approximately 50–70 m long, the majority of buildings within a radius of about 200 m are 1–3-story family houses and urban villas with slanted roofs. Patches of grass and rather high trees (~10 m) are also found in the same area. The arrangement of buildings, trees and grass is irregular. Approximately 100 m westward of the measuring site there is a N-S oriented road that has weak to moderate traffic, and an E-W oriented road that has very weak traffic is located approximately 50 m southward.

The Model 8520 DustTrak™ Aerosol Monitor (TSI, Inc., Shoreview, MN, USA) was placed 3 m above the roof terrace (i.e., 15.8 m above the ground) of the 3-story DG building. It was equipped with a waterproof environmental enclosure designed for outdoor measurements. An automatic meteorological station, META 2000 (AMES, Brezovica, Slovenia), measured meteorological data. Horizontal and vertical wind, and global radiation sensors were in the vicinity of the aerosol monitor, 4.5, 4.3 and 1.8 m above the roof terrace (17.3, 17.1 and 14.6 m above the ground), respectively. Air temperature and relative humidity sensors were in a standard meteorological shelter (2 m above the ground) located in a grassy area on campus, while an air pressure sensor was inside the DG building at a height of 7.5 m above the ground (187.55 m above the mean sea level). Both PM1.0 mass concentrations and meteorological data were measured every second, and 1-min averages were stored. Measurements were made from 17 March to 12 May, 2010. In total, 80054 1-min records were collected.

14.3   Data Analyses and Discussion of Results

Table 14.1 shows the basic statistics of the measured data. During the period of investigation, the average 1-min PM1.0 mass concentration was about 28 $\mu g\ m^{-3}$, which is about 2.5 times higher than regional background concentrations obtained for the Mediterranean Basin (10–11 $\mu g\ m^{-3}$) [18].

Diurnal variations in PM1.0 mass concentrations (not shown here) strongly suggest the importance of traffic in the greater town area. Namely, two maximums do not exactly coincide with the morning and evening rush hours, but occur 1–2 h later. This implies that elevated mass concentrations do not primarily arise because of traffic emissions from the two roads closest to the measuring site. Instead, assuming that the wind speed is within an order of magnitude of 1 $m\,s^{-1}$ (Table 14.1), the elevated concentrations correspond to particles transported from a distance of several kilometers.

To account for the relative position of the measuring site with respect to various urban sources, 16 wind directions were analyzed separately. Winds having a northern component (specifically N, NNW, NW, WNW and NNE) were expected to transport clean air from the nearby mountain, Medvednica, toward the measuring site. Thus, they should have generally had low PM1.0 concentrations. Conversely, southeastern winds (ESE, SE and SSE) favored transport of polluted air from the industrial zone; therefore, they should have had elevated PM1.0 concentrations.

Table 14.1 Basic statistics of 1-min data measured data during 17 March – 12 May, 2010 period

Quantity	Mean	Standard deviation	Minimum	Maximum
PM1.0 mass concentration (mg m^{-3})	0.0279	0.0216	0.000	0.844
Air temperature (°C)	13.12	4.48	2.8	28.2
Relative humidity (%)	66.09	17.68	21	97
Global radiation (J cm^{-2})	331.68	282.52	1	1344
Pressure (hPa)	994.21	6.07	979.6	1008.7
Horizontal wind speed (m s^{-1})	1.87	1.35	0.0	14.3
Vertical wind (cm s^{-1})	13.85	21.12	−105.9	148.1
Absolute vertical wind (cm s^{-1})	17.40	18.30	0.0	148.1

Global radiation statistics is given only for the daytime data

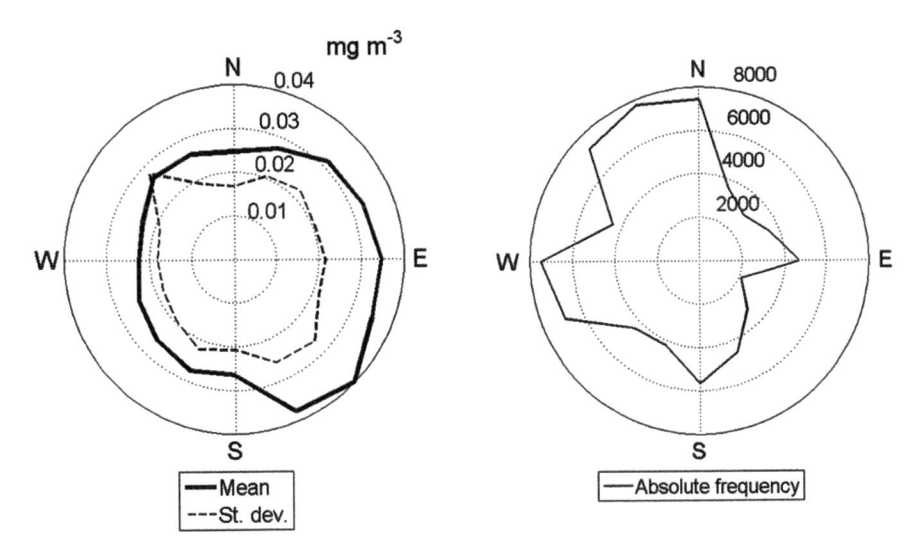

Fig. 14.1 Mean PM1.0 concentrations and standard deviations of concentrations with respect to the wind direction (*left*). Absolute frequencies of wind directions during investigated period (*right*)

As expected, the highest average concentrations were related to SE (40 µg m^{-3}), SSE (38 µg m^{-3}) and ESE (35 µg m^{-3}) winds (Fig. 14.1, left), which suggested the importance of the industrial zone. Similar results were obtained for the daily means of PM2.5 and PM10 concentrations at another location in the northern part of Zagreb [19, 20]. The lowest average concentrations were measured when there were W (22 µg m^{-3}) and WNW (23 µg m^{-3}) winds. As seen from the frequency distribution of wind directions (Fig. 14.1, right), airflows related to the highest average mass concentrations were rare during this investigation, with relative frequencies of 4% (SE), 6% (SSE) and 3% (ESE). Several other multiyear studies of daily averages also showed that southeastern airflows over the greater Zagreb area are the most rare, with a relative frequency of 8% at most [19–21]. In contrast, NNW, W and NW airflows, which were accompanied by relatively low average concentrations (26, 22 and 27 µg m^{-3}, respectively), where the most frequent (10%, 9% and 9%, respectively).

Inspection of the relationship between PM1.0 mass concentrations and relative humidity of all airflows together, and separately for each of 16 wind directions, clearly showed an increase in concentration with increasing relative humidity, except at the highest values of relative humidity. Only the results for SE, NNW and all wind directions together are shown here (Fig. 14.2), where SE and NNW flows favored the advection of polluted air from the industrial zone and clean air from nearby mountain, respectively. Decreased PM1.0 concentrations at the highest values of relative humidity (>90%) can probably be attributed to removal of the particles by rainout and/or washout processes. In the former, particles serve as cloud condensation nuclei or are captured by cloud water; in the latter, below-cloud particles are removed by raindrops.

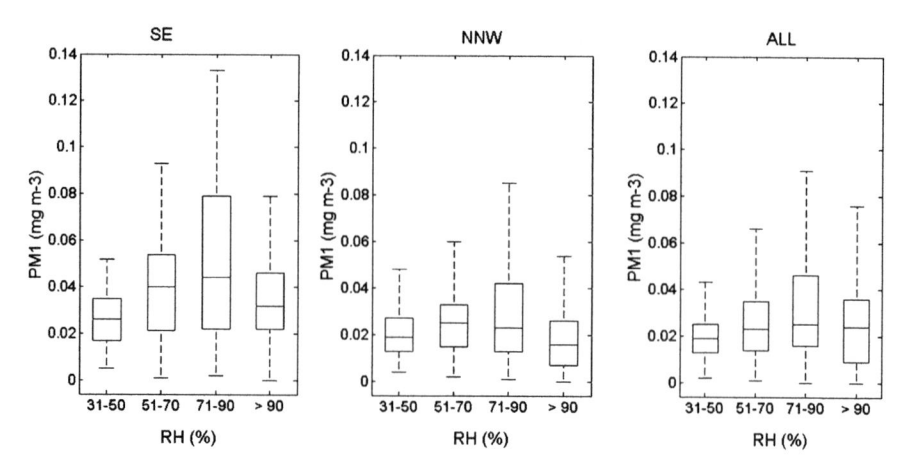

Fig. 14.2 Box-plots of PM1.0 mass concentrations vs. relative humidity for southeastern (*left*) and north-northwestern (*centre*) and all flows (*right*). The former and latter correspond to advection of polluted and clean air, respectively. On each box, the central mark is the median and the edges of the box are the 25th and 75th percentiles. The whiskers extend to the most extreme data points not considered outliers, where the outliers (not shown here) are values that are more than 1.5 times the interquartile range away from the *top* or *bottom* of the box

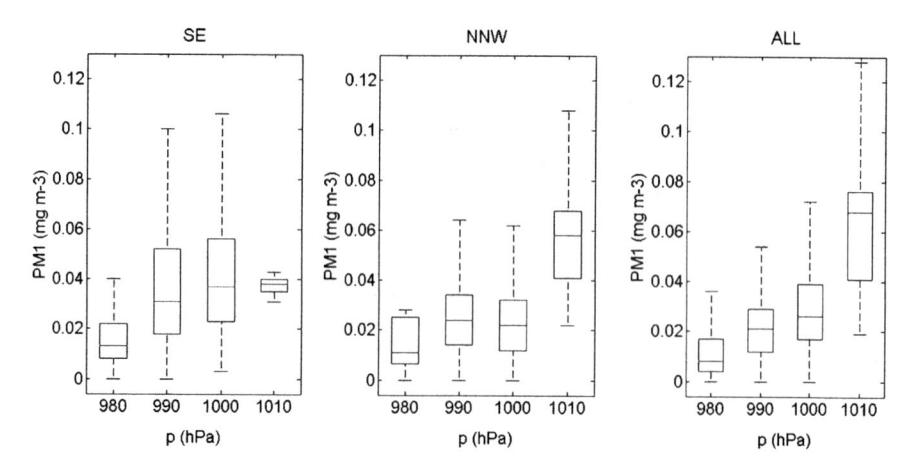

Fig. 14.3 Same as Fig. 14.2 but for the air pressure

Figure 14.3 illustrates the relationship between PM1.0 levels and the air pressure of polluted, clean and all airflows. It shows that PM1.0 concentrations increased with increasing air pressure. Similar patterns were also observed with other wind directions. Low-pressure fields usually accompany the passage of fronts or cyclones. Typically, these are accompanied by higher wind speeds (i.e., more efficient ventilation) and precipitation (removal of pollutants from the atmosphere), which result in low pollutant concentrations. Conversely, high-pressure fields are associated with

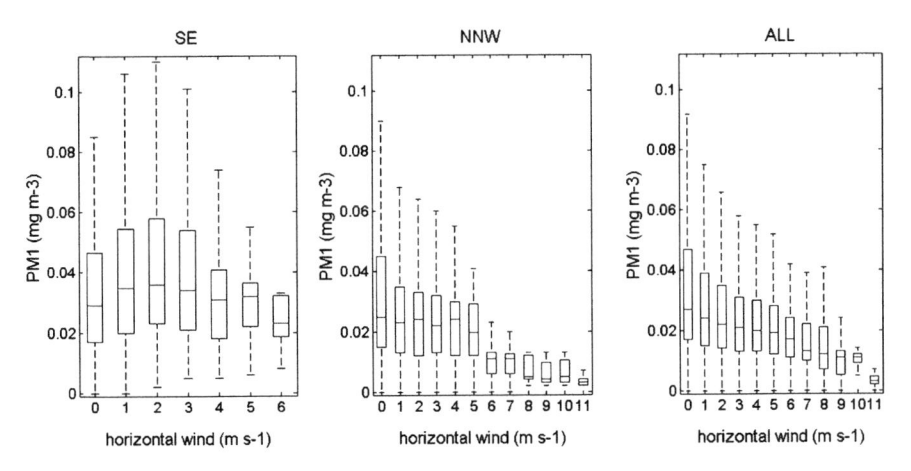

Fig. 14.4 Same as Fig. 14.2 but for the horizontal wind speed

stable, anticyclonic weather conditions and weak winds (i.e., inefficient ventilation), and accordingly, high pollutant concentrations.

The effects of ventilation and consequent dilution of pollutant concentrations are illustrated in Fig. 14.4. For all wind directions together, concentrations decreased with increasing horizontal wind speeds. The same relationship was also found for all individual wind directions (e.g., NNW in Fig. 14.4), except for those favoring the transport of air from the industrial zone to the measuring site (ESE, SE and SSE) and the flows with a pronounced western component (W, WSW and SW). For winds that favored the transport of pollutants from the industrial zone (e.g., SE in Fig. 14.4), concentrations increased with increasing wind speed if the wind speed was weak enough (speeds below 2.5 m s^{-1}). This means that for weak winds, the effects of the advection of polluted air dominated the effects of ventilation. Conversely, for stronger winds (speeds above 2.5 m s^{-1}), the effects of ventilation dominated the advection of pollution, and thus, concentrations decreased with increasing wind speeds.

The dependence of PM1.0 on the horizontal wind speed was U-shaped for W, WSW and SW winds (Fig. 14.5). For weak winds (speeds below 2.5–3.5 m s^{-1}) ventilation was efficient, that is, concentrations decreased with increasing wind speed. On the contrary, for higher speeds (above 2.5–3.5 m s^{-1}) concentrations increased with wind strengthening, implying the domination of the advection of pollutants. Because the average horizontal wind speeds at the measuring site were somewhat below 2 m s^{-1} (Table 14.1), it was reasonable to assume that stronger winds (e.g., 5 m s^{-1} or more) were the result of larger scale (synoptic) forcing and consequently, larger scale advection. Thus, the increase in concentrations at high wind speeds might have been due to the larger-scale, meaning regional or long-range, transport of pollutants. The more synoptic-scale western and west-southwestern airflows may have transported the air from industrialized regions of northern Italy towards the measuring site. Although one cannot quantitatively distinguish between

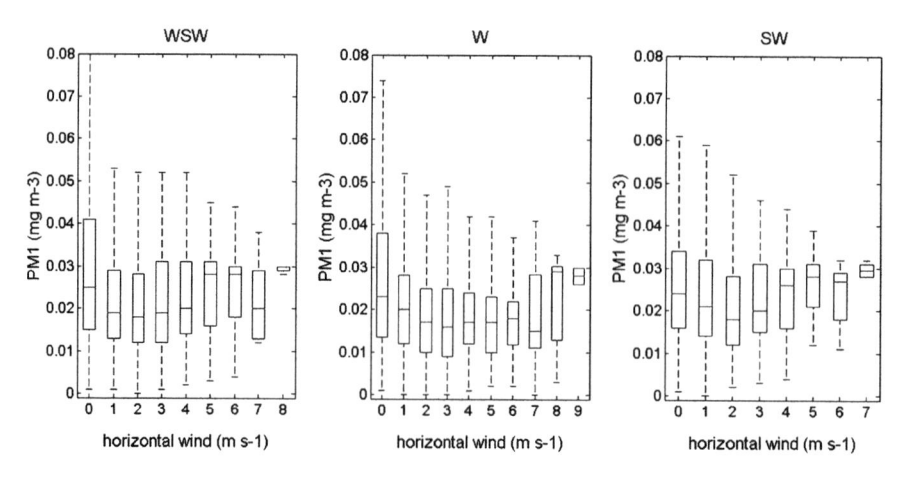

Fig. 14.5 2 Box-plots of PM1.0 mass concentrations vs. horizontal wind speed for WSW, W and SW flows

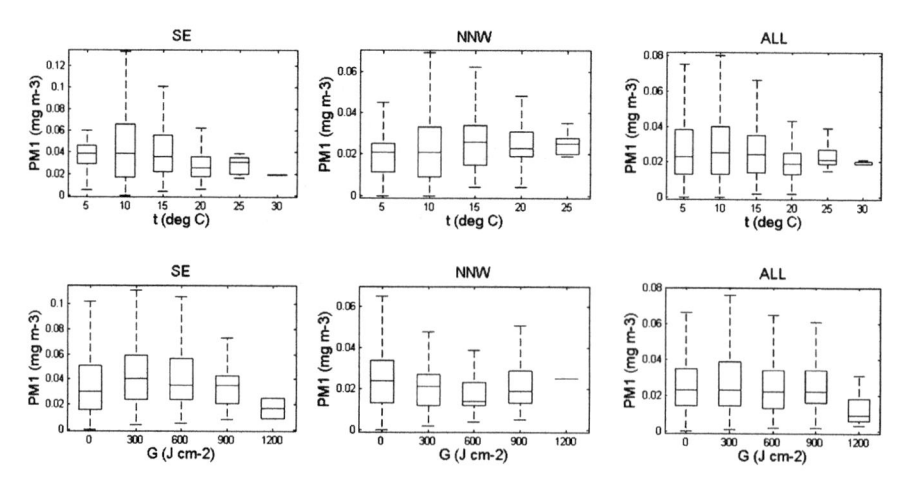

Fig. 14.6 Same as Fig. 14.2 but for the temperature and global radiation (*upper* and *lower* panels, respectively)

the contributions from urban and distant sources without an extensive air quality modeling study (which is beyond the scope of this investigation), it is very likely that distant sources contributed to the elevated concentrations recorded at high W, WSW and SW winds, and thus, they could not be neglected. Other recent urban atmosphere studies [8, 22] also point to an important role of the long-range transport in the formation of the accumulation mode.

Figure 14.6 shows PM1.0 mass concentrations versus temperature and global radiation. For both meteorological parameters, patterns were not straightforward.

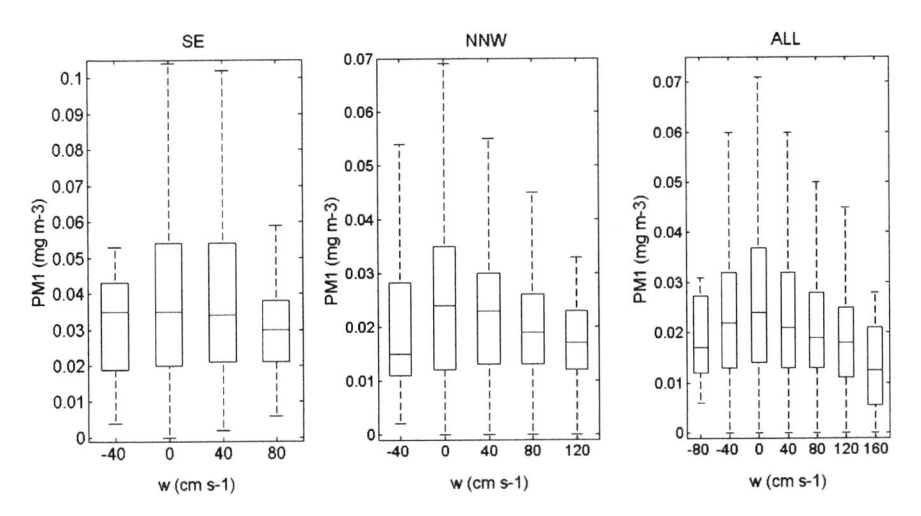

Fig. 14.7 Same as Fig. 14.2 but for the vertical wind speed

For temperature, for 8 out of 16 directions (namely WNW, NW, NNW, N, NNE, NE, ENE and E), concentrations generally increased with the increasing temperatures (e.g., NNW flow in Fig. 14.6). Meanwhile, for the remaining flows and for all flows together, they decreased (e.g., SE flow in Fig. 14.6). For global radiation (values over 150 J cm^{-2}), concentrations decreased with increasing radiation for all flows together and for 10 out of 16 directions (e.g., SE flows in Fig. 14.7). Exceptions with more irregular patterns arose with N, NNE, NE, ENE, E and NNW winds. Such complicated behavior regarding both temperature and global radiation was probably the result of (photo) chemically intricate nature of measured particles, which are both unknown and temporally variable.

Figure 14.7 shows box-plots for PM1.0 mass concentrations versus vertical wind speed. For all directions analyzed together, concentrations were the highest at vertical wind speeds around a value of zero, and they decreased with both upward and downward motions. Concentrations were the lowest for the most intense upward/downward motions, meaning the highest absolute values of vertical wind speed. Apart from S, WSW and W flows, similar behavior was also observed with other wind directions (e.g., SE and NNW winds in Fig. 14.7), although maximum concentrations frequently shifted at negative vertical wind speeds (not shown here). Minimum concentrations, which were observed at the strongest upward motions, most likely occurred because of efficient pollution dilution due to convection and/or transport of particles to heights above the instrument inlet (almost 16 m above the ground). Conversely, the strongest downward motions may have captured particles within a thin layer next to the ground, i.e., below the instrument inlet, which also resulted in low measured concentrations. We note however, that the above patterns might have been different if measurements were performed at different heights above the ground.

14.4   Conclusions

An investigation of the relationships between 1-min mean PM1.0 mass concentrations and meteorological parameters revealed the dependence of PM1.0 levels on horizontal and vertical wind speeds, air pressure and relative humidity. The role of temperature and global radiation in PM1.0 levels remains unclear probably because PM1.0 is a complex and time-varying mixture of various species having different (photo) chemical characteristics. Thus, further investigation of the association between PM1.0 levels and these two parameters is desirable. However, it would also require collection of data allowing the investigation of the chemical composition of the PM.

If climate is indeed changing, we can expect there to be some effects on PM pollution, and consequently, at least local influence on human health because of the dependence of PM1.0 levels on meteorological conditions. Additionally, possible climatically-induced changes in PM pollution may cause feedback processes that will affect the future state of the atmosphere.

Finally, the observed pattern in the diurnal variation of PM1.0 mass concentrations, together with the low PM1.0 concentrations with western winds, suggest that weak to moderate road traffic approximately 100 m away from the measuring site did not substantially affect PM1.0 levels. Thus, measured concentrations mainly originated from other urban sources that were several kilometers away.

Acknowledgements The study was supported by the Croatian Ministry of Science, Education and Sport (grant No. 119-1193086-1323).

References

1. John W (2001) Size distribution characteristics of aerosols. In: Baron PA, Willeke K (eds) Aerosol measurement: principles, techniques, and applications, 2nd edn. Wiley-InterScience Inc, New York, pp 99–116
2. Engelbrecht JP, Swanepoel L, Chow JC, Watson JG, Egami RT (2002) The comparison of source contributions from residential coal and low-smoke fuels, using CMB modeling, in South Africa. Environ Sci Policy 5:157–167
3. Watson JG, Zhu T, Chow JC, Engelbrecht J, Fujita EM, Wilson WE (2002) Receptor modeling application framework for particle source apportionment. Chemosphere 49:1093–1136
4. Kampa M, Castanas E (2008) Human health effects of air pollution. Environ Pollut 151:362–367
5. Whitby KT (1978) The physical characteristics of sulphur aerosols. Atmos Environ 12:135–159
6. Buseck PR, Adachi K (2008) Nanoparticles in the atmosphere. Elements 4:389–394. doi:10.2113/gselements.4.6.389
7. Monks PS, Granier C, Fuzzi S, Stohl A, Williams ML et al (2009) Atmospheric composition change – global and regional air quality. Atmos Environ 43:5268–5350. doi:10.1016/j.atmosenv.2009.08.021

8. Costabile F, Birmili W, Klose S, Tuch T, Wehner B, Wiedensohler A, Franck U, König K, Sonntag A (2009) Spatio-temporal variability and principal components of the particle number size distribution in an urban atmosphere. Atmos Chem Phys 9:3163–3195

9. Sun J, Zhang Q, Canagaratna MR, Zhang Y, Ng NL, Sun Y, Jayne JT, Zhang X, Zhang X, Worsnop DR (2009) Highly time- and size-resolved characterization of submicron aerosol particles in Beijing using an aerodyne aerosol mass spectrometer. Atmos Environ. doi:10.1016/j.atmosenv.2009.03.020

10. Birmili W, Alaviippola B, Hinneburg D, Knoth O, Tuch T, Borken-Kleefeld J, Schacht A (2009) Dispersion of traffic-related exhaust particles near the Berlin urban motorway – estimation of fleet emission factors. Atmos Chem Phys 9:2355–2374

11. Ehrlich C, Noll G, Kalkoff W-D, Baumbach G, Dreiseidler A (2007) PM_{10}, $PM_{2.5}$ and $PM_{1.0}$ – emissions from industrial plants – Results from measurement programmes in Germany. Atmos Envrion 41:6236–6254

12. Anderson HR (2009) Air pollution and mortality: a history. Atmos Environ 43:142–152

13. Choi J, Fuentes M, Reich BJ (2009) Spatial-temporal association between fine particulate matter and daily mortality. Comput Stat Data An 53:2989–3000

14. Kim JJ, Smorodinsky S, Lipsett M, Singer BC, Hodgson AT, Ostro B (2004) Traffic-related air pollution near busy roads. The east bay children's respiratory health study. Am J Respir Crit Care Med 170:520–526

15. Chalupa DC, Morrow PE, Oberdörster G, Utell MJ, Frampton MW (2004) Ultrafine particle deposition in subjects with asthma. Environ Health Persp 112:879–882

16. Ranft U, Schikowski T, Sugiri D, Krutmann J, Krämer U (2009) Long-term exposure to traffic-related particulate matter impairs cognitive function in the elderly. Environ Res 109:1004–1011. doi:10.1016/j.envres.2009.08.003

17. Metzger S, Lelieveld J (2007) Reformulating atmospheric aerosol thermodynamics and hygroscopic growth into fog, haze and clouds. Atmos Chem Phys 7:3163–3193

18. Querol X, Alastuey A, Pey J, Cusack M, Pérez N, Mihalopoulos N, Theodosi C, Gerasopoulos E, Kubilay N, Koçak M (2009) Variability in regional background aerosols within the Mediterranean. Atmos Chem Phys 9:4575–4591

19. Bešlić I, Šega K, Čačković M, Klaić ZB, Vučetić V (2007) Influence of weather types on concentrations of metallic components in airborne PM10 in Zagreb, Croatia. Geofizika 24:93–107

20. Bešlić I, Šega K, Čačković M, Klaić ZB, Bajić A (2008) Relationship between 4-day air mass back trajectories and metallic components in PM_{10} and $PM_{2.5}$ particle fractions iz Zagreb air, Croatia. Bull Environ Contam Toxicol 80:270–273. doi:10.1007/s00128-008-9360-6

21. Čanić KŠ, Vidič S, Klaić ZB (2009) Precipitation chemistry in Croatia during the period 1981–2006. J Environ Monit 11:839–851. doi:10.1039/b816432k

22. Timonen H, Saarikoski S, Tolonen-Kivimäki O, Aurela M, Saarnio K, Petäjä T, Aalto PP, Kulmala M, Pakkanen T, Hillamo R (2008) Size distributions, sources and source areas of water-soluble organic carbon in urban background air. Atmos Chem Phys 8:5635–5647

Chapter 15
Modelling of Heavy Metals: Study of Impacts Due to Climate Change

Amela. Jeričević, I. Ilyin, and S. Vidič

Abstract Heavy metals are a category of pollutants recognized as dangerous to human health and human exposure occurs through all environmental media. Since metals are naturally occurring chemicals that do not break down in the environment and can accumulate in soils, water and the sediments of lakes and rivers, it is important to evaluate the contribution of natural emission sources in the environment. Owing to climate change, the water content in soil is decreased while evapotranspiration is increased as a consequence the higher resuspension of soil dust particles. In this work, a modelling study of heavy metals was performed in order to assess the levels of heavy metals pollution, particularly lead, in Croatia and to estimate the effects of an increase in lead natural emissions due to climate change.

Heavy metals are emitted into environment mainly as a result of anthropogenic activities, complemented by naturally occurring chemicals in the environment; therefore it is important to evaluate the contribution and patterns of their natural emissions. The main paths for heavy metals through the atmosphere and water are dispersion and deposition processes leading to the accumulation in soils and water sediments, which, consequently, become reservoirs for secondary, semi-natural release of heavy metals back to the atmosphere and other media. Both the strength and spatial patterns of this release naturally depend on climate conditions and change accordingly. A rise in temperature causes soil water content to decrease while evapotranspiration increases, and thus impacts resuspension of soil dust particles. In this study, modelling of heavy metals, particularly lead, was performed in order to assess the influence of climate-sensitive variables and resuspension of heavy metals to the levels and their distribution in Croatia.

Keywords Heavy metals • Climate change • Health impacts • Resuspension

A. Jeričević (✉) • S. Vidič
Meteorological and Hydrological Service of Croatia, Zagreb, Croatia
e-mail: jericevic@cirus.dhz.hr

I. Ilyin
Meteorological Synthesizing Centre-East, Moscow, Russia

H.J.S. Fernando et al. (eds.), *National Security and Human Health Implications of Climate Change*, NATO Science for Peace and Security Series C: Environmental Security, DOI 10.1007/978-94-007-2430-3_15, © Springer Science+Business Media B.V. 2012

15.1   Introduction

Increased exposure to air pollution coming from industrial activities, traffic and energy production is one of the consequences of the current stage of industrialization. Several decades ago, convincing scientific data emerged that related specific air pollutants to health effects, and the results of such studies provided, inter alia, arguments for setting limit values for specific pollutants in ambient air. Heavy metals are one category of pollutants recognized as dangerous to human health. Human exposure to heavy metals occurs through all environmental media [1]. The heavy metals cadmium, lead and mercury are common air pollutants, being emitted mainly as a result of various industrial activities. Although the atmospheric levels are low, they contribute to the deposition and build-up in soils. Heavy metals are persistent in the environment and are subject to bioaccumulation in food chains.

The modelling of heavy metals is an important tool for the assessment of air pollution. The MSCE-HM (Meteorological Synthesizing Centre East-Heavy Metal) chemical transport model, which was developed for the assessment of heavy metal airborne pollution in Europe and in support of the review, extension and implementation of the U.N. Protocol on Heavy Metals, was used in this work. The investigation of different model resolutions, improved emission inventories and incorporation of the observed lead concentrations in soil is provided.

Since the industrial revolution, a considerable amount of heavy metals has been deposited in soils [2]. The resuspension of surface material could give rise to a significant source term contributing to the natural emissions of lead and considerably influencing the level of pollution. Furthermore, according to the climate change projections, water content in soil is expected to decrease due to a temperature rise while evapotranspiration increases. As a consequence, the higher resuspension of soil dust particles is envisaged.

The goal of this paper is to use the model in order to assess levels of heavy metals pollution, particularly lead, in Croatia, and to estimate the effects of potential increase in natural emissions of lead due to climate change.

15.1.1   Heavy Metals and Health Impacts

Metals are naturally occurring chemicals that do not break down in the environment and accumulate in soils, water and the sediments of lakes and rivers. Some metals are essential for human health at very low levels but toxic at higher concentrations. Others such as lead, cadmium and mercury, defined as 'heavy metals', have no known benefits and can be harmful if ingested. These metals are common air pollutants produced largely by industrial processes and fossil fuel combustion. They persist in the environment, move from the air into soil and migrate into the food chain.

Heavy metals have been linked to variety of health problems, such as kidney and bone disease, and developmental and behavioural disorders. At higher doses some metals may also contribute to other health problems including cardiovascular disease and cancers. Lead exposures have developmental and neurobehavioral effects on fetuses, infants and children, and elevate blood pressure in adults. Cadmium exposures are associated with kidney and bone damage as well as potential human carcinogen, causing lung cancer. Mercury is toxic in its elemental and inorganic forms, but the main concern is associated with the organic compounds, especially methylmercury, that accumulate in the food chain, i.e., in predatory fish in lakes and seas, as these are the main routes of human exposure. In addition, when exposed to high levels of mercury, people can suffer from many diseases. The major source of exposure is through the consumption of fish and other seafood, which accumulates mercury in their tissues. Mercury is a significant airborne pollutant because it can travel thousands of kilometers and pose threats to the environment and health far from its source. It poses additional threats to small-scale miners who use the metal to extract gold. The United Nations Industrial Development Organization estimates that this use of mercury threatens the health of 15 million people.

Long-range transboundary air pollution is just one source of exposure to these metals but, because of their persistence and potential for global atmospheric transfer, atmospheric emissions affect even the most remote regions.

15.2 Climate Change and Heavy Metals

Since the Industrial Revolution (Fig. 15.1), the production of heavy metals such as lead, copper, and zinc has increased exponentially (e.g., [2, 3]). Heavy metals have been used in a variety of ways for at least two millennia. For example, lead has been used in plumbing, and lead arsenate has been used to control insects in apple orchards. The Romans added lead to wine to improve its taste, and mercury was used as a salve to alleviate teething pain in infants. Although emission of lead has been drastically decreased, there are still significant amounts of lead deposited in the environment. Regarding the health impacts, the contamination by heavy metals is considerably more efficient through nutrition than by direct atmospheric exposure [1]. However, the resuspension of surface material could give rise to a significant source term contributing to the natural emissions of lead and influencing the level of pollution considerably.

Analysis of observed trends in Croatia has shown that average annual soil temperatures have increased by $1^\circ C$ in the period 1991–2009 (e.g., [4]). The highest increase $\approx 2^\circ C$ was observed in the south of Croatia, in the Dubrovnik area. An increase of soil temperature and potential evapotranspiration together with a decrease and unfavourable distribution of precipitation has been noted at available Croatian stations for two climatological periods 1981–2009 and 1961–1990 (e.g., [5]).

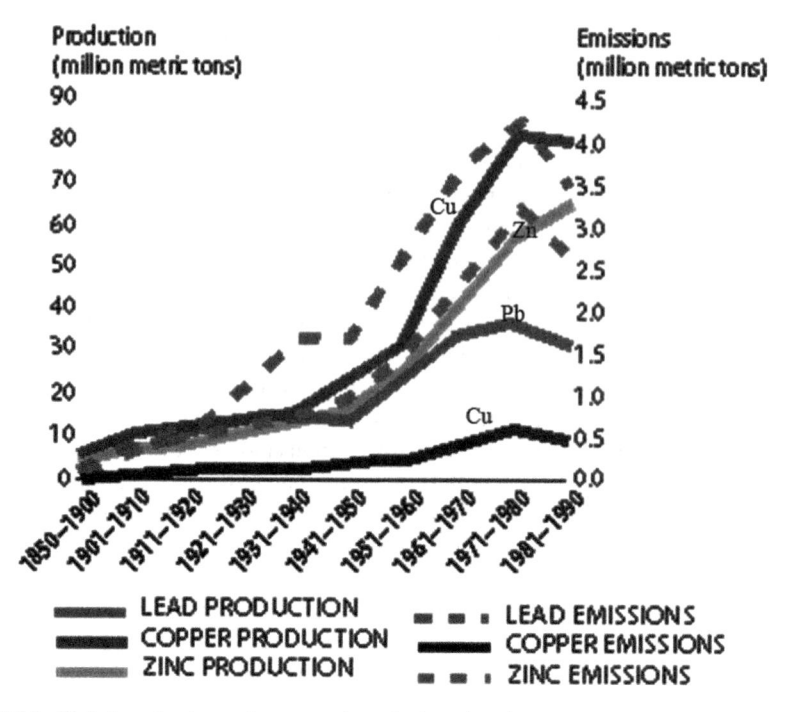

Fig. 15.1 Global production and consumption of selected toxic metals, 1850–1990 (Source: [3])

Decreased soil water content and increased evapotranspiration due to climate change is exacerbating resuspension of the soil dust particles that comprise suspended heavy metals. Climate change has also increased the frequency of extreme events such as floods, and measurements have shown that river basins contain larger amounts of lead concentrations that are brought to the surface by flooding.

15.3   Modeling of Heavy Metals

15.3.1   The MSCE-HM Chemical Transport Model

The MSCE-HM (Meteorological Synthesizing Centre East-Heavy Metal) chemical transport model has been developed for assessment of heavy metal airborne pollution in Europe and in support of the review, extension and implementation of the U.N. Protocol on Heavy Metals (e.g., [6–8]). MSCE-HM is a three-dimensional Eulerian-type chemical transport model driven by off-line meteorological data. It was developed to evaluate atmospheric transport and deposition of such heavy metals as lead, cadmium and mercury. Additionally, pilot parameterisations for some other toxic

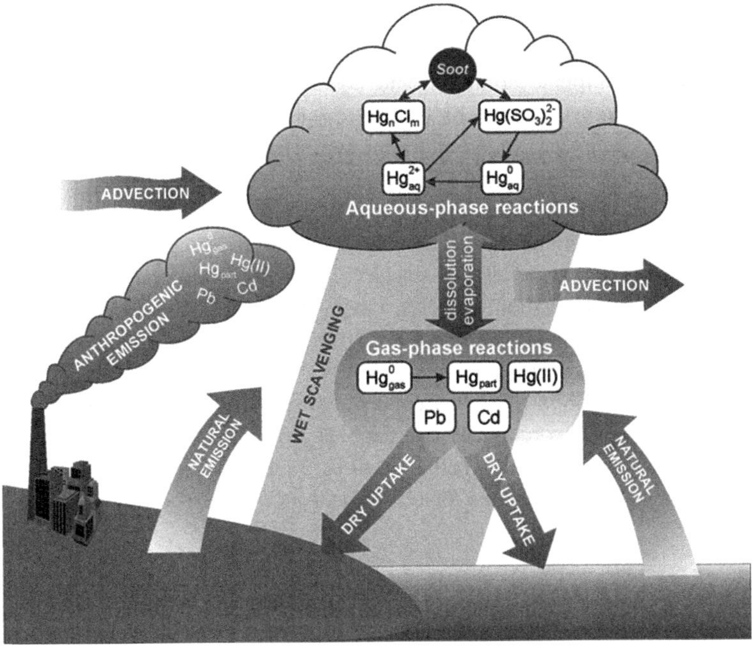

Fig. 15.2 The model scheme of heavy metal behaviour in the atmosphere (Source [7])

metals and metalloids like chromium, nickel and arsenic were developed. The model domain covers the EMEP region (Europe, part of Northern Africa and Middle East, the north-eastern Atlantic and part of the Arctic) with spatial resolution 50×50 km^2. The vertical structure of the model is formulated in the sigma-pressure (σ-p) coordinate system. The model domain consists of 15 irregular σ-layers and has a maximum height at a pressure level equal to 100 hPa. The layers are confined by surfaces of constant σ and do not intersect the ground topography. The midlevel of the lowest σ-layer approximately corresponds to 37 m. The top of the model domain can be roughly estimated at 15 km.

The model takes into account key processes governing the behaviour of heavy metals in the atmosphere and their deposition to the ground. These processes include anthropogenic and natural emissions, advective transport, turbulent mixing, wet and dry removal, and mercury chemical transformations both in gaseous and aqueous phases. Schematically these processes are depicted in Fig. 15.2.

Advective and vertical transport are evaluated by a Bott scheme. Turbulent mixing is approximated by a second-order implicit numerical scheme. Lead and cadmium are assumed to be transported in the atmosphere only as a part of aerosol particles. Besides, chemical transformations of these metals do not change the removal properties of their particles-carriers. Physical and chemical transformations of mercury include dissolution of gaseous elemental mercury in cloud droplets, gas-phase and aqueous-phase oxidation by ozone and chlorine, aqueous-phase formation of chloride complexes, reactions of mercury ion reduction through the decomposition of sulphate complex, and adsorption by soot particles in droplet water. A dry deposition scheme is based on a resistance-analogy approach. Modelled dry deposition velocity depends on surface type (forests, arable lands, water, etc.) and atmospheric conditions (atmospheric stability, wind velocity, etc.). At present the model is capable of calculating dry deposition fluxes to 18 categories of land cover. The model distinguishes in-cloud and sub-cloud wet scavenging. Boundary concentrations of heavy metals are set along the outer boundaries of the EMEP region and updated once a month. Mercury concentrations at the domain boundaries are derived from hemispheric-scale model. The concentrations of lead and cadmium are based on monitoring data.

The Meteorological Synthesizing Centre East uses MM5 as a meteorological pre-processor to prepare meteorological data for heavy metal and POP regional transport models. The MM5 system is described in detail in [9]. The configuration of system used by MSC-E, and its input and output meteorological are overviewed in MSCE report [8].

15.3.2   Modelling Case Study

Assessment of heavy metal airborne pollution of the environment encompasses various aspects, including estimation of atmospheric emissions, monitoring of pollution levels and application of chemical transport models. However, there are a number of factors constraining the quality of assessment on the European scale. In particular, measurements of heavy metals are not regular and only partly cover the EMEP domain. Furthermore, estimates of heavy metal emissions are often characterized by significant uncertainty and unaccounted emission sources while modelling brings in parameterisation of resuspension with wind-blown dust which requires detailed information on land-cover, soil concentrations, etc., and inserts a considerable factor of uncertainty. In order to facilitate the assessment process, several European countries, including Croatia, are participating in case studies initiated with the aim to collect and evaluate all available information on heavy metals on European/national and local scales and to minimise sources of uncertainty.

The proposed study aims at a complex analysis of factors affecting the quality of the assessment of heavy metal pollution levels in Croatia using a variety of available information (detailed emissions, monitoring and modelling results). Priority metals targeted by the study are lead, cadmium and mercury. Information on emissions, monitoring, and modelling is particularly for the year 2007.

The general scheme of the case study covers several steps. Firstly, emissions, monitoring, meteorological and geophysical information are collected and used as input to the atmospheric transport model. Model calculations will be performed at two spatial scales (resolutions): 50×50 km^2 resolutions and 10×10 km^2. For investigation of discrepancies between modelled and measured concentrations, and for the assessment of overall model performance a comparison of measured and modelled values as well as back trajectory analysis will be used.

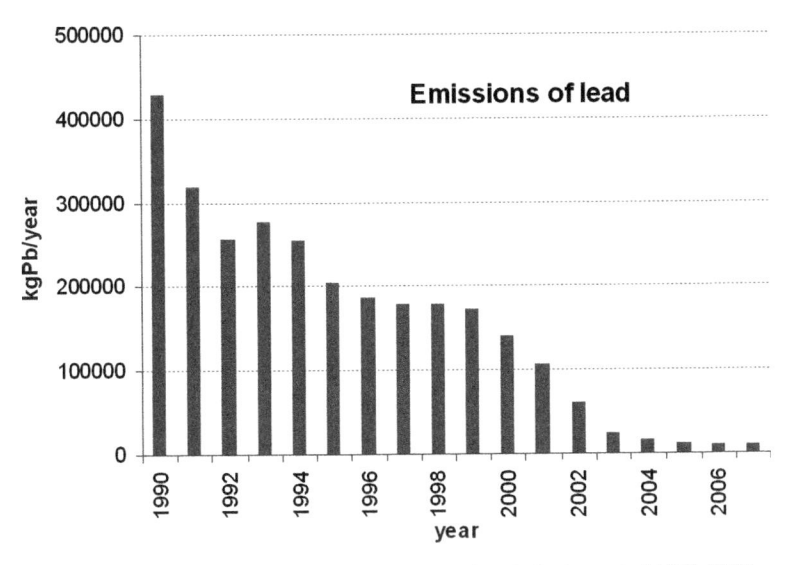

Fig. 15.3 Time series of lead emissions (kg Pb/year) in Croatia in the period 1990–2007

15.3.3 Anthropogenic Emissions of Lead

Air emissions of heavy metals arise from a wide range of different sources. Trace quantities are found in fossil fuels, and there are a number of industrial processes which give rise to emissions of specific metals. The relative contributions of these sources to the emissions total varies by metal, for example, lead emissions used to be almost completely from road transport but are now dominated by processes in the iron and steel sector. The observed decreases in lead emissions in Croatia (Fig. 15.3) are primarily a result of the decreasing lead content of petrol beginning in the late 1990s when unleaded petrol was introduced.

UN/ECE Heavy Metals Protocol from 1998 (www.unece.org/env/lrtap/) targets the emissions of three metals: lead, cadmium and mercury. The Protocol demands that countries reduce their emissions of heavy metals to levels below the values in 1990. The protocol also specifies nine limit values for emissions from stationary sources and requires that the best available technology be used for emission reduction from these sources. The protocol also required countries to phase out leaded petrol as well as to lower heavy metal emissions from other products e.g., mercury in batteries and introduce measures for other mercury-containing products.

Since model results are highly dominated by emissions, the quality of emission data, including their spatial distribution, is of great importance for final results. For this study, spatially distributed (gridded) emissions are used as input. However, it must be pointed out that the model is sensitive to grid resolution

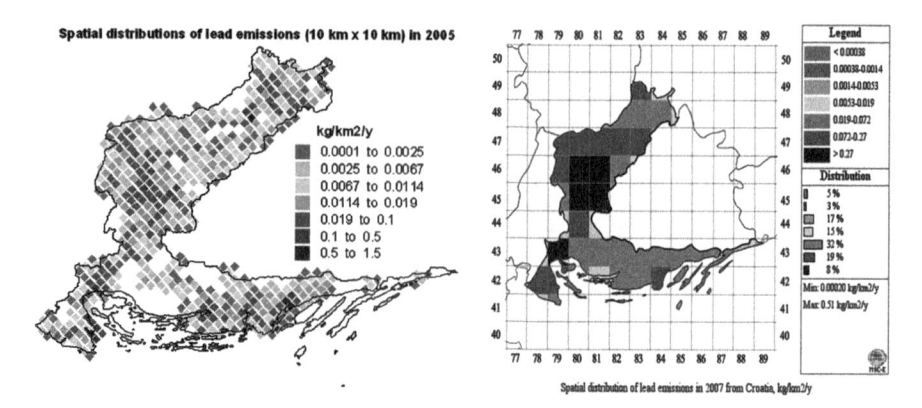

Fig. 15.4 Spatial distribution of lead emissions on different resolutions: 10 km × 10 km (*left*) and 50 km × 50 km (*right*)

which also affects the results. In Fig. 15.4, the spatial distribution of lead emissions on two different resolutions, 10 km × 10 km and 50 km × 50 km, is presented.

15.3.4 Heavy Metal Resuspension

There are a number of natural mechanisms responsible for emission of aerosol-bound heavy metals to the atmosphere. In particular, they include emissions with wind-blown dust and sea-salt aerosol. Since human activity has led to significant increases of concentrations of heavy metals in soils, compared to pre-industrial times, the meteorologically-driven emissions include both natural component and re-emission of previously deposited matter from anthropogenic sources called "natural and historical emission."

It is important to recognise that the emissions inventory (Fig. 15.4) is a "primary" emissions inventory. It does not typically include emissions from resuspension. Significant levels of all metals are expected to have been emitted since the start of the industrial revolution, and deposited to the land and sea. As a result, it is possible that the resuspension of surface material could give rise to a significant source term. Currently, the relative importance of this resuspension component at the national scale is not well known, although there are modelling studies which have attempted to include a resuspension component.

Wind resuspension of particle-bound heavy metals (like lead and cadmium) from soil and seawater appears to be an important process affecting the ambient concentration and deposition of these pollutants, particularly in areas with low direct anthropogenic emissions. Pilot parameterization of heavy metal wind resuspension was included in the MSCE-HM model ([6] and [7]). The parameterization is based on the approaches widely applied in contemporary mineral dust production models (e.g., [11–13]). Particularly, suspension of dust aerosol from soil is considered as a combination of two major processes – saltation and sandblasting – presenting horizontal movement of large soil aggregates driven by wind stress and ejection of fine dust particles, respectively. The dust suspension is estimated for non-vegetated surfaces (deserts and bare soils, agricultural soils during the cultivation period, and urban areas). The generation of sea-salt and wind suspension of heavy metals from the sea surface was also considered based on the empirical Gong-Monahan parameterization [13].

The model performance has been evaluated using different emission inventories [10] and it was shown that natural emissions, i.e., resuspension processes, have an important part of heavy metal concentrations in air.

15.4 Heavy Metal Concentration in Soils

Spatial distribution of heavy metal concentrations in soils is an input to the model, providing the quantity to be driven by the resuspension processes. The measurements of heavy metal compositions in a high spatial resolution, 5 km × 5 km have been conducted by Croatian Geological Survey and published in the Geochemical Atlas of the Republic of Croatia [14]. In Fig. 15.5, the spatial distribution of the measured lead concentrations aggregated to 10 km × 10 km resolution over Croatia

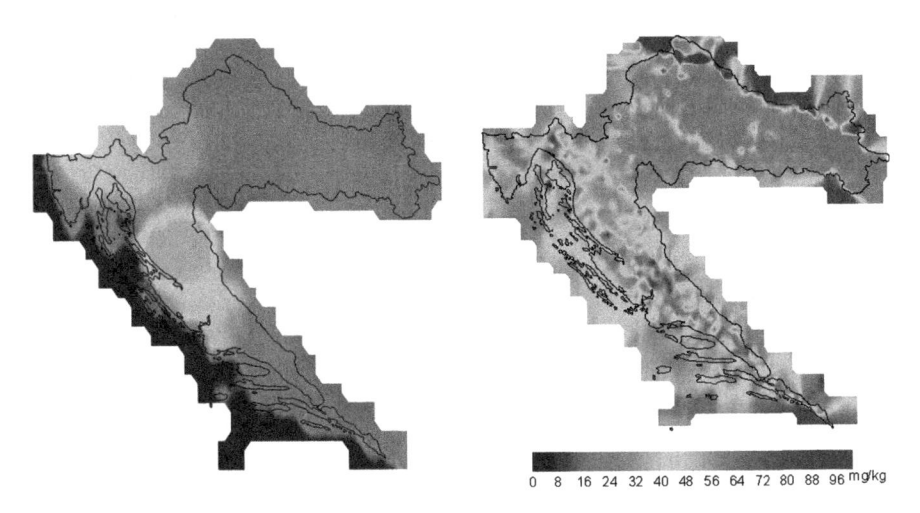

0 8 16 24 32 40 48 56 64 72 80 88 96 mg/kg

Fig. 15.5 Soil concentrations of lead in Croatia used in the model: concentrations from the FOREGS (Forum of European Geological Surveys) database and concentrations in soils (mg/kg) from the geochemical atlas of Croatia with the samples taken in 5 km × 5 km horizontal resolution (*right*)

are shown together with previously used heavy metals concentration in topsoil from the Geochemical Atlas of Europe developed under the auspices of the Forum of European Geological Surveys (FOREGS) [http://www.gsf.fi/foregs/geochem/]. The highest concentrations of lead are found in North Croatia and vary between 15 and 699 mg/kg. In this region, anomalous concentrations have been recorded in the Drava River valley and in soils above the alluvial and flood plain deposits. Increased lead concentrations have also been noticed on the Mura River alluvium. Those high concentrations are a consequence of the lead ore deposits situated upstream in Austria and Slovenia, where intense mining activity has existed for the last two centuries (Bleiberg, Mrežica and others). Coastal Croatia has high lead concentration in its soils with an extremely high regional median value of 48.7 mg/kg, which can be contributed to atmospheric pollution and the composition of red soils. Mountainous Croatia also has high lead concentrations with the highest values in the mountainous belts of Gorski Kotar (Risnjak) and Lika (Velebit Mnt) indicating an origination from atmospheric pollution.

15.5 Results

15.5.1 The Impact of Model Resolution on Pollution Levels

The EMEP-HM model is applied on different horizontal resolutions for the year 2007 and the annual concentrations and depositions of lead are calculated. The measured soil concentrations of lead are used as an input in the model contributing to the resuspension of lead concentrations in the air. The impact of the model resolution on pollution levels is evident (Fig. 15.6). The higher concentrations of lead in the air are simulated when 10 km × 10 km horizontal resolution is used, especially in the coastal area. However, the highest values are confined to urban areas where high emission sources are dominant. Similarly, for the deposition levels, higher values are simulated when higher horizontal resolution is used in the model (Fig. 15.7).

The distribution of lead depositions is highly correlated with the amount of precipitation. Therefore, modelled mean annual precipitation (mm) over Croatia calculated with the MM5 model on different horizontal resolutions was modelled for comparison (Fig. 15.8). There is a considerable difference in the model performance when different resolutions are used which is the mainly consequence of orography representation in the model. Model with the lower horizontal resolution tends to flatten mountains and it can be noted in Fig. 15.8 that high precipitation fields are at the Alps area on the west and above the Dinarides Mountains in Bosnia and Herzegovina. Mean annual precipitation (mm) calculated from the measurements for the period 1961–1990 over Croatia [15] is shown in Fig. 15.9.

The MM5 model with the higher horizontal resolution (Fig. 15.8) was able to capture the spatial distribution of precipitation more realistically.

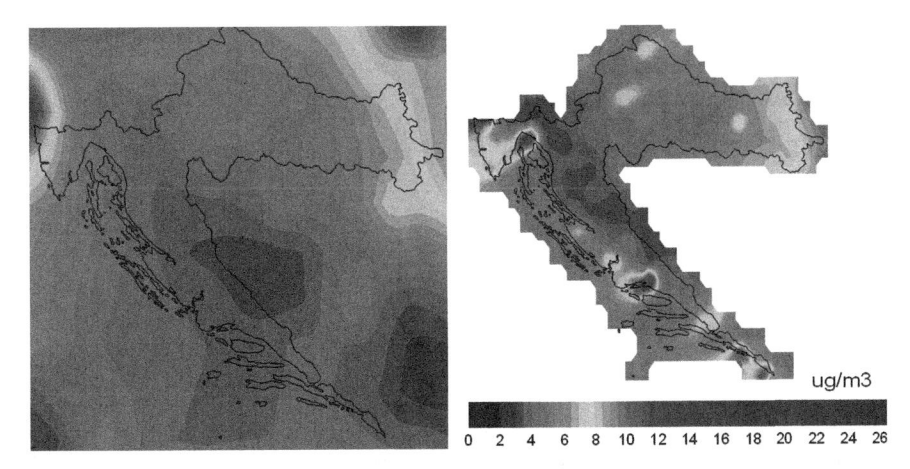

Fig. 15.6 The spatial distribution of average modelled yearly lead concentrations in air (ug/m³) over Croatia calculated with the EMEP-HM model on different horizontal resolutions: 50 km×50 km (*left*) and 10 km×10 km (*right*) for the year 2007

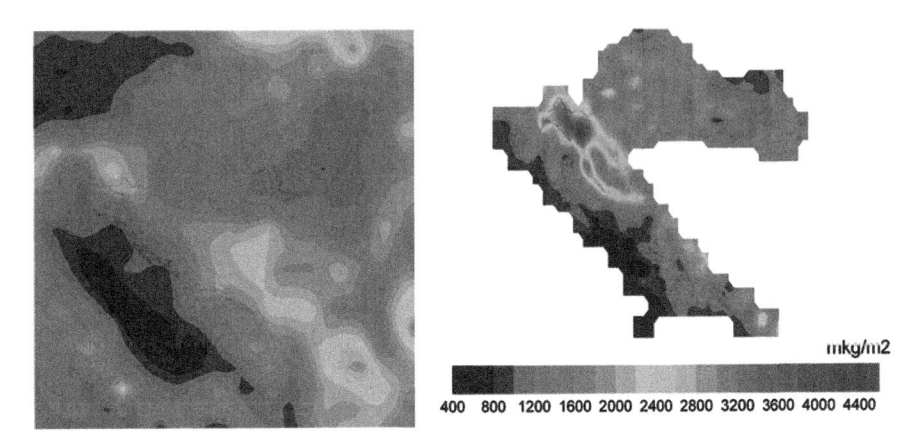

Fig. 15.7 The spatial distribution of average modelled yearly lead depositions (mkg/m²) over Croatia calculated with the EMEP-HM model on different horizontal resolutions: 50 km×50 km (*left*) and 10 km×10 km (*right*) for the year 2007

15.5.2   The Effects of Climate Change on Lead Concentrations

It has been shown that climate change yields drier soils and enhances the accumulation of lead levels in the atmosphere by increasing the natural emissions of lead and other heavy metals into the atmosphere. In order to estimate the increase in natural lead emissions due to climate change a test case run was performed with the EMEP-HM model with the higher horizontal resolution employed. In the idealized

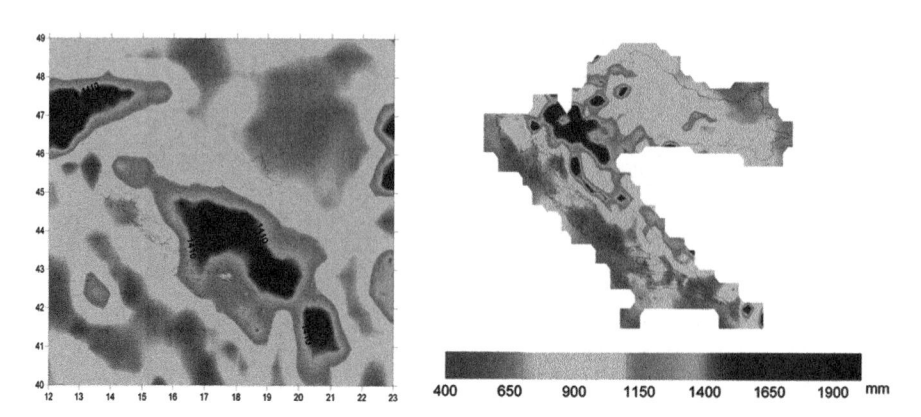

Fig. 15.8 Modelled mean annual precipitation (mm) over Croatia calculated with the MM5 model on different horizontal resolutions: 50 km×50 km (*left*) and 10 km×10 km (*right*) for the year 2007

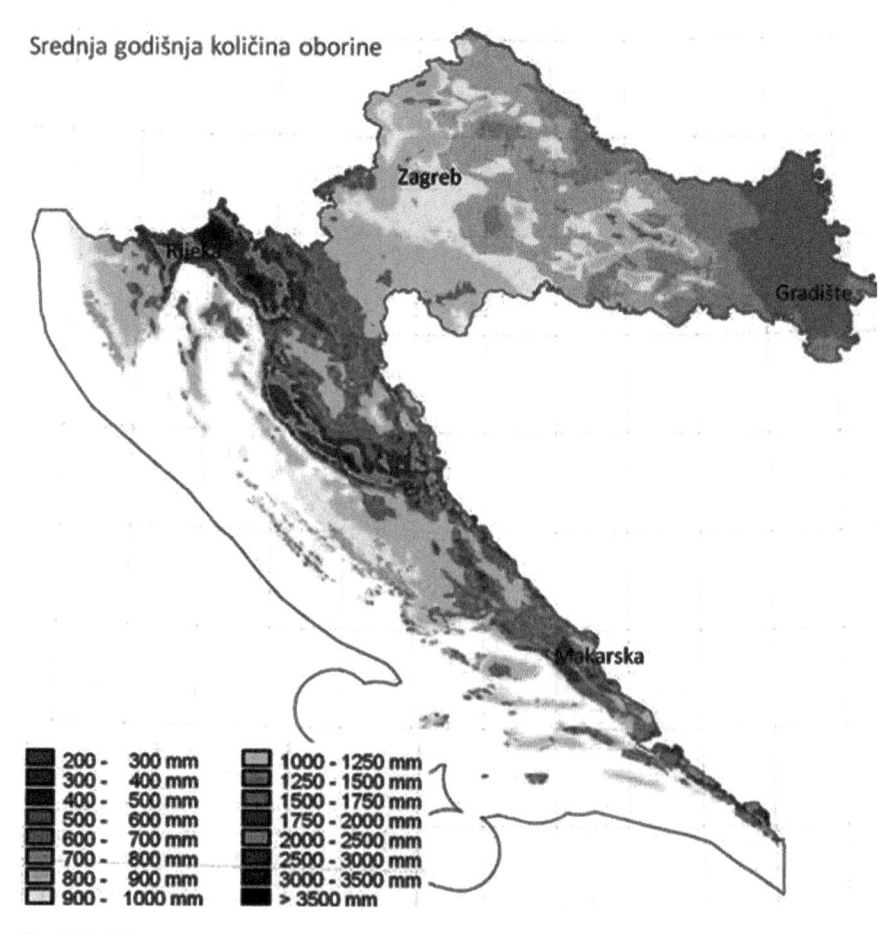

Fig. 15.9 Mean annual precipitation (mm) for the period 1961–1990 over Croatia (Source [14])

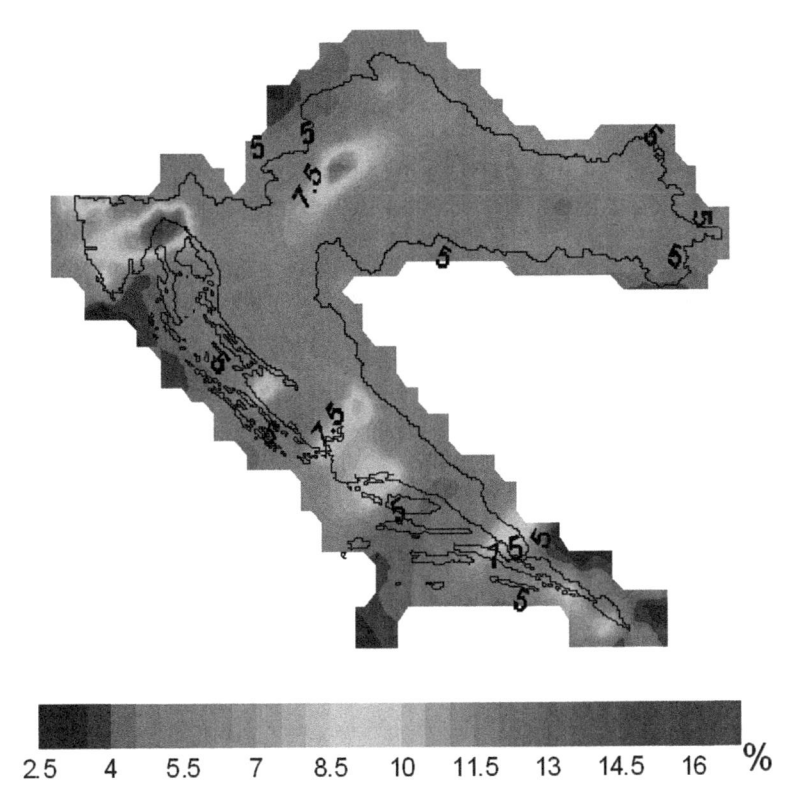

Fig. 15.10 Spatial distribution of bias differences determined between the modelled yearly lead concentrations in air calculated with the EMEP-HM with the normal model run and the test case run

test case run, parameterizations of resuspension processes were conducted in order to simulate the effects of climate change. Results are shown in Fig. 15.10. The largest impacts of the climate change expressed in bias values up to 17% are found for urban areas while for the rest of the country an increase of $\approx 7\%$ in modelled lead concentrations is found.

15.6 Conclusions

Climate change effects shift the risk and contamination pathways of heavy metals due to evident reduction in ground water contamination and an increase in contaminant concentration in soil dust particles. Assessment of heavy metal airborne pollution of the environment involves different aspects including estimates of atmospheric emissions, monitoring of pollution levels and application of chemical transport models. In order to estimate the increase of heavy metals natural emissions due to climate change, a modelling study was conducted as a case study on a complex assessment of heavy metal pollution at the national level organized by MSC-E and EMEP/TFMM. The EMEP-HM model was used with varying modelling resolutions, with an improved national emissions inventory that provided concentrations of lead in topsoil. The high impact of the model resolution as well as of emissions and soil concentrations on simulated pollution levels is shown. Furthermore, in the idealized test case run, parameterizations of resuspension processes were performed, which resulted in 7–18% higher spatial concentrations of lead over Croatia. The developed parameterization of wind resuspension from soil and seawater is to be improved and refined further in the future.

Due to the complex interactions between climate parameters and soil properties, the best response to these contamination conditions in view of impending climate change conditions would be site specific, and determined by perceived contamination pathways which would be influenced by the end use purposes for the sites, both at the present and in the foreseeable future.

References

1. Wilson B, Pyatt FB (2007) Heavy metal dispersion, persistence, and bioaccumulation around an ancient copper mine situated in Anglesey, UK. Ecotoxicol Environ Saf 66:224–231
2. Järup L (2003) Hazards of heavy metal contamination. Br Med Bull 68:167–182
3. Nriagu JO (1996) History of global metal pollution. Science 272:223–224
4. Derežić D, Vučetić V (2010) Tendency of increase in average soil temperature in Croatia', the present-day challenges in meteorology – conference of Croatian meteorological society, Zagreb, Croatia, 8–9 Nov 2010
5. Ferina J (2011) The spatial distribution of water balance components in Croatia', graduation thesis, University of Zagreb, Zagreb, pp 52
6. Gusev A, Ilyin I, Mantseva L, Rozovskaya O, Shatalov V, Travnikov O (2006) Progress in further development in MSCE-HM and MSCE-POP models (implementation of the model review recommendations). EMEP/MSC-E Technical Report 4/2006, p 114
7. Ilyin I, Rozovskaya O, Travnikov O, Aas W (2007) Heavy metals: transboundary pollution of the environment. EMEP Status Report 2/2007, p 5
8. Ilyin I, Travnikov O (2005) Modelling of heavy metal airborne pollution in Europe: evaluation of the model performance. EMEP/MSC-E Technical Report 8/2005, June 2005
9. Grell GA, Dudhia J, Stauffer DR (1995) A description of the fifth-generation Penn State/NCAR Mesoscale Model (MM5). NCAR/TN-398+STR. NCAR Technical Note. Mesoscale and Microscale Meteorology Division. National Center for Atmospheric Research, Boulder, pp 122
10. Gussev A, Ilyin I, Travnikov O, Shatalov V, Rozovskaya O (2008) 'Atmospheric deposition of selected heavy metals and POPs to the OSPAR maritime area (1990–2005)', Report to OSPAR Commission, pp 99
11. Gomes L, Rajot JL, Alfaro SC, Gaudichet A (2003) Validation of a dust production model from measurements performed in semi-arid agricultural areas of Spain and Niger. Catena 52:257–271
12. Zender CS, Bian H, Newman D (2003) Mineral dust entrainment and deposition (DEAD) model: description and 1990s dust climatology. J Geophys Res 108(D14):4416
13. Gong SL (2003) A parameterization of sea-salt aerosol source function for sub- and super-micron particles. Global Biogeochem Cycles 17(4):1097

14. Halamić J, Miko S (eds) (2009) Geochemical atlas of the republic of the Croatia. Croatian Geological Survey, Zagreb, 87 pp
15. Zaninović K, Gajić-Čapka M, Perčec Tadić M et al (2008) Climate atlas of Croatia 1961–1990, 1971–2000. Državni hidrometeorološki zavod, Zagreb, 200 pp

Chapter 16
New Atmospheric Pollution Indicators and Tools to Support Policy for Environmental Sustainable Development

Maria Cristina Mammarella, Giovanni Grandoni, Pasquale Fedele, Harindra J.S. Fernando, Silvana Di Sabatino, Laura S. Leo, Marco Cacciani, Giampietro Casasanta, and Ann Dallman

Abstract Air pollution, in particular high air pollution events, increase the effects of climate change on human health. It is necessary to undertake actions to prevent and minimize these adverse events with the aim of supporting a policy for environmental sustainable development.

For this purpose science, industry and institutions became allies through an applied research project in the valley of Biferno on the Adriatic Sea shore in central Italy. The purpose of this project is to search for new atmospheric pollution indicators and tools to support a policy for environmental sustainable development, useful to tackle climate change. The EART (ENEA Atmosphere Research Team) of ENEA (Italian National Agency for New Technologies, Energy and Sustainable Economic Development) has, with the cooperation of American, European and Russian research groups leading at

M.C. Mammarella (✉) • G. Grandoni • P. Fedele
Italian National Agency for New Technologies, Energy and Sustainable Economic Development, Technical Unit of Environmental Technologies, ENEA, EART Group, Rome, Italy
e-mail: mammarella@enea.it; giovanni.grandoni@enea.it

H.J.S. Fernando
Department of Civil Engineering and Geological Sciences, Wayne and Diana Murdy Endowed Professor of Engineering and Geosciences, University of Notre Dame, South Bend, IN, USA

S. Di Sabatino • L.S. Leo
Material Science Department, Climatological And Meteorological Laboratory, Salento University (Lecce), Italy

M. Cacciani • G. Casasanta
Physics Department, Laboratory of Environmental Physics, Rome University, Sapienza, Italy

A. Dallman
CEFD (Center of Environmental Fluid-Dynamics), Arizona State University (ASU), Phoenix, AZ, USA

H.J.S. Fernando et al. (eds.), *National Security and Human Health Implications of Climate Change*, NATO Science for Peace and Security Series C: Environmental Security, DOI 10.1007/978-94-007-2430-3_16, © Springer Science+Business Media B.V. 2012

the environmental level, investigated an industrial site through the concept of a "meteodiffusivity scenario". Meteodiffusivity is based on a new way to think of air quality, as the result of a strong interaction between emissions and local meteorological climatic factors. The micrometeorological parameters, and especially the Planetary Boundary Layer (PBL) depth, modulate the airborne concentration of emissions, causing their build up or dispersion, depending on the atmospheric turbulence.

Generally high air pollution events are affected by a low PBL height that prevents pollutant dispersion. High air pollution events frequently occur in anthropogenically polluted areas despite the fact that atmospheric emissions are checked and do not exceed legal limits. The meteodiffusivity method of analysis enables a more accurate interpretation of how the air quality of a place reacts to the pressure caused by anthropogenic activities, in order to try and mitigate the impact on the environment and human health. In fact, the control of local energy flows can reduce the negative effects of air quality on the land and climate. This innovative approach is based on new meteorological indicators and information tools in order to contribute to a development shift, from uncontrolled expansion to sustainability. The case study presented here is a project which was realized in lower Molise in the Biferno Valley, near Termoli (Molise Region, Italy), where a large manufacturing district coexists with a former fishing village that is now a well known tourist resort. The project execution and results were published in a volume entitled "Research on Environmental Management in a Coastal Industrial Area: new indicators and tools for air quality and river investigations" ISBN 9788860818997 and edited by Armando publisher*s*. The book was presented for the first time at the conference NATO ARW "Climate Change, Human Health and National Security" in Dubrovnik from 28 to 30 April 2011. It is also available as a CD ROM.

Keywords Air quality • Air pollution indicator • Planetary boundary layer • Meteodiffusivity • Sustainability • Climate change

16.1 Introduction

The ENEA (Italian National Agency for New Technologies, Energy and Sustainable Economic Development), with the cooperation of leading national and international environmental scientific groups, collectively, aimed at understanding, monitoring and checking the quality of the air everybody breathes in the valley and its vicinity. The project is a response to concerns expressed by local entities and dwellers vis a vis the environment and any findings are communicated and made available to the general public and local companies. The resulting solutions enable better management of air quality issues in industrial areas, promote the arrangement of interventions and ultimately provide the data required to plan the potential development of this area.

Research activities were characterized by an original analytical methodology aimed at accurately exploring any close relations existing between air quality and micrometeorology in an environmental context like the Biferno Valley. The valley is both geographically well defined and complex and it is characterized by the simultaneous presence of local and other than local atmospheric elements which overlap with sea and valley breezes. The focal of the study is an in-depth understanding of air quality aspects with special reference to improving the air quality of the Biferno Valley. This valley covers a portion of land near the city of Termoli, one of the most densely inhabited cities of Molise with approximately 30,000 dwellers. The valley, about 3 km wide, ends near the Adriatic Sea and exhibits a nearly east-west alignment. The 96 km long Biferno River runs across the valley. Along its course, the river is environmentally variable and thus conducive to the preservation of biodiversity. In fact, a number of such areas specifically account for the preservation of biological diversity by protecting habitats and animal and vegetable species. These areas fall under the protection scope of Sites of Community Importance (SCI) and Special Protection Areas (SPA).

Inside the Valley and near the sea, the Industrial Development Consortium of the Biferno Valley occupies a surface of approximately 1,000 Ha; its borders are adjacent to the left side of the Biferno River and, to the south, to the city of Termoli. The industrial pole of Termoli, a key reference for the economy in Lower Molise, experienced its utmost expansion and success in the 1970s, during the greatest Italian economic boom. Today it is home to about 100 companies, mostly in the manufacturing, mechanical and also chemical industries.

16.2   Approach and Methods

According to the needs expressed in Lower Molise, the EART team designed and planned an applied research project to air quality by including a group of researchers from Europe, Russia and America who joined forces to work together locally. The international team continues working beyond the project, and is ready to assist and participate in other areas using the most innovative set of skills, tools and methodologies aimed at improving air quality management, even in the most complex meteorological-environmental conditions. Working with the ENEA EART (Italy) project planner and coordinator, the experts are an international pool including the University of Helsinki, the Finnish Meteorological Institute (FMI), Russian State Hydrometeorological University of St. Petersburg (RSHU), Arizona State University (ASU), University of Notre Dame (Indiana), University of Salento (Lecce) and Sapienza University (Rome).

The research activities carried out in the Biferno Valley followed a new approach: investigating and analyzing air pollution levels as a profound interaction between airborne anthropogenic emissions and local climatic factors. From the above, a specific micrometeorology characterizes each site and influences any atmospheric release from emission sources so that the micrometeorology determines the quality of the air that the people breathe. The activities included a detailed analysis of typical meteorological diffusivity scenarios to identify the atmospheric capacity to concentrate or disperse the gaseous substances released into it. This observation explains why it is necessary to know and describe correlations between emissions and local/non local meteorological factors, by carefully measuring quite important meteorological dispersive parameters like PBL height. Throughout the project and in its field tests, the concept of meteodiffusivity was explored and better defined.

The study area has become an interesting "open sky laboratory" where research groups, according to their skills and methods, performed and shared studies and atmospheric experimental tests during two measurement meteorological campaigns, under the coordination of the EART team. Significant results were obtained, fulfilling the project objectives. During the experimental atmospheric measurement campaigns, an innovative set of equipment as well as previously existing and available meteorological instruments were employed, including remote sensing equipment like LIDAR and Vaisala CL 31 CEILOMETER. While the ceilometer generally measures the cloud height, as a pilot equipment it could continuously monitor PBL height. Moreover, four ultrasonic ANEMOMETERS aligned with the valley axis were deployed. According to in-field testing it became clear that the time variation of the PBL height is fundamentally important to characterize air quality and apply simulation models more realistically.

After gathering information during the experimental campaigns, airborne pollutant dispersion models were developed and applied. Through numerical analysis, concentration levels of pollutants (gas and aerosol) from industrial plants inside the Biferno Valley were simulated to determine the contribution of factories to local pollution. The Advanced Dispersion Modelling System ADMS 4.0 (CERC Ltd, www.cerc.co.uk) has been used with emissions from industrial plants coupled with meteorological variables acquired during campaigns as an input, and, later, with data collected by the IAF meteorological station of Termoli. The implementation of mathematical models also allowed highlighting the significance of proper positioning of meteorological measurement equipment, to better represent measures and estimates of modeling applications. During the project, it became increasingly clear that in order to investigate and analyze air quality at a given site, the monitoring of atmospheric patterns and of meteodiffusivity is crucial. For this reason, the Meteodiffusivity Indicator (MDI) was developed; it is expressed as a linear combination of meteorological and environmental parameters. The analytical nature of this indicator, as confirmed by tests conducted in the course of the project, brings an additional added value to basic science investigating the environment: it can easily provide any time changes of meteodiffusivity in a given site.

MDI is a new tool, essential to improve air quality management; it can be defined as a "potential pollution" indicator, kind of a sentinel of air quality. PBL height and MDI values are communicated to local authorities. They are dynamically calculated to obtain meteodiffusivity time changes. On an appropriate working platform as the one delivered to the Biferno Valley, this new integrated information along with conventional meteorological and air quality data, allowed definition of the overall local picture in order to design the best environmental policies and strategies, opening up a new path to an improved management of air quality.

16.3 Main Project Findings

16.3.1 *Flow Patterns*

The micrometeorology of a complex land like the Biferno area is characterized by local air flows which interact with the general wind circulation producing multiple, heterogeneous and mutual interactions. An overall picture of atmospheric motions highlights an increasing complexity caused by the effect of sea and land breeze. The presence of hill slopes around the valley, towards the sea, and higher mountains towards the inland, promote a channeling of air masses along the valley.

During the winter-spring and summer experimental campaigns in the Biferno Valley, the broad range of meteorological phenomena to observe and measure generated a huge amount of extremely detailed data. During the two campaigns both synoptic and weak synoptic forcing gave rise to two main distinctive flow regimes. The first regime (synoptic) was clearly dominated by large scale pressure gradient flows and unstable weather. The second (weak synoptic) resulted from the effect of local temperature gradients (thermally-driven flows). Despite the presence of a relatively high pressure system over central Italy for the experimental period few clear sky conditions were observed. During both campaigns, daily flow patterns in the valley usually showed some deviations from the daily cycle of a pure thermally-driven valley flow, mainly in the diurnal progression of wind direction. Nevertheless, friction velocity, turbulent kinetic energy and heat fluxes showed a diurnal periodicity with an increase during the middle of day and the night.

Well-defined diurnal flow patterns made up of a succession of morning transition, up-valley flow, evening transition followed by down-valley drainage flow were detected during the summer campaign and some days of the winter-spring campaign.

16.3.2 *ABLh Investigation*

During the summer campaign, a stronger differentiation between the maximum and minimum heights reached by the ABL, observed by Lidar data, was probably due to more frequent variations in the atmospheric stability, while the really low and almost constant ABLH values often observed during the daytime, in the presence of stable stratification, can be attributed to the effect of winds originated from sea-breeze conditions affecting the Lidar site and limiting the ABL growth. Lidar data suggest an ABLH almost always lower than 1 km, and in most cases lower than 300 m, in good agreement with the behavior shown by Ceilometer CL31 data. The Ceilometer is a state of the art nephoipsometer, which includes a module to measure the planetary boundary layer depth with a specific, dedicated algorithm offering an excellent performance even at low altitude. This instrument is located in the Consortium's area and continuously collects ABL height data. The Ceilometer data are related to the meteodiffusivity indicator MDI in order to improve the air quality in the industrial area of Biferno Valley.

16.3.3 Dispersion Modelling Application

The dispersion model ADMS 4.0 is able to calculate concentrations by using as input emissions from industries and meteorological variables. ADMS 4.0 was used to numerically assess the impact on air quality associated with emissions related to some of the biggest industries in the Biferno Valley study area. Due to the complex topography of the study area, the meteorological characterization of the site was comprised of a series of the meteorological data at different spatial and temporal scales, allowing differentiation between the actual contribution of the local meteorology and diffusion transport of pollutants in the valley and in the surrounding areas. The relation between the pollutant concentrations observed at the monitoring stations and those estimated by the numerical model, shows that the contribution of the industries to the pollution in the Biferno valley varies from 20% to 50% depending on the type of pollutant.

16.3.4 MeteoDiffusivity Indicator (MDI) and MeteoDiffusivity Detector (MDD)

In order to describe air quality in a site it is important to understand and analyze the atmospheric dynamic patterns and meteodiffusivity monitoring over time. Meteorological conditions affect the atmospheric capacity to concentrate or disperse substances released into the atmosphere, as a function of local and non-local phenomena. Special attention is required in outlining typical air mass motions and atmospheric turbulence of a given site, including PBL height as one fundamental parameter. The meteodiffusivity indicator (MDI) is described by a linear combination of meteorological and environmental parameters and represents a new methodology. After gathering locally acquired data, MDI can assess the atmospheric dispersion capacity of a given site.

Finally, to make MDI operational in the Biferno Valley, in addition to air quality data, a new information instrument was designed, realized and installed in the Cosib industrial area. This instrument is a monitoring, computing and management tool called MeteoDiffusivity Detector (MDD), which detects PBL height and other parameters while providing an hourly MDI pattern across the valley. MDD is the project technological novelty; an innovative environmental diagnostic tool which can dynamically provide information on meteorological scenarios, with special reference to meteodiffusivity in the Biferno Valley.

16.4 Policy Implications

Overall, the project reflected a joint effort to support both environmental resource and manufacturing operations. At the same time, it offered a means suitable to recommend the most effective air quality improvement strategies. This project is a practical example of how land governance, scientific resources and manufacturing plants may jointly help the cause of environmental sustainability and this experience can be replicated in any industrial area expressing the same needs and sensitivity. These findings are an asset not only for the project participants, but for all stakeholders. Meteodiffusivity and MDI outline a new way toward an improved air quality management, as key elements for environmental qualification and climate change mitigation. The programs, plans and actions aimed at environmental sustainability and reducing environmental vulnerability of a site, help to maintain the overall energy balance of the climate system.

In light of the above, it can be concluded that for a Country or institutions in charge of environmental monitoring and management, it is strategically important to incentivize meteodiffusivity investigations, both in their *ex ante* and *ex post* evaluations and in the authorization process provided for new industrial settlements. To make the best possible use of meteodiffusivity and MDI in the management of air quality in industrial and urban critical areas as well as in any emergency situation, MDD provides the hourly measurement of PBL height, indicator local patterns and potential pollution values.

MDD, an easy to install and operate instrument, is an information tool providing users with a constantly up-to-date meteodiffusivity pattern and it allows the evaluation of meteorological scenarios affecting air quality even in the presence of critical events. Furthermore, it contributes to improved air quality management and supports appropriate mitigation measures for a local ecologically sustainable development.

In order to apply these instruments, new professions are needed that are capable of managing air quality not only through obligatory checks but also by using diagnostic tools, elaborations and pollution prevention. It would be useful to have:

- an international agreement and methodology standardization for air quality control that uses the PBL height as an indicator (like MDI) in order to mitigate and prevent high pollution levels.
- a specific educational program dedicated to health care professionals in order to acquire and adopt new air quality indicators and tools useful for prevention of pathologies connected to air pollution.

These innovative methodologies and equipment, a legacy for dwellers and local administrations alike, are quite valuable. In fact, newly trained professionals with an expertise in specific sectors are now available for the proper operation of such novel equipment. For the Biferno Valley, there was a divide between "before" and "after" the project and the methodologies and technologies developed during the project resulted in a growth opportunity for the entire area.

Finally, the project paid special attention to communication of information to local entities (inhabitants, associations, schools), institutions and the scientific community in order to open up a new era of conscious growth. To disseminate information dedicated communications media were used such as meetings and workshops, as well as a video narrated by project participants that highlighted the most important experimental activities and technical-scientific solutions adopted by the Project.

Chapter 17
Numerical and Experimental Simulations of Local Winds

Franco Catalano, Antonio Cenedese, Serena Falasca, and Monica Moroni

Abstract Local circulation dynamics have a strong impact on the climate evolution as they contribute to the redistribution of energy and scalars from the regional to the global scale. Mesoscale phenomena are driven by surface heat, momentum and moisture fluxes; the intensity and distribution of these forcings can be significantly modified by the urbanization. The present work describes numerical and experimental investigations of the flow over an urban area. The circulation arises from the temperature difference between the city and the suburbs, called the Urban Heat Island (UHI) phenomenon. The three-dimensional non-hydrostatic meteorological model WRF has been used to perform Large Eddy Simulations of the UHI flow and its evolution during the complete day-night cycle. The domain is assumed to be planar in the cross-flow direction and periodic lateral boundary conditions are imposed. The laboratory experiments are conducted in a thermally controlled water tank to simulate an initially stably stratified environment and an electric heater solidal with the bottom of the tank mimics the urban site. Image analysis techniques have been used to reconstruct the velocity fields, while temperatures are acquired by multiple thermocouple arrays. The high resolution of both the numerical and laboratory experiments allows a detailed characterization of both mean and turbulent properties of the UHI circulation. Present numerical and laboratory results, normalized by similarity theory scaling parameters, compare well with literature data.

Keywords UHI • LES • Water tank model • Local climate • PBL turbulence

F. Catalano (✉) • A. Cenedese • S. Falasca • M. Moroni
Department of Civil, Construction and Environmental Engineering, Sapienza University
of Rome, via Eudossiana 18, Rome 00184, Italy
e-mail: franco.catalano@uniroma1.it; antonio.cenedese@uniroma1.it

H.J.S. Fernando et al. (eds.), *National Security and Human Health Implications of Climate Change*, NATO Science for Peace and Security Series C: Environmental Security, DOI 10.1007/978-94-007-2430-3_17, © Springer Science+Business Media B.V. 2012

17.1 Introduction

Local winds originate in the atmospheric boundary layer (ABL) as a result of non-uniform ground heating. They play a fundamental role in the evolution of the local climate, especially in absence or under negligible geostrophic winds; this condition often occurs in the Mediterranean region.

Local winds can be classified into:

- sea and land breezes which arise from the temperature difference between the sea and the land; in fact, the sea temperature is constant during the diurnal cycle, whereas the ground temperature may vary by more than 10 K in the same time frame
- valley and slope currents originate in mountainous regions because of the differential heating between the air in a valley and that over an adjacent plane (valley winds) or between the air adjacent to the slope and the ambient air at the same altitude (slope flows); during daytime, the air adjacent to the slope is warmer than the ambient air, the flow is then upslope (anabatic); during nighttime the temperature of the air adjacent to the slope cools faster than the surrounding air, so the flow is downslope (katabatic)
- urban heat island circulation which generates from the temperature difference between urban (or industrial) areas and the neighboring rural zones

Local winds have been investigated through laboratory experiments ([8], hereinafter CM, [12, 22]) and numerical simulations [5–7]. To allow a better characterization of the single circulation typologies, simplified controlled conditions have been imposed: absence of geostrophic winds and negligible effects of the Coriolis force, i.e. high values for Rossby and Ekman numbers. It should be pointed out here that, even if the Ekman number (ratio between viscous and Coriolis forces) is large, the vorticity of local winds is not due to viscosity but to baroclinicity.

For its relevance to the local climate change and adaptation scenarios, this work will be focused on the Urban Heat Island phenomenon.

Urbanization determines significant changes to the Earth's surface with important alterations of the local climate. Two kinds of modifications of the land surface which affect the energy budget in the Urban Boundary Layer (UBL) can be identified [23]: the first is connected with the architectural and morphological/geometrical aspects of the city (thermal and optical properties of the construction materials, soil roughness, presence of street canyons and building-barriers), the other one is connected with the energetic aspect of the human activities (transportations, air-conditioning, any process involving energy transformation). The connection between the Urban Heat Island (UHI) phenomenon and weather conditions, like precipitations and cloud cover, is largely recognized. Moreover, Chen et al. [9] proposed the UHI as an important contributor to global warming.

The UHI is defined as the temperature anomaly observed over urban areas with respect to the suburbs, the intensity ($\Delta\theta_m$) depending on weather conditions, soil moisture, geographical setting, latitude and city size. $\Delta\theta_m$ can be more than 10 K for big cities (CM).

The UHI circulation is characterized by a strong updraft motion at the city center, a divergent flow at the elevated levels, a horizontal convergent flow near the surface and a weak downdraft far from the city that closes the circulation (CM). This circulation pattern is more evident in the absence or weakness of synoptic winds, especially in the nighttime.

Early studies of the UHI date back to the nineteenth century [17].

Atkinson [1], by means of a mesoscale model, analyzed the effects of surface characteristics on the UHI intensity, finding that roughness length and surface resistance to evaporation are the most important parameters during daytime, while anthropogenic heat dominates in nighttime. The characteristic dimension of the city did not appear to be a dominant parameter. Taha [27] found that surface albedo and evapotranspiration have a more effective impact on the local climate than the anthropogenic heat. The UHI influences pollutant dispersion and hence air quality, meteorological conditions, energetic demand. Hinkel et al. [16] found that there is a mutual influence between aerosol and gaseous pollutants' concentration in the urban canopy layer and the radiation exchange between the surface and the atmosphere.

Buechley et al. [4] observed a correlation between the UHI phenomenon and the occurrence of human health problems, due to the combination of pollution and high temperatures.

Simplified linear and weakly nonlinear solutions of the equations describing the UHI circulation, proposed by Baik and Chun [2] and Baik et al. [3], evidenced the damping effect of stratification on the thermal plume and the fast and intense development of the circulation under neutral conditions.

The UHI circulation has been widely investigated by mesoscale models [15, 24, 29] in real scale. Computational Fluid Dynamics models have been used to reproduce the UHI phenomenon in laboratory scale [18, 19] under initially stably stratified conditions.

Despite the rich literature, the UHI phenomenon is nowadays not completely understood. Furthermore, real-scale, high-resolution numerical simulations of the UHI have never been performed. The results can be used to evaluate the impact of the UHI on the local climate. Only a few laboratory investigations of the UHI have been published ([20a] and [20b], hereinafter LUa and LUb, CM).

The present work describes a high resolution numerical and laboratory analysis of the phenomenon. Large-Eddy Simulations (LESs) are performed with the three dimensional non–hydrostatic meteorological model WRF (Weather Research and Forecast). The laboratory measurements are conducted in a temperature-controlled water tank where the UHI is simulated by an electric heater. The LES and laboratory results, normalized according to the similarity theory proposed by LUa, are compared with numerical, experimental and field data.

Section 17.2 briefly presents the numerical code and settings used for the LESs, together with the laboratory setup. Results are discussed in Sect. 17.3; conclusions and remarks are given in Sect. 17.4.

17.2 Numerical and Experimental Setup

The Large Eddy Simulations are performed using the three-dimensional non-hydrostatic fully compressible model WRF. For a detailed description of the code refer to Skamarock et al. [25]. A non-uniform terrain-following hydrostatic-pressure vertical coordinate is employed with a higher resolution ($\cong 2$ m) close to the ground. A 3rd order Runge-Kutta (RK) scheme for the time integration, 5th order advection scheme for the horizontal integration and 3rd order for the vertical integration are used. The time-splitting scheme allows the integration of acoustic and gravity waves with a smaller time step, into the RK loop, keeping larger time steps for the RK integration. Planetary Boundary Layer (PBL) parameterizations are switched off for the LESs. A modified subgrid-scale (SGS) model [5], which solves a prognostic equation for the SGS turbulent kinetic energy (TKE), is introduced to take into account the effects of the anisotropy of the grid. The model has been shown to be a valid tool for LESs of local winds over flat terrain [7] and in presence of orography [6].

Two different conditions are tested to introduce the heat forcing at the bottom boundary of the domain (ground and UHI site): in one case the surface heat flux is directly imposed in the thermal energy equation. A more realistic forcing is then employed by imposing sinusoidal time dependence for the surface temperature anomaly and using a surface layer parameterization based on the Monin-Obukhov similarity theory to compute the surface heat flux and the friction velocity necessary to drive the first layer of the grid. Periodic lateral boundary conditions (LBCs) are imposed to mimic an infinite succession of UHIs on the x direction and an infinitely long urban site along the y direction. To test the influence of the boundary effects on the solution, different domain dimensions are analyzed. The top boundary of the domain is assumed to be at constant pressure with zero vertical velocity. Five numerical simulations have been conducted varying the domain's x dimension, the UHI's forcing, the urban size D, the Brunt-Väisälä frequency N and the ground heat forcing (Table 17.1).

The laboratory model consists of a parallelepiped tank containing initially stably stratified water, heated from below. The experimental set-up is described in CM. Details of the experimental apparatus and measuring techniques are reported in Dore et al. [12].

The laboratory experiments were performed in a rectangular tank (Fig. 17.1) of length 1.8 m, height 0.2 m and width of 0.6 m (along the x-, z- and y-axes respectively), filled with distilled water, open at the top and with a horizontal aluminum surface at the bottom. Some of the experiments have been published in CM, others belong to a novel measurement campaign which differs from the past one for the improved monitoring apparatus and for the shape of the UHI. Distilled water is used as working fluid, to allow both a large heating rate and sufficient time to take measurements of the evolving thermal structures. The bottom of the tank is divided into two sections, the sea (S) side and the land (L) side, to make the test section suitable to run sea- and land-breeze experiments as well. Both sides are kept at the same temperature by means of two heat exchangers, connected to two thermostats, consisting of counter-flow channels. The surplus of surface heat flux H_0 between the

Table 17.1 Parameters for the numerical simulations and the laboratory experiments

	N_x, N_y, N_z	UHI width D (m)	Ground heat forcing	UHI forcing	N (s^{-1})	z_i (m)	$\Delta\theta_m$ (K)	Re	Fr	u_L (m s^{-1})
Sim 1	200×100×58	2,000 (rectangular)	–	$H_0 = 0.03$ K m s^{-1}	0.0128	390	1.1	1.5·10^8	0.05	1.25
Sim 2	400×100×58	2,000 (rectangular)	–	$H_0 = 0.03$ K m s^{-1}	0.0128	304	1.3	1.5·10^8	0.05	1.25
Sim 3	400×100×58	2,000 (rectangular)	Sinusoidal: $\Delta\theta_{s,max} = 5$ K	$\Delta\theta_m = 5$ K	0.0128	1,150	5	2.9·10^8	0.09	2.39
Sim 4	400×100×58	4,000 (rectangular)	Sinusoidal: $\Delta\theta_{s,max} = 5$ K	$\Delta\theta_m = 5$ K	0.0128	1,281	5	7.3·10^8	0.06	3.06
Sim 5	400×100×58	4,000 (rectangular)	Sinusoidal: $\Delta\theta_{s,max} = 5$ K	$\Delta\theta_m = 5$ K	0.0256	475	5	6.3·10^8	0.02	2.63
Exp 1	–	0.05 (rectangular)	–	$H_0 = 4.8·10^{-5}$ K m s^{-1}	0.45	0.013	1.1	90	0.080	0.0018
Exp 2	–	0.05 (rectangular)	–	$H_0 = 7.2·10^{-5}$ K m s^{-1}	0.43	0.016	1.3	103	0.097	0.0021
Exp 3	–	0.05 (rectangular)	Constant: 7.2·10^{-5} K m s^{-1}	$H_0 = 4.8·10^{-5}$ K m s^{-1}	0.43	0.017	1.4	122	0.115	0.0024
Exp 4	–	0.05 (rectangular)	Constant: 7.2·10^{-5} K m s^{-1}	$H_0 = 7.2·10^{-5}$ K m s^{-1}	0.43	0.027	2.2	130	0.121	0.0026
Exp 1CM	–	0.1 (circular)	–	$H_0 = 7.2·10^{-5}$ K m s^{-1}	0.7	0.009	2.0	270	0.038	0.0027
Exp 2CM	–	0.1 (circular)	Constant: 7.2·10^{-5} K m s^{-1}	$H_0 = 7.2·10^{-5}$ K m s^{-1}	0.7	0.013	1.9	340	0.049	0.0034
Exp 4CM	–	0.1 (circular)	–	$H_0 = 1.2·10^{-4}$ K m s^{-1}	0.7	0.012	2.7	320	0.045	0.0032
Exp 5CM	–	0.1 (circular)	Constant: 7.2·10^{-5} K m s^{-1}	$H_0 = 1.2·10^{-4}$ K m s^{-1}	0.7	0.018	2.2	370	0.052	0.0037
Exp 7CM	–	0.1 (circular)	–	$H_0 = 1.8·10^{-4}$ K m s^{-1}	0.7	0.015	3.5	360	0.051	0.0036
Exp 8CM	–	0.1 (circular)	Constant: 7.2·10^{-5} K m s^{-1}	$H_0 = 1.8·10^{-4}$ K m s^{-1}	0.7	0.021	3.0	410	0.058	0.0041

(continued)

Table 17.1 (continued)

	N_x, N_y, N_z	UHI width D (m)	Ground heat forcing	UHI forcing	N (s^{-1})	z_i (m)	$\Delta\theta_m$ (K)	Re	Fr	u_L (m s^{-1})
Exp 10CM	–	0.1 (circular)	–	$H_0 = 3.0 \cdot 10^{-4}$ K m s^{-1}	0.7	0.020	5.5	430	0.061	0.0043
Exp 11CM	–	0.1 (circular)	Constant: $7.2 \cdot 10^{-5}$ K m s^{-1}	$H_0 = 3.0 \cdot 10^{-4}$ K m s^{-1}	0.7	0.024	4.0	460	0.067	0.0046
Exp 13CM	–	0.1 (circular)	–	$H_0 = 4.8 \cdot 10^{-4}$ K m s^{-1}	0.7	0.024	6.5	500	0.073	0.0050
Exp 14CM	–	0.1 (circular)	Constant: $7.2 \cdot 10^{-5}$ K m s^{-1}	$H_0 = 4.8 \cdot 10^{-4}$ K m s^{-1}	0.7	0.028	6.7	530	0.076	0.0053

N is the Brunt-Väisälä frequency. LBCs for the numerical cases are periodic. Horizontal resolution is $\Delta x = \Delta y = 50$ m, vertical grid is stretched with $\Delta z \cong 2$ m near the ground and $\cong 90$ m at domain top. Domain dimensions are: $L_x = 20{,}000$ m, $L_y = 5{,}000$ m and $L_z = 2{,}400$ m for all cases except for Sim 1 where $L_x = 10{,}000$ m. The time step is $\Delta t = 1$ s and the surface roughness is $z_0 = 0.1$ m

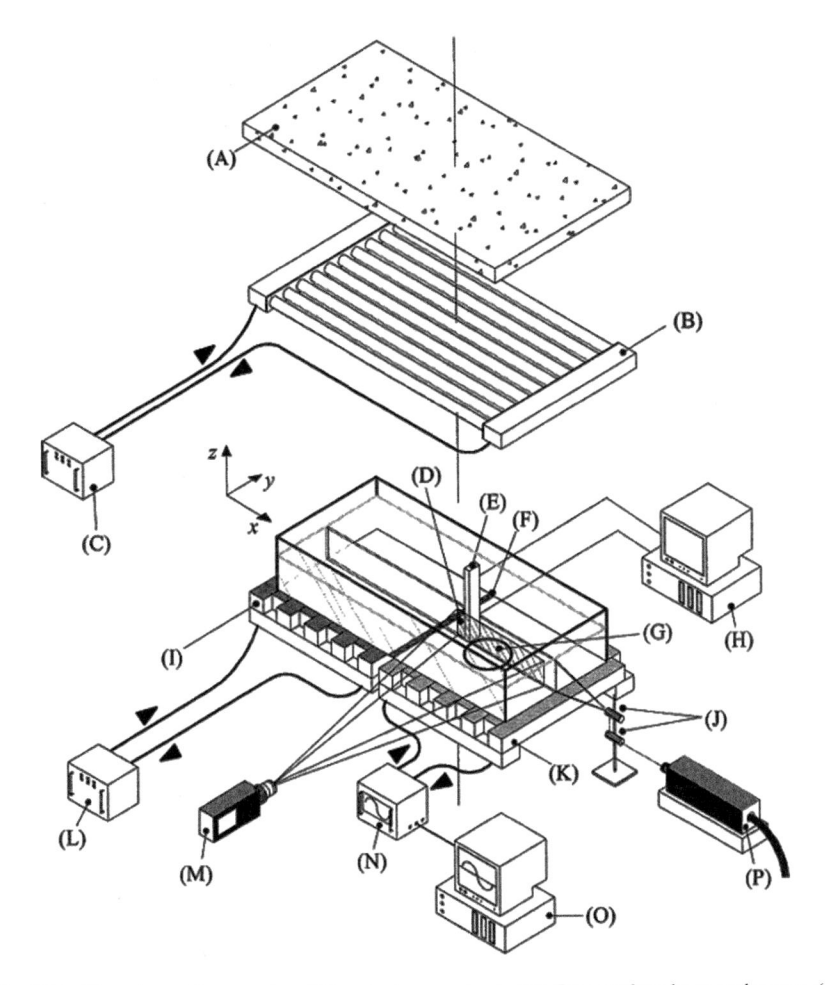

Fig. 17.1 Experimental apparatus: (**A**) polystyrene sheet, (**B**) Free surface heat exchanger, (**C**) Free surface thermostat, (**D**) framed area, (**E**) thermocouple array, (**F**) coastline, (**G**) heating disk, (**H**) personal computer (thermocouple controller), (**I**) heat exchanger (sea side), (**J**) optics (mirrors, lens), (**K**) heat exchanger (land side), (**L**) thermostat (land side), (**M**) video camera, (**N**) thermocryostat (land side), (**O**) personal computer (land temperature controller), and (**P**) laser

city and the rural environment is simulated by means of a thin ($2 \cdot 10$-4), circular-shaped electric heater (diameter $D = 0.1$ m) for CM experiments and a rectangular-shaped (side $D = 0.05$ m) electric heater for the novel experiments, both connected to a suitable power supply. The diurnal case is reproduced starting simultaneously heating the electric resistance and increasing the temperature of the bottom of the tank from the initial stably stratified conditions. The test section is illuminated through a planar light sheet obtained through a laser (CM) or a high power lamp. Images are acquired with a 764×576 pixels black and white camera at 25 Hz (CM) and a high resolution CMOS camera ($1{,}732 \times 2{,}532$ pixels) at 10 fps; due to the low velocity, images have been occasionally under-sampled. The area under investigation is rectangular, lying in the vertical x-z plane and passing through the center of the resistance, corresponding to the origin of the reference system. The framed area is 0.12 m long (x axis) and 0.040 m high (z axis). The velocity field has been determined through Particle Tracking Velocimetry (PTV) and Feature Tracking (FT), which both provide a Lagrangian description of the flow field. The sparse velocity vectors are interpolated on a regular grid to gather an Eulerian description of the flow field. Temperature is detected through T-type (copper-constantan) thermocouples characterized by uncertainty of 0.1 K and sample frequency of 1 Hz. Thermocouples are placed within the test section along vertical arrays to measure vertical profiles at locations of interest and on the lower boundary to test horizontal homogeneity in supplying heat. Thermocouples are placed about 0.15 m from the illuminated plane on the y-axis in order not to disturb the flow field. Four new laboratory cases are analysed, varying the UHI and the rural heat flux (Table 17.1).

17.3 Results

The numerical and experimental results are scaled according to the bulk model of LUa developed for nighttime low (<1) aspect ratio (defined as the ratio z_i/D between the mixing height z_i and the horizontal size D) UHIs. The phenomenon is completely described by three parameters: D, N giving the ambient thermal stratification and the surface heat flux H_0. The scale quantities are the mixing height of the UHI z_i, the horizontal velocity scale $u_L = (g\beta H_0 D)^{1/3}$, the vertical velocity scale $w_L = u_L^2 / (ND)$, the mean temperature scale given by the UHI intensity $\Delta\theta_m$. The convective temperature scale $\theta_L = H_0/uL$ is used to normalize the turbulent fluxes. z_i is estimated following the method of the maximum vertical gradient of the potential temperature [26]. The values of the scale quantities u_L, w_L and z_i, together with the Froude (Fr $= u_L / (ND)$) and Reynolds (Re $= u_L D/v$) numbers are reported in Table 17.1. The expressions for Fr and Re are taken from LUa.

Experiments, started right after the heater is turned on, evidence an initial transient phase when the UHI rapidly grows and develops until it reaches an equilibrium depth. Figure 17.2 shows the trajectories of the seeding particle reconstructed over 64 frames (~6 s) at time t=650 s for Exp 2. The picture shows the flow in a quasi steady state regime, i.e. when the circulation is fully established and sidewalls effects are still negligible. The classical shape of an axisymmetrical convective cell is clearly visible, with the thermal plume axis at the center of the heater (x/D=0), the converging flow near the bottom, and the diverging flow at upper levels. Note that the diameter of the thermal plume becomes narrower with height (a feature of area-source plumes) and that the plume contraction ratio, namely the ratio of minimum plume diameter to UHI diameter, is ~0.25.

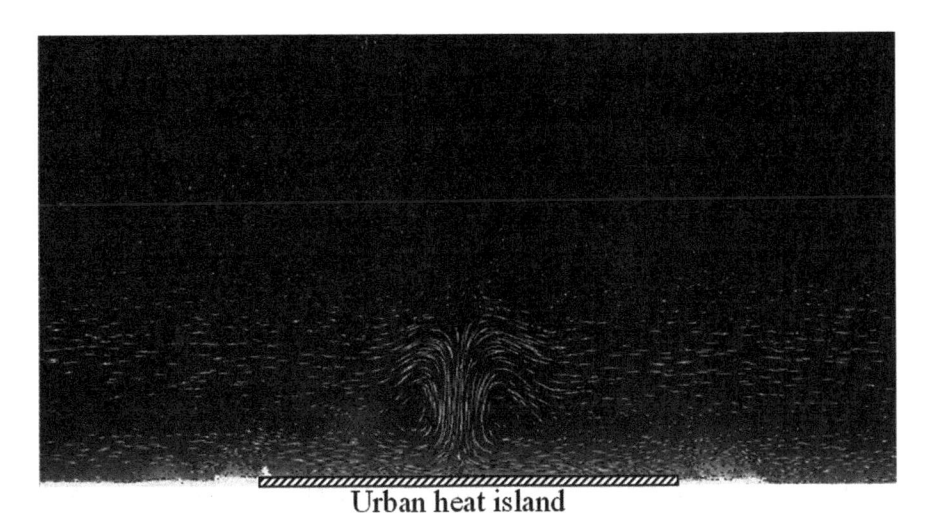

Urban heat island

Fig. 17.2 Trajectories reconstructed by FT at $t=650$ s over ~6 s for Exp 2

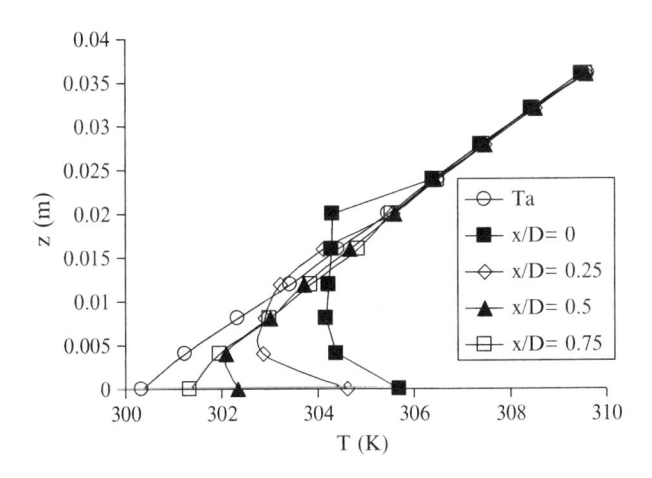

Fig. 17.3 Mean temperature profiles as a function of the radial distance from the UHI center for Exp 8CM ($z_i=0.020$ m; $z_e=0.015$ m)

For a given $d\theta_a/dz$ and D, the island characteristic dimension, the time needed to reach nearly steady-state flow conditions depends on the horizontal velocity scale u_L. For all cases considered, the UHI circulation reaches a quasi steady-state a few minutes after the heating is started. This condition persists for about 20 min, when the induced circulation arrives at the sidewalls and starts to be influenced by them.

Figure 17.3 shows the mean temperature profiles at various radial distances from the UHI center for Exp 11CM. The figure illustrates that, except for the superadiabatic surface layer adjacent to the bottom, inside the middle portion of the UHI ($x/D=0$ and 0.25), temperature does not vary appreciably with height.

Conversely, outside the heater ($x/D=0.75$), the boundary layer remains stably stratified with a temperature profile similar to that imposed for the ambient temperature θ_a. Temperature profiles suggest that the resulting UHI circulation is dome shaped and characterized by well-mixed conditions within its central region. Its maximum depth is at $x/D=0$. In particular, at $x/D=0$, temperature is nearly constant in the vertical range $0 < z < z_i = 0.020$ m. Furthermore, one can observe that for $z_e = 0.016$ m $< z < z_i$ (where z_e is the height of the base of the entrainment layer), the fluid inside the plume is cooler than that outside the plume at the same height.

Figure 17.4 shows the velocity field associated with the UHI circulation for Exp 2, 3 and 4. Nearly 500 velocity samples belonging to the averaging-time interval $\Delta t_{aver} = 20$ s are employed for each of the 130×74 grid cells. The shape of the flow patterns for all cases is similar to the trajectories shown in Fig. 17.2 and is consistent with the mean temperature profiles previously discussed. Figure 17.4b, c displays the averaged velocity field carried out for diurnal cases after the UHI had become well established. The UHI shape is similar to that observed for the corresponding nocturnal UHI (Exp 2, Fig. 17.4a), even though in the daytime it is wider the area with significant vertical velocities (plume), relative to its nocturnal counterpart. In particular, the increased UHI depth detected for the daytime case Exp 4 ($z_i = 0.027$ m), as compared with that observed for the nocturnal one Exp 2 ($z_i = 0.016$ m), is related to the presence of the daytime Convective Boundary Layer (CBL), which makes the vertical development of the thermal plume easier.

In the following the numerical results corresponding to the simulation time $t = 6$ h from the initial state at rest will be discussed; at that time the UHI circulation and the turbulence are fully developed. In Sim 1 and 2 the ground heat forcing is absent (nighttime conditions), while in Sim $3-5$ $t = 6$ h corresponds to the maximum amplitude of the surface temperature (daytime conditions). In Sim 1 and 2 $\Delta\theta_m$ evolves with time and at $t = 6$ h is, respectively 1.1 and 1.3 K, while it is 5 K for the other cases at all times. z_i is smaller for nocturnal cases, in particular for Sim 2 ($\cong 304$ m) where the resultant UHI intensity is minimum, and larger for diurnal cases, attaining its maximum value in Sim 4 ($\cong 1{,}281$ m) characterized by a UHI dimension larger than Sim 3 and by a smaller N than Sim 5.

Two convective cells are clearly visible in Fig. 17.5a–e, symmetrical with respect to the center of the UHI ($x/D=0$). The convergent flow depth is about $0.3z_i$ for all cases, except for Sim 3 (Fig. 17.5c) where it is $\cong 0.4z_i$. An upper divergent flow is observed from the top of the convergent flow up to $z/z_i \cong 0.8$.

Figure 17.5a evidences a slight influence of the horizontal dimension of the domain on the flow field; in fact in Sim 1 ($L_x = 10{,}000$ m) there is a residual circulation above the mixing zone (at z about $1.2z_i$) whereas this is not observed for the other simulations ($L_x = 20{,}000$ m).

The effect of the forcing is evident in the thermal structure (Fig. 17.6a–e). At $t = 6$ h in Sim $3-5$ the UHI interacts with the developing CBL, while Sim 1 and 2 represent a nighttime PBL. Above the UHI the normalized temperature anomaly $(\theta - \theta_a)/\Delta\theta_m$ is close to zero in all the cases. Outside the UHI region, scalar transport is due only to advection in Sim 1 and 2, whereas for diurnal cases it is intensified by the CBL growth. Furthermore, in Sim 1 a thin stratified surface layer is observed

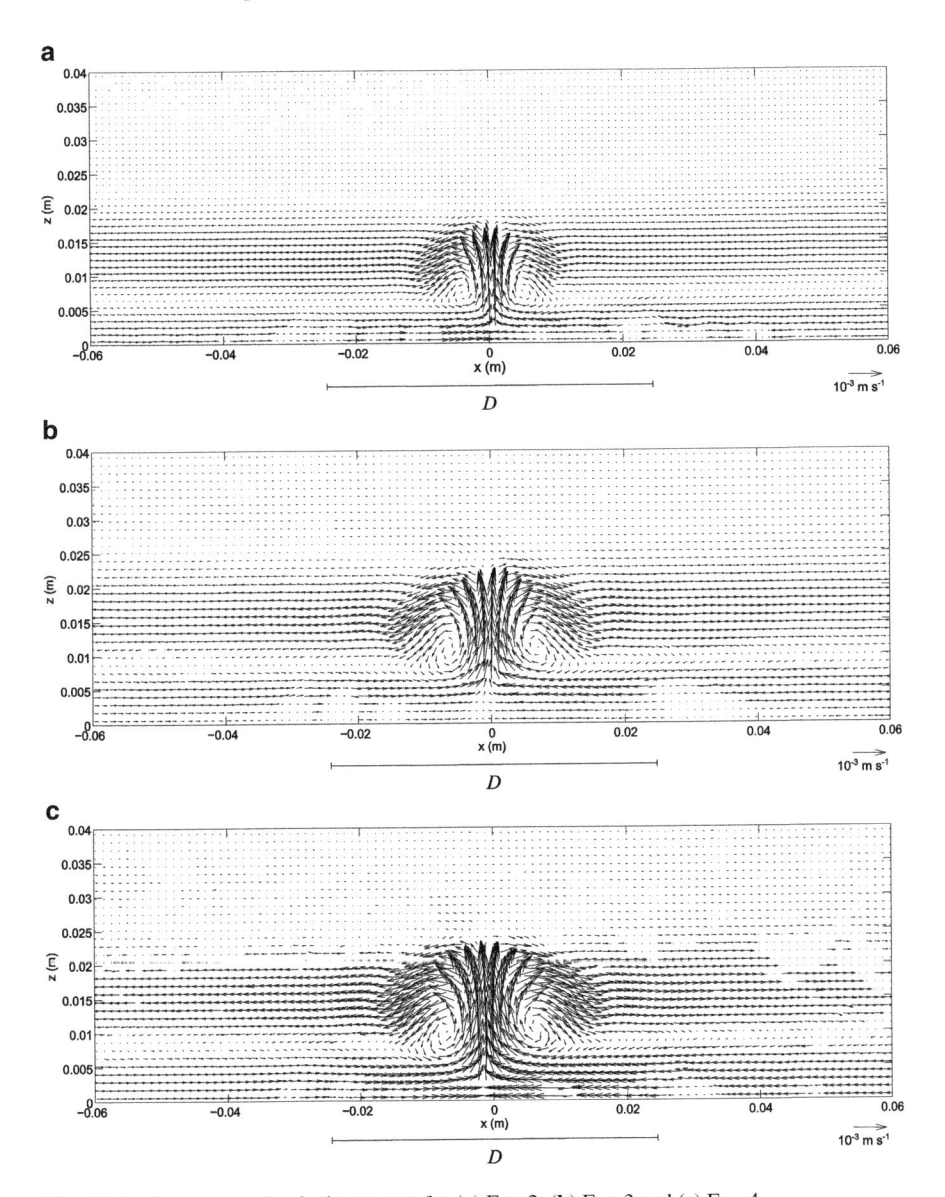

Fig. 17.4 Time-averaged velocity vectors for (**a**) Exp 2, (**b**) Exp 3 and (**c**) Exp 4

outside the UHI. In all cases there is a significant entrainment zone, as evidenced by the negative values of the temperature anomaly in the upper part of the PBL, i.e. thermal inversion.

Figure 17.7 show a vertical cross section of the normalized heat flux. Again, the entrainment zone can be identified above the thermal plume ($1 < z/z_i < 1.3$). The horizontal extent of the inversion region appears to be larger in Sim 5. In diurnal cases

Fig. 17.5 Vertical
cross-sections of the
y-averaged normalized
horizontal velocity
component for Sim 1 (**a**), Sim
2 (**b**), Sim 3 (**c**), Sim 4 (**d**),
Sim 5 (**e**) at $t = 6$ h

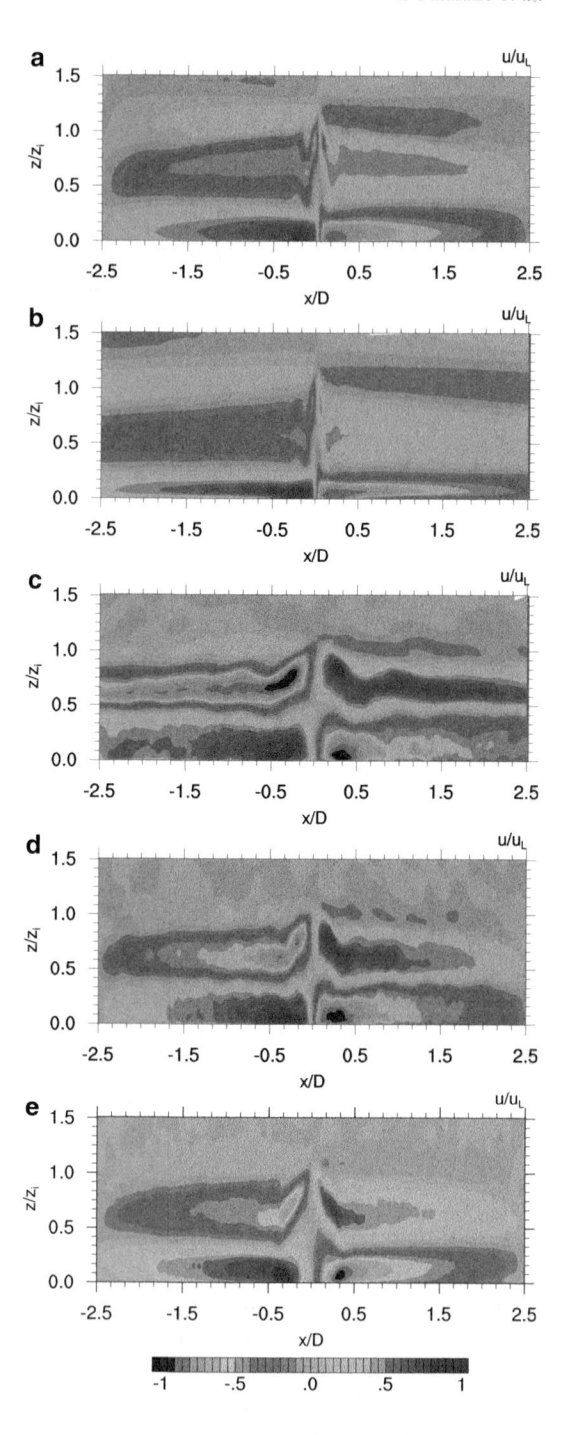

Fig. 17.6 y-averaged vertical cross-sections of the normalized temperature anomaly at $t = 6$ h for Sim 1 (**a**), Sim 2 (**b**), Sim 3 (**c**), Sim 4 (**d**), Sim 5 (**e**) at $t = 6$ h

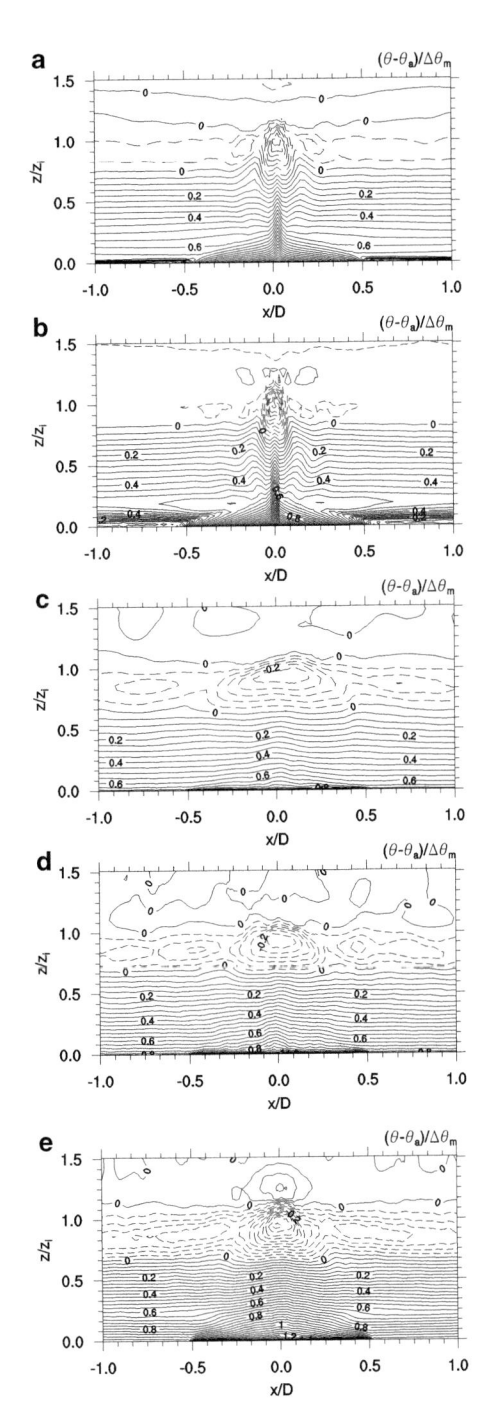

Fig. 17.7 Vertical cross-sections of the y-averaged normalized heat flux for Sim 1 (**a**), Sim 2 (**b**), Sim 3 (**c**), Sim 4 (**d**), Sim 5 (**e**) at $t = 6$ h

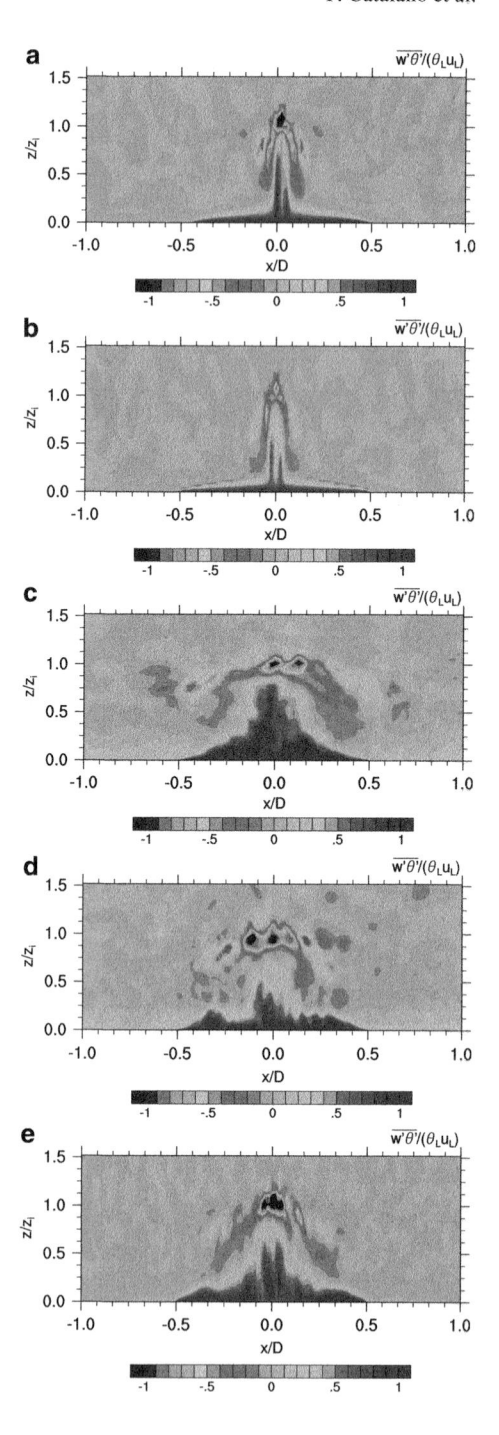

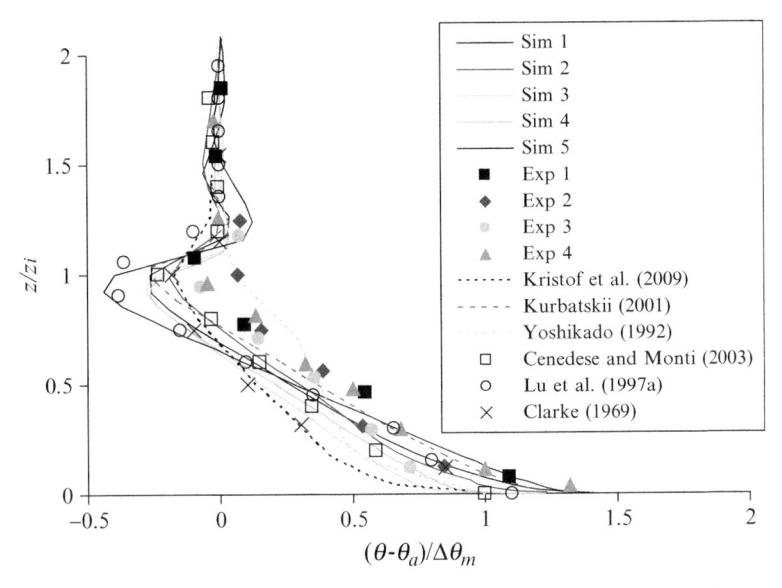

Fig. 17.8 Vertical profiles of the non-dimensional temperature anomaly at $x/D=0$ for the numerical simulations and the laboratory experiments, compared with literature data: numerical results by Kristof et al. [18], Kurbatskii [19], Yoshikado [29]; laboratory experiments of Cenedese and Monti [8], Lu et al. [20]; field observations by Clarke [10]

the area characterized by positive heat flux is more irregular than for the nocturnal ones, especially near the surface, thanks to the coupling of heat and momentum fluxes which causes the non-homogeneous distribution of the surface heat flux. In Sim 1 and 2, where a constant surface heat flux is imposed, the values of the normalized $\overline{w'\theta'}$ are more homogeneous close to the ground (Fig. 17.7a, b). Initial thermal stratification also plays a role in the homogenization of the surface heat flux, suppressing vertical motions and the development of the thermal plumes, as evidenced by Sim 5 (Fig. 17.7e).

To verify the similarity between present numerical and laboratory results and literature data, the normalized vertical profiles of the temperature anomaly and the horizontal and vertical velocity components are compared (Figs. 17.8–17.10). Figure 17.8 compares non-dimensional temperature profiles in correspondence to the UHI's centre ($x/D=0$) with laboratory results by LUa for Re=2,920 and Fr=0.089, data from field experiments in Cincinnati, Ohio (Re=$1.2\cdot10^9$ and Fr=0.013; [10]), and numerical results by Yoshikado [29]. For the latter case, values of the surface heat flux, Froude and Reynolds numbers are not reported and have therefore been estimated. Given $D=25,000$ m and $d\theta_a/dz=0.007$ K m^{-1}, the city radius and the ambient temperature gradient selected by Yoshikado, and assuming $H_0 \cong 9.6\cdot10^{-6}$ K m s^{-1} as a typical value for nocturnal heat fluxes, the velocity scale, Froude number, and Reynolds number are $u_L \cong 3$ m s^{-1}, Fr=0.009, and Re=$4.8\cdot10^9$. Despite the large differences in Re and Fr, the agreement among the curves is rea-

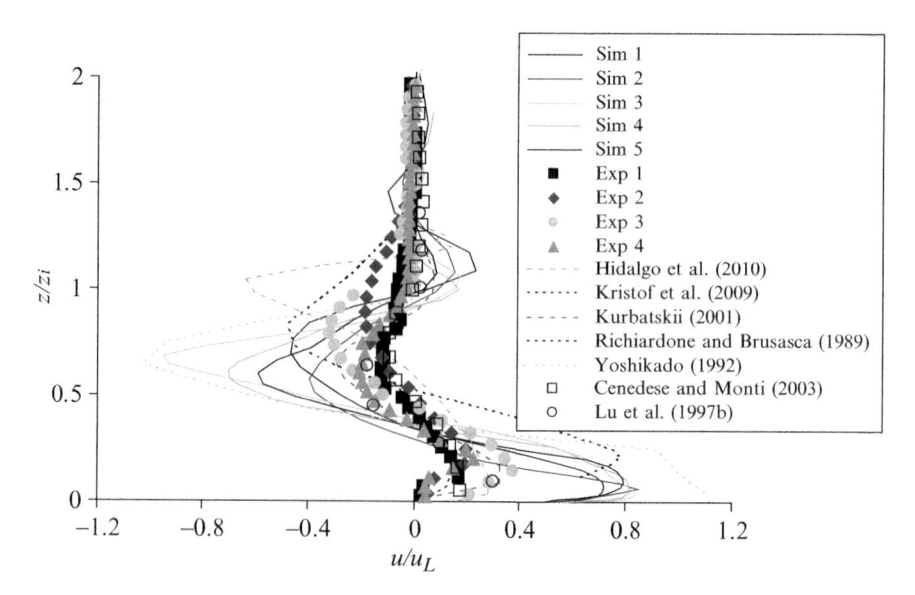

Fig. 17.9 Normalized horizontal velocity component vertical profiles at $x/D = 0.5$ for the numerical simulations and the laboratory experiments, compared with literature data: numerical results by Hidalgo et al. [15], Kristof et al. [18], Kurbatskii [19], Richiardone and Brusasca [24], Yoshikado [29]; laboratory experiments by Cenedese and Monti [8], Lu et al. [21]

sonable. This confirms that the shape of the non-dimensional mean temperature profiles of low-aspect-ratio UHIs depends on x/D rather than on Reynolds and Froude numbers (LUa). Figure 17.9 compares the profiles of the horizontal velocity component, normalized by u_L, at $x/D = 0.5$. The agreement among the curves is reasonable; this supports the possibility to extend the LUa theory for the daytime regime.

The oscillation over the mixing zone for Sim 1 can be attributed to sidewalls effects. It is remarkable that the maximum intensities for real scale numerical results, including literature data are larger than those of the laboratory scale data, evidencing a dependence on the fluid type (air, water). This suggests that the velocity scale u_L could be not an appropriate scaling parameter. Further investigation is needed to address this theoretical aspect. Figure 17.10 reports the profiles for the vertical velocity, normalized by w_L. Present numerical and laboratory results show a fair agreement with literature data. Maximum vertical velocity is attained at $0.5z_i$, except for the nighttime cases Sim 1 and 2 where it is located at about $0.3z_i$.

Detailed information on the turbulence structure is required by many environmental and wind assessment applications. Non dimensional vertical profiles of the standard deviation for the two components of the velocity are shown in Figs. 17.11 and 17.12 and compared with literature data. It is interesting to note that the high resolution of present LESs allowed a better characterization of the lower layer of the PBL, particularly important for σ_u, which attains its maximum near the surface. The profiles of σ_u show a general fair agreement with other data, but for Sim 3 and 4 which evidence much stronger values near the ground. The vertical velocity

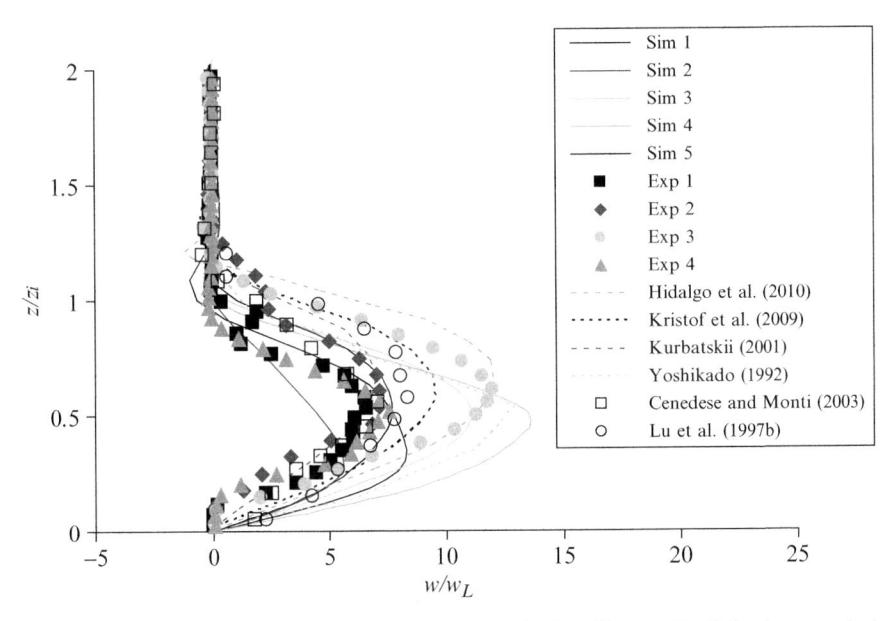

Fig. 17.10 Normalized vertical velocity component vertical profiles at $x/D=0$ for the numerical simulations and the laboratory experiments, compared with literature data: numerical results by Hidalgo et al. [15], Kristof et al. [18], Kurbatskii [19], Yoshikado [29]; laboratory experiments by Cenedese and Monti [8], Lu et al. [21]

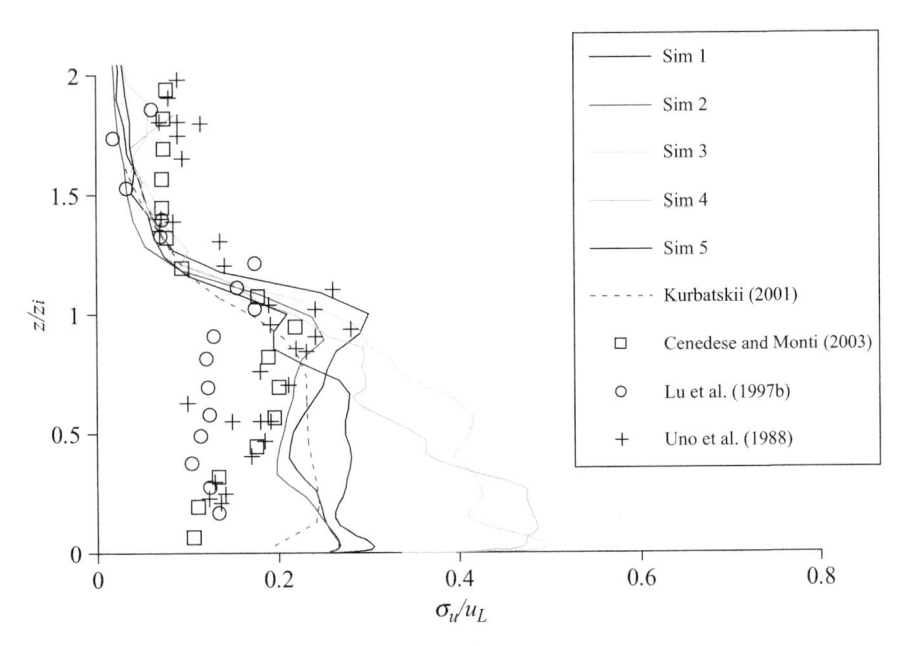

Fig. 17.11 Normalized horizontal velocity component standard deviation vertical profiles at $x/D=0$ for the numerical simulations, compared with literature data: numerical results by Kurbatskii [19]; laboratory experiments by Cenedese and Monti [8], Lu et al. [21]; field observations by Uno et al. [28]

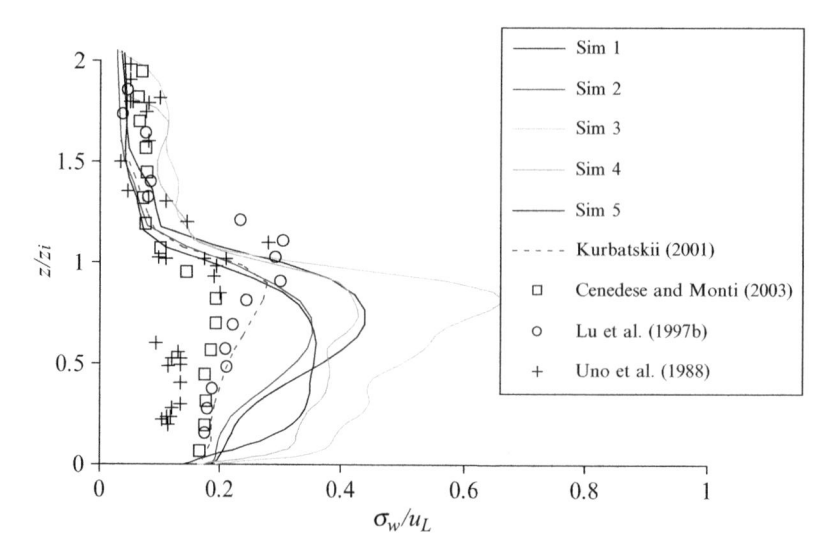

Fig. 17.12 Normalized vertical velocity component standard deviation vertical profiles at $x/D=0$ for the numerical simulations, compared with literature data: numerical results by Kurbatskii [19]; laboratory experiments by Cenedese and Monti [8], Lu et al. [21]; field observations by Uno et al. [28]

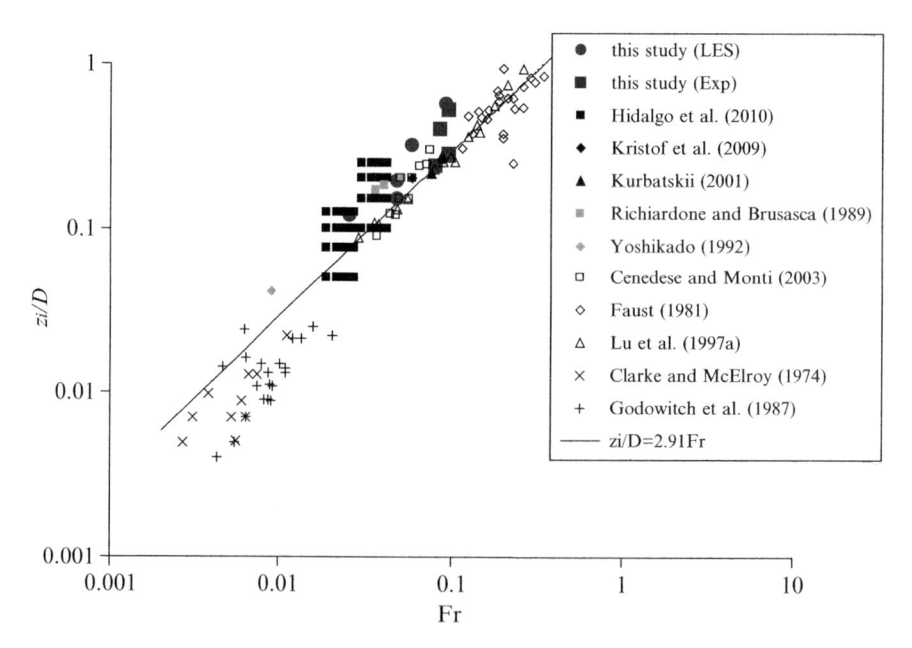

Fig. 17.13 Ratio of UHI aspect ratio z_i/D versus Froude number for the numerical cases and the laboratory experiments, compared with literature data: numerical results by Hidalgo et al. [15], Kristof et al. [18], Kurbatskii [19], Richiardone and Brusasca [24], Yoshikado [29]; laboratory experiments by Cenedese and Monti [8], Faust [13], Lu et al. [20]; field observations by Clarke and McElroy [11], Godowitch et al. [14]. For a better visualization, both axes are in logarithmic scale. The regression line, characterized by a slope of 2.91, is shown in solid

standard deviations (Fig. 17.12) present a structure similar to that of the penetrative free convection (PFC) but the height where the maximum is attained is shifted upward to about $0.7z_i$, whereas it is about $0.4z_i$ in PFC [7]. This could be attributed to the effect of the shear associated to the UHI circulation. A similar feature was found for other baroclinic local wind systems [6].

Figure 17.13 reports the UHI aspect ratio versus the Froude number for present LES and laboratory results, compared with literature data. The ensemble of data appears to collapse on a line with slope 2.91, slightly larger than the coefficient found by LUa and CM (2.86).

17.4 Conclusions

The characteristic circulation associated to the UHI phenomenon has been reproduced by means of LES and laboratory experiments. This allowed a detailed characterization of the PBL structure in terms of mean and turbulent fields. The presence of a significant inversion layer, associated to an intense entrainment over the urban site has been evidenced. This feature is of particular relevance to environmental aspects, since it strongly affects the dispersion properties of the atmosphere. Furthermore, the characteristic thermal structure plays a fundamental role in the redistribution of heat fluxes to the larger scales and hence on the regional climate.

The results, normalized following the LUa theory, show a fairly good agreement with literature data with respect to the temperature field. Differences between numerical and laboratory results are evidenced for the velocity maxima, which suggest a possible limitation of the theory.

Acknowledgements We thank Arianna Ferrari and Marco Giorgilli for their assistance in taking the laboratory measurements.

References

1. Atkinson BW (2003) Numerical modelling of urban heat-island intensity. Bound-Lay Meteorol 109:285–310
2. Baik J-J, Chun H-Y (1997) A dynamical model for urban heat islands. Bound-Lay Meteorol 83:463–477
3. Baik J-J, Kim Y-H, Kim J-J, Han J-Y (2007) Effects of boundary-layer stability on urban heat island-induced circulation. Theor Appl Climatol 89:73–81
4. Buechley RW, Truppi LE, Van Brugg J (1972) Heat island=death island? Environ Res 5:85–92
5. Catalano F, Cenedese A (2010) High-Resolution numerical modeling of thermally driven solpe winds in a valley with strong capping. J Appl Meteorol Climatol 49:1859–1880
6. Catalano F, Moeng C-H (2010) Large-eddy simulation of the daytime boundary layer in an idealized valley using the weather research and forecasting numerical model. Bound-Lay Meteorol 137:49–75

7. Catalano F, Moroni M, Dore V, Cenedese A (2011) An alternative scaling for unsteady penetrative free convection. Submitted to J Atmos Sci
8. Cenedese A, Monti P (2003) Interaction between an inland urban heat island and a sea-breeze flow: a laboratory study. J Appl Meteorol 42:1569–1583
9. Chen X-L, Zhao H-M, Li P-X, Yin Z-Y (2006) Remote sensing image-based analysis of the relationship between urban heat island and land use/cover changes. Remote Sens Environ 104:133–146
10. Clarke JF (1969) Nocturnal urban boundary layer over Cincinnati, Ohio. Mon Weather Rev 97:582–589
11. Clarke JF, McElroy JL (1974) Effects of ambient meteorology and urban morphological features on the vertical temperature structure over cities. 67th annual meeting of air pollution control association, Denver, CO
12. Dore V, Moroni M, Le Menach M, Cenedese A (2009) Investigation of penetrative convection in stratified fluids through 3D-PTV. Exp Fluids 47:811–825
13. Faust KM (1981) Modelldarstellung von Wärmeinselströmungen durch konvektionsstrahlen. SFB 80/ET/201 PhD Dissertation Universität Karlsruhe 144 pp
14. Godowitch JM, Ching JKS, Clarke JF (1987) Spatial variation of the evolution and structure of the urban boundary layer. Boundary-Layer Meteorol 38:249–272
15. Hidalgo J, Masson V, Gimeno L (2010) Scaling the daytime urban heat island and urban-breeze circulation. J Appl Meteorol Climatol 49:889–901
16. Hinkel KM, Nelson FE, Klene AE, Bell JH (2003) The urban heat island in winter at Barrow, Alaska. Int J Climatol 23:1889–1905
17. Howard L (1833) Climate of London deduced from meteorological observations, vol 1–3. Harvey and Darton, London
18. Kristof G, Rácz N, Balogh M (2009) Adaptation of pressure based CFD solvers for mesoscale atmospheric problems. Bound-Lay Meteorol 131:85–103
19. Kurbatskii AF (2001) Computational modeling of the turbulent penetrative convection above the urban heat island in a stably stratified environment. J Appl Meteorol 40:1748–1761
20. Lu J, Arya SP, Snyder WH, Lawson RE Jr (1997) A laboratory study if the urban heat island in a calm and stably stratified environment. Part I: temperature field. J Appl Meteorol 36:1377–1391
21. Lu J, Arya SP, Snyder WH, Lawson RE Jr (1997) A laboratory study if the urban heat island in a calm and stably stratified environment. Part II: velocity field. J Appl Meteorol 36:1392–1402
22. Moroni M, Cenedese A (2006) Penetrative convection in stratified fluids: velocity measurements by image analysis techniques. Nonlinear Proc Geophy 13:353–363
23. Oke TR (1982) The energetic basis of the urban heat island. Q J R Meteorol Soc 108:1–24
24. Richiardone R, Brusasca G (1989) Numerical experiments on urban heat island intensity. Q J R Meteorol Soc 115:983–995
25. Skamarock WC, Klemp JB, Dudhia I, Gill DO, Barker DM, Duda MG, Huang X-Y, Wang W, Powers JG (2008) A description of the advanced research WRF version 3. NCAR/TN-475, 113 pp
26. Sullivan PP, Moeng C-H, Stevens B, Lenschow D, Mayor SD (1998) Structure of the entrainment zone capping the convective atmospheric boundary layer. J Atmos Sci 55:3042–3064
27. Taha H (1997) Urban climates and heat islands: albedo, evapotranspiration, and anthropogenic heat. Energ Buildings 25:99–103
28. Uno I, Wakamatsu S, Ueda H, Nakamura A (1988) An observational study of the structure of the nocturnal urban boundary layer. Bound-Lay Meteorol 45:59–82
29. Yoshikado H (1992) Numerical study of the daytime urban effect and its interaction with the sea breeze. J Appl Meteorol 31:1146–1164

Chapter 18
Wind Effects on Man-Made Structures in a World with a Changing Climate

Hrvoje Kozmar and Zvjezdana Bencetić Klaić

Abstract A number of epidemiologic studies reported correlations between ambient concentrations of air pollution and adverse health effects, such as respiratory and heart diseases, premature mortality, premature delivery and low birth weight. Apart from indirect effects of the wind on health, humans can experience 'mechanical' wind-induced injuries due to collapsing engineering structures, windborne debris, and wind-induced traffic accidents. In this study, basic features of the wind/structure interaction were briefly addressed and some effects of a changing climate on local wind characteristics were reported. Therefore, wind-tunnel simulations of the atmospheric boundary layer flow indicate the applicability of truncated vortex generators in reproducing the wind characteristics in the lower atmosphere. A loading of a vehicle exposed to cross-wind gusting gives evidence about the aerodynamics significantly different than on vehicles exposed to 'standard' atmospheric turbulence.

Keywords Human health • Wind/Structure interaction • Climate change • Wind-tunnel simulations • Atmospheric boundary layer flow • Transient winds

H. Kozmar (✉)
Faculty of Mechanical Engineering and Naval Architecture, University of Zagreb, Ivana Lučića 5, Zagreb 10000, Croatia
e-mail: hrvoje.kozmar@fsb.hr

Z.B. Klaić
Andrija Mohorovičić Geophysical Institute, Department of Geophysics, Faculty of Science, University of Zagreb, Zagreb 10000, Croatia
e-mail: zklaic@gfz.hr

H.J.S. Fernando et al. (eds.), *National Security and Human Health Implications of Climate Change*, NATO Science for Peace and Security Series C: Environmental Security, DOI 10.1007/978-94-007-2430-3_18, © Springer Science+Business Media B.V. 2012

18.1   Winds Can Affect Human Health in Different Ways

Winds can harm human health in different ways, chronically and/or acute. A number of epidemiologic studies reported correlations between ambient concentrations of air pollution, and adverse health effects, such as respiratory and heart diseases, premature mortality, premature delivery and low birth weight (e.g. [1, 2, 16, 22, 23]), even at relatively low concentrations [6]. While modern diesel engines produce less toxic gases, the number of harmful toxic nanoparticles continues to increase. Thus, substantial scientific effort is currently devoted to addressing nanoparticles (e.g. [21]). In addition, it is well known that pollution levels can be strongly affected by atmospheric conditions (e.g. [4, 5, 15, 25]). Thus, a number of studies deal with the influence of airborne pollutant concentrations, weather conditions, population characteristics and public health policies.

Within the wind-related phenomena in particular, advection and turbulence are among the most important atmospheric factors affecting the fate of atmospheric pollutants, and consequently, in influencing human health. Apart from these indirect effects of the wind on health, humans can experience 'mechanical' wind-induced injuries (e.g. [3]) due to collapsing engineering structures (Fig. 18.1), windborne debris (Fig. 18.2), and wind-induced traffic accidents (Fig. 18.3). In general, knowing the exact characteristics of winds and atmospheric turbulence can (a) improve measures to prevent humans from inhaling harmful gases; (b) lead to an improved design of structures to make them able to withstand extreme weather conditions; and, (c) enable the development of up-to-date warning and sheltering systems.

Fig. 18.1   The Tacoma Narrows Bridge collapse on November 7, 1940 (Photo taken by Barney Elliott, presented according to fair-use principles)

Fig. 18.2 Hurricane Andrew, August 1992, east US coast. Winds were strong enough to shoot a piece of plywood through a tree trunk in Homestead, FL (Photo courtesy of NOAA)

Fig. 18.3 Tow truck workers hook a tractor-trailer close to San Bernandino, CA, overturned by strong cross-wind gust on February 2nd, 2011 (Photo courtesy of Jennifer Cappuccio Maher)

18.2   The Atmospheric Boundary Layer Structure Determines Wind-Structure Interaction

The aspects of the atmospheric boundary layer flow that are of interest in structural design are mean wind profiles, wind speeds in different roughness regimes, and the structure of atmospheric turbulence [30]. Incompressibility may be assumed, as wind speeds are considerably lower than the speed of sound. Since the structural engineer is primarily concerned with the effects of strong winds, in most cases it is assumed that the flow is neutrally stratified and, as in strong wind conditions, mechanical turbulence dominates the heat convection and thus, consequent intense turbulent mixing tends to produce neutral stratification. However, for slender structures like chimneys, vortex shedding may produce the greatest wind load on structures [11]. Therefore, the Canadian code prescribes a considerable thermal influence on the load due to vortex shedding.

In general, the wind loads on structures can be divided into steady and unsteady. For many structures, the wind-induced resonant vibrations are negligible and the fluctuating wind responses can be calculated using procedures applicable for static loads. Since the majority of buildings belong to this category, the so-called static wind load is very important in connection with stress calculations and design. The Newton's hypothesis that the load on a fixed body in a flow is proportional to the flow velocity squared is still considered to be correct for sharp-edged bodies, and also for curved structures in certain intervals of Reynolds number values.

The main sources of the fluctuating pressures and forces on engineering structures are (a) atmospheric turbulence in the free stream flow, also called buffeting; (b) unsteady flow generated by the body itself, by phenomena such as separations, reattachments and vortex shedding; (c) fluctuating forces due to movement of the body itself, e.g. aerodynamic damping; and (d) buffeting forces from the wakes of other structures upwind of the structure of interest. Therefore, the turbulence in the approaching flow determines the size of the reattachment zones behind the leading edges of the building, and a magnitude of suction acting on building facades in these zones.

Owing to the turbulent nature of winds there is the potential to excite resonant dynamic responses of structures, or parts of structures, with natural frequencies less than about 1 Hz. When a structure experiences resonant dynamic responses, counteracting structural forces come into play to balance the wind forces. These are: (a) inertial forces proportional to the mass of the structure; (b) damping or energy-absorbing forces; and (c) elastic or stiffness forces proportional to the deflections or displacements.

The spectral approach [8], based on random vibration theory, has been commonly applied by wind engineers to calculate unsteady wind loads on structures (Fig. 18.4). In this approach, it is assumed that unsteady wind forces acting on structures cannot be predicted deterministically due to complexities of atmospheric turbulence. Several types of structures are particularly sensitive to atmospheric winds, for example, tall buildings, roofs of large buildings, and slender structures and bridges, as reported in Holmes [13].

Resonant dynamic response in along-wind, cross-wind and torsional modes characterize the overall structural loads experienced by tall buildings, whereas extreme local cladding pressures may be experienced on their side walls. Major problems are the vulnerability of glazed cladding to direct wind pressures and wind-borne debris, as well as serviceability problems due to excessive motion near the top of these buildings.

For large buildings, such as convention centers, stadiums and aircraft hangars two facts are important: (a) the quasi-steady approach, although appropriate for small buildings, is not applicable for large roofs, and (b) resonant effects, although not dominant, can be significant.

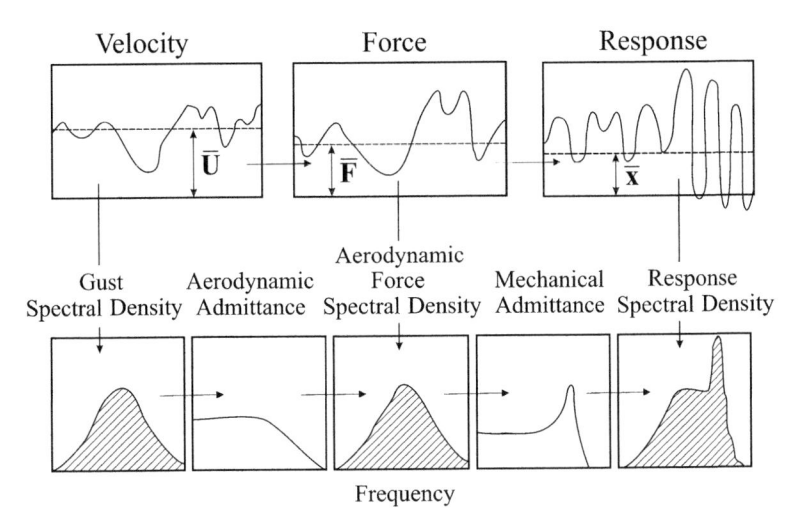

Fig. 18.4 The random vibration approach to resonant dynamic response by Davenport [8]

For slender structures, such as chimneys of circular cross-sections, poles carrying lighting arrays or mobile telephone antennas, and guyed masts: (a) the fundamental mode shapes are generally non-linear; (b) higher modes are more likely to be significant in the resonant dynamic response; (c) since the aspect ratio is higher, i.e. the width is much less than the height, the aerodynamic 'strip' theory can be applied. That is, the total aerodynamic coefficients for the cross-section can be used with the wind properties upstream, at the same height; (d) if the mass per unit height is low, aerodynamic damping is significant; (e) as for tall buildings, the cross-wind response can be significant (except for lattice structures). However, due to the smaller cross-wind width, the velocity at which the vortex shedding frequency (or the maximum frequency of the cross-wind force spectrum) coincides with the first mode vibration frequency is usually much lower than for tall buildings.

In bridge design, as the spans increase, wind actions become more critical. For the longest suspension or cable-stayed bridges, the dynamic wind forces excite resonant responses, often in several modes, and equally important are the aeroelastic forces, in which the motion of the structure itself generates force. Therefore, a large number of cases with vibrations of bridge decks induced by vortex shedding have been reported as, e.g., Fujino and Yoshida [12].

18.3 Winds Are Affected by Climate Change

An insight into scientific publications gives solid evidence for a changing structure of local winds due to the climate change. The first study that comprehensively explored the effect of climate change on wind speeds in the U.S. [26]

indicated that global warming lowers wind speeds enough to handicap the wind energy industry. The wind speeds acquired at hundreds of locations across the U.S. appear to be waning, in many locations by more than 1% a year. On the other hand, there are signs that climate change is leading to higher wind speeds in the English Channel [31]. In particular, wind speeds have been rising so much that wind farms could generate 50% more electricity than expected a decade ago. As reported in Science Daily from January 24th [29], Dartmouth researchers have learned that the prevailing winds in the mid latitudes of North America, which now blow from the west, once blew from the east. They reached this conclusion by analyzing wood samples from areas in the mid-latitudes of North America, north of Denver and Philadelphia and south of Winnipeg and Vancouver. Therefore, Sanchez et al. [28] presented a description of the atmospheric boundary layer characteristics for current and future climate periods. Present climate shows an annual cycle for vertically integrated turbulent kinetic energy with a clear summer maximum for southern regions, while northern regions of Europe exhibit a smoother or even a lack of cycle. Future climate conditions indicate that changes in the turbulent transport from the atmospheric boundary layer to the free troposphere can affect atmospheric circulations. Deepthi and Deo [10] reported effects of climate change on design wind at the Indian offshore locations. Their study showed that the magnitude of the long term wind speed would increase if the effect of global climate change is incorporated in the analysis. For the two locations considered, the increase in the 100-year wind was found to be varying from 44% to 74%. Debernard et al. [9] considered a possible change in future wind, wave, and storm surge climate for the regional seas northeast of the Atlantic. The authors report an expected increase in all variables in the Barents Sea and a significant reduction in wind and waves north and west of Iceland. Also, there is a significant increase in wind speed in the northern North Sea and westwards in the Atlantic Ocean, and a comparable reduction southwest of the British Isles in the autumn. Finally, in the review article on climate change impacts on wind energy, Pryor and Barthelmie [27] conclude that global climate change may redistribute the regions favorable for wind exploitation, resulting thus in 'winners' and 'losers', that is regions where the wind energy industry may benefit and region where they may suffer losses due to the climate change.

18.4 Small-Scale Experiments Can Predict Wind Environment and Wind Loading of Structures

The ABL wind-tunnel simulations have been commonly created using well-established methods [7, 14], as well as recently modified methodology using the truncated vortex generators (e.g. [17–19]), with some representative results reported in Fig. 18.5.

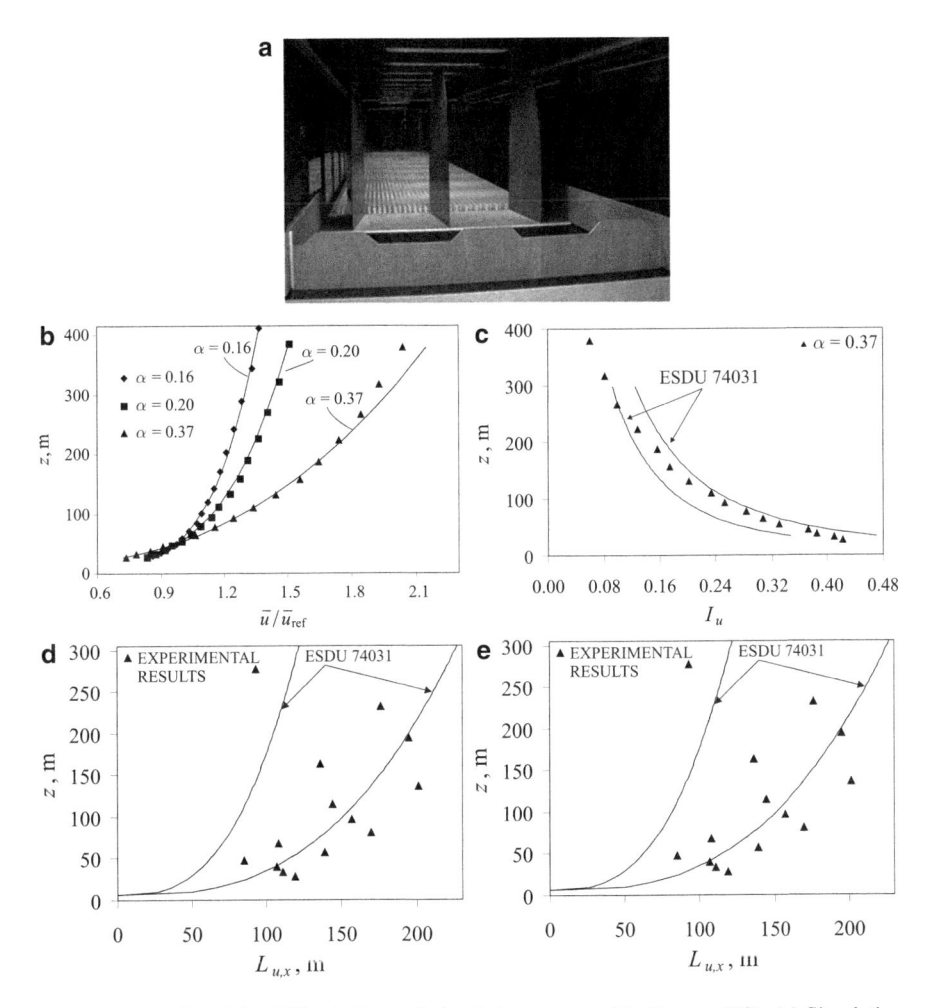

Fig. 18.5 Results of the ABL wind-tunnel simulations reported in Kozmar [18]: (**a**) Simulation hardware in the boundary layer wind tunnel at the Technische Universität München, (**b**) Mean velocity profiles in comparison with the empirical power law for the rural, suburban, and urban terrain exposures, (**c**) Turbulence intensity profile for urban terrain exposure in comparison with ESDU 74031 data sheets, (**d**) Turbulence length scales for urban terrain exposure in comparison with ESDU 74031 data sheets, (**e**) Power spectral density of longitudinal velocity fluctuations for urban terrain exposure in comparison with the Kolmogorov inertial subrange and von Kármán design curve (© 2010 Elsevier Ltd.)

In the past decade, several transient flow field simulators (e.g. at Miyazaki University, Iowa State University, University of Notre Dame) have been built in order to enable experimental simulation of transient winds as well, e.g., tornado, downburst, bora winds and others. Results from the Notre Dame facility indicate that the aerodynamics of a vehicle on a bridge exposed to cross-wind gusting can significantly differ from the aerodynamics of a vehicle exposed to 'standard' atmospheric turbulence (Fig. 18.6).

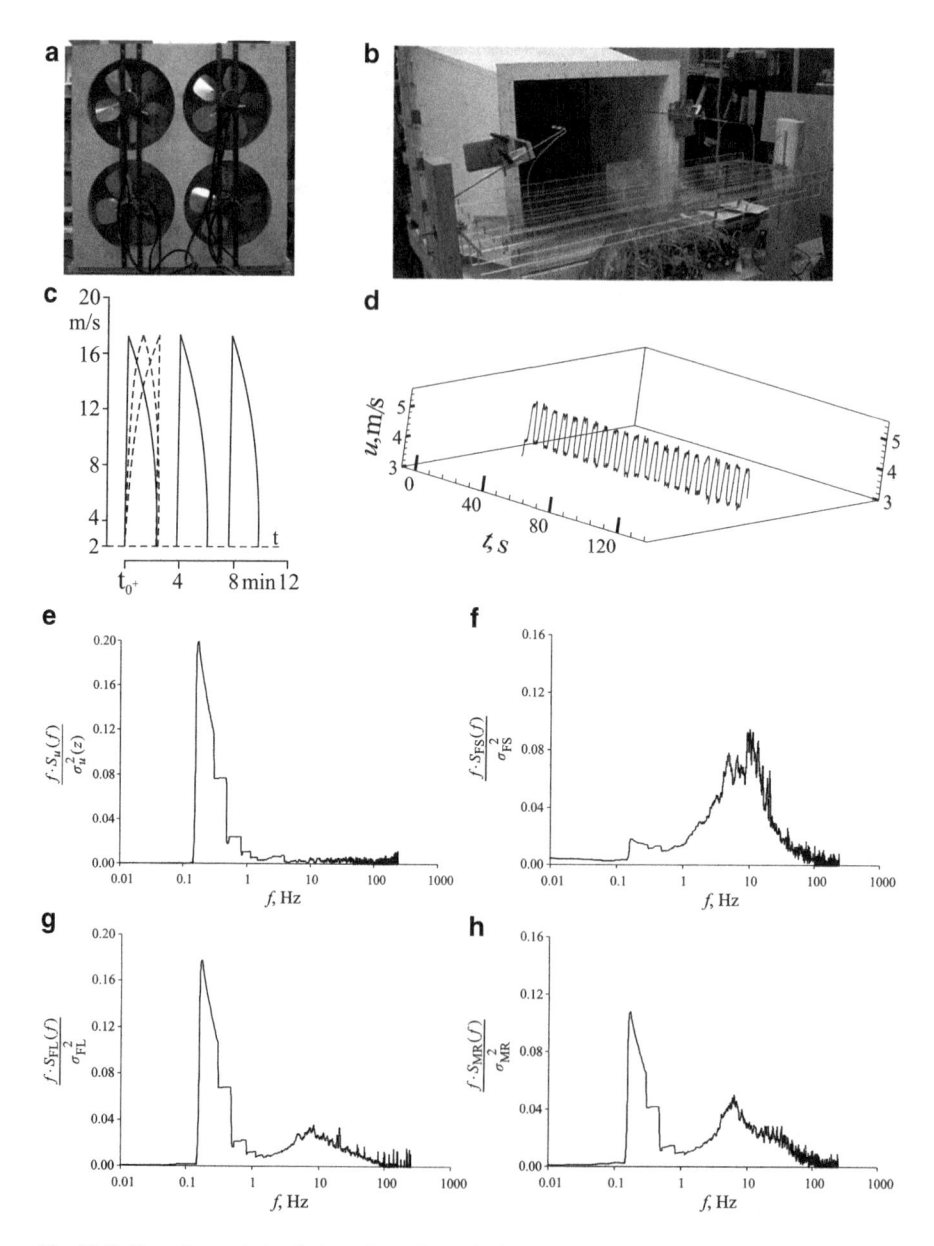

Fig. 18.6 Experimental simulation of transient winds. (**a**) Wind blowers in the NatHaz Transient Flow Field Simulator (TFFS), (**b**) Vehicle model on the bridge section model in the TFFS test section, (**c**) Time series of full-scale bora wind gusting after Petkovšek [24], (**d**) Time series of simulated wind gusting in the TFFS. Power spectral density of: (**e**) velocity fluctuations, (**f**) side force fluctuations, (**g**) lift force fluctuations, (**h**) overturning moment fluctuations (Adapted from Kozmar et al. [20])

In general, existing experimental facilities proved to be capable of successfully reproducing the characteristics of winds. This makes one confident that they would be capable of reproducing the modified winds in the decades to come as well. In addition, numerical methods for these purposes have been developing at a tremendous pace over the past few years providing a valuable complement to small-scale experiments and full-scale measurements.

18.5   Concluding Remarks

Winds can influence human health indirectly through the effects on air pollution levels, as well as directly, by mechanical wind-induced injuries. Namely, strong winds can cause traffic accidents, and they can induce wind-sensitive structures to collapse or become severely damaged. Additionally, they affect energy production, and consequently economy. While experimental data and prediction models give evidence about considerable changes in the structure of local winds, in the future it would be necessary to address several important questions in regard to the effects of winds on human health and national security in the light of a changing climate. These are as follows:

- Are the calculation methods currently employed in structural design of wind-sensitive structures going to be acceptable for the newly developed structures exposed to winds modified by climate change?
- International wind loading standards and codes of practice are strongly dependent on the structure of local winds. Are they going to be realistic in the future as well?
- Would it be necessary to re-evaluate wind loading of already existing structures?
- Is it necessary to expect wind-induced traffic accidents at new sites?

Acknowledgements The first author acknowledges the support of the Fulbright Foundation for experimental simulations of transient winds, which were carried out at the University of Notre Dame, USA. The Croatian Ministry of Science and Technology, the German Academic Exchange Service (DAAD), and the Croatian Academy of Sciences and Arts (HAZU) supported a project on wind-tunnel simulations of the atmospheric boundary layer flow, which was conducted at the Technische Universität München (TUM), Germany.

References

1. Alebić-Juretić A, Cvitaš T, Kezele N, Klasinc L, Pehnec G, Šorgo G (2007) Atmospheric particulate matter and ozone under heat-wave conditions: do they cause an increase of mortality in Croatia? Bull Environ Contam Toxicol 79:468–471. doi:10.1007/s00128-007-9235-2
2. Anderson HR (2009) Air pollution and mortality: a history. Atmos Environ 43:142–152
3. Belušić D, Klaić ZB (2006) Mesoscale dynamics, structure and predictability of a severe Adriatic bora case. Meteorol Z 15:157–168
4. Bešlić I, Šega K, Čačković M, Klaić ZB, Vučetić V (2007) Influence of weather types on concentrations of metallic components in airborne PM10 in Zagreb, Croatia. Geofizika 24:93–107

5. Bešlić I, Šega K, Čačković M, Klaić ZB, Bajić A (2008) Relationship between 4-day air mass back trajectories and metallic components in PM10 and PM2.5 particle fractions iz Zagreb air, Croatia. Bull Environ Contam Toxicol 80:270–273. doi:10.1007/s00128-008-9360-6
6. Costigliola V (2010) 10th EMS annual meeting, 10th European Conference on Applications of Meteorology (ECAM) abstracts. Boundary-Layer & health, Zürich, 2010
7. Counihan J (1969) An improved method of simulating an atmospheric boundary layer in a wind tunnel. Atmos Environ 3:197–214
8. Davenport AG (1963) The buffeting of structures by gusts. In: Proceedings of the international conference on wind effects on buildings and structures, Teddington, 1963
9. Debernard J, Saetra O, Roed LP (2002) Future wind, wave and storm surge climate in the northern North Atlantic. Clim Res 23(1):39–49
10. Deepthi R, Deo MC (2010) Effect of climate change on design wind at the Indian offshore locations. Ocean Eng 37:1061–1069
11. Dyrbye C, Hansen SO (1997) Wind loads on structures. Wiley, New York
12. Fujino Y, Yoshida Y (2002) Wind-induced vibration and control of Trans-Tokyo Bay crossing bridge. J Struct Eng ASCE 128(8):1012–1025
13. Holmes JD (2001) Wind loading of structures. Spon Press, London
14. Irwin HPAH (1981) The design of spires for wind simulation. J Wind Eng Ind Aerod 7:361–366
15. Jelić D, Klaić ZB (2010) Air quality in Rijeka. Geofizika 27:147–167
16. Kampa M, Castanas E (2008) Human health effects of air pollution. Environ Pollut 151:362–367
17. Kozmar H (2010) Scale effects in wind tunnel modeling of an urban atmospheric boundary layer. Theor Appl Climatol 100(1–2):153–162
18. Kozmar H (2011) Truncated vortex generators for part-depth wind-tunnel simulations of the atmospheric boundary layer flow. J Wind Eng Ind Aerod 99(2–3):130–136
19. Kozmar H (2011) Wind-tunnel simulations of the suburban ABL and comparison with international standards. Wind Struct 14(1):15–34
20. Kozmar H, Butler K, Kareem A (2009) Aerodynamic loads on a vehicle exposed to cross-wind gusts: an experimental study. In: 7th Asia-Pacific conference on wind engineering, Taipei, 2009
21. Kumar P, Robins A, Vardoulakis S, Britter R (2010) A review of the characteristics of nano-particles in the urban atmosphere and the prospects for developing regulatory controls. Atmos Environ 44:5035–5052
22. Mohorović L (2004) First two months of pregnancy – critical time for preterm delivery and low birthweight caused by adverse effects of coal combustion toxics. Early Hum Dev 80:115–123
23. Monks PS, Granier C, Fuzzi S, Stohl A, Williams ML et al (2009) Atmospheric composition change – global and regional air quality. Atmos Environ 43:5268–5350. doi:10.1016/j.atmosenv.2009.08.021
24. Petkovšek Z (1987) Main bora gusts – a model explanation. Geofizika 4:41–50
25. Prtenjak MT, Jeričević A, Kraljević L, Bulić IH, Nitis T, Klaić ZB (2009) Exploring atmospheric boundary layer characteristics in a severe SO_2 episode in the north-eastern Adriatic. Atmos Chem Phys 9:4467–4483
26. Pryor SC, Barthelmie RJ, Young DT, Takle ES, Arritt RW, Flory D, Gutowski WJ Jr, Nunes A, Roads J (2009) Wind speed trends over the contiguous United States. J Geophys Res 114:D14105. doi:10.1029/2008JD011416
27. Pryor SC, Barthelmie RJ (2010) Climate change impacts on wind energy: a review. Renew Sust Energ Rev 14:430–437
28. Sánchez E, Yagüe C, Gaertner MA (2007) Planetary boundary layer energetics simulated from a regional climate model over Europe for present climate and climate change conditions. Geophys Res Lett 34:L01709. doi:10.1029/2006GL028340
29. Science Daily (2007) Winds of change: North America's wind patterns have shifted significantly in the past 30,000 years
30. Simiu E, Scanlan RH (1996) Wind effects on structures, 3rd edn. Wiley, New York
31. Webb T (2009) Winds of change blow for offshore power operators, The Observer, published on April 26, 2009

Chapter 19
Remote Sensing and Public Health Issues in a Changing Climate and Environment: The Rift Valley Fever Case

Y.M. Tourre, J.-P. Lacaux, C. Vignolles, and M. Lafaye

Abstract Climate and environment are changing rapidly. We must then cope with new challenges posed by new and re-emerging diseases, innovate beyond benches and bedsides, i.e., using high resolution technology, and re-invent health politics and multidisciplinary management, all in a climate change context. The new concept of *tele-epidemiology* is presented. The detailed conceptual approach (CA) associated with Rift Valley Fever (RVF) epidemics in Senegal (monitored from space) is given. Ponds were detected by using high-resolution SPOT-5 satellite images. Data on rainfall events obtained from the Tropical Rainfall Measuring Mission (NASA/JAXA) were combined with in-situ data. Localization of vulnerable and parked hosts (from QuickBird satellite) is also used. The dynamic spatio-temporal distribution and aggressiveness of one of the main RVF vectors, *Aedes vexans*, were based on total rainfall amounts, pond dynamics and entomological observations. Detailed risks zones (hazards and vulnerability) are expressed in percentages of parks where animals are submitted to mosquitoes' bites. This CA, which simply relies upon rainfall distribution evaluated from space, is meant to contribute to the implementation of an operational early warning system (EWS) for RVF or RVFews. It is based on environmental risks associated with climatic and environmental

Y.M. Tourre (✉) • C. Vignolles • M. Lafaye
Lamont Doherty Earth Observatory (LDEO), Centre National d'Etudes Spatiales (CNES), LDEO of Columbia University, Palisades, NY, USA

Météo-France, Toulouse, France
e-mail: Yvestourre@aol.com

J.-P. Lacaux
OMP, Université Paul Sabatier (UPS), Toulouse, France

H.J.S. Fernando et al. (eds.), *National Security and Human Health Implications of Climate Change*, NATO Science for Peace and Security Series C: Environmental Security, DOI 10.1007/978-94-007-2430-3_19, © Springer Science+Business Media B.V. 2012

changing conditions: natural and anthropogenic. It is to be applied to other diseases and elsewhere. This is particularly true in new places where vectors have been rapidly adapting recently whilst viruses circulate from an ever moving and increasing population.

Keywords Climate change • Public health • Remote sensing • Risk mapping • Infectious diseases • Rift valley fever • Early warning systems • Health information system

19.1 Rationale

19.1.1 The Changing Climate

Climate changes and varies on all sort of temporal scales. Natural climate variability is essentially due to: (i) the relative position of the sun with regard to the earth, and its activity (i.e., sunspots, irradiance, magnetism, eruption…); (ii) the interactions and feedbacks between the components of the climate system; and (iii) the Milankovitch cycles. Key climate signal fluctuations have been identified from the diurnal to multi-decadal (MD) periods, along with seasonal, quasi-biennial (QB), El-Niño-Southern Oscillation (ENSO), quasi-decadal (QD) and inter-decadal (ID) oscillations at least [10]. The natural variability of the global climate during the twentieth century is reproduced in Fig. 19.1 [3]. Adding to these natural fluctuations is the anthropogenic component, from population increases and energetic needs leading to global pollution of the Earth System. All natural fluctuations are interacting with and are modulated by the anthropogenic effects, with direct impacts on public health, ecosystems healthiness, and socio-economic activity.

19.1.2 Climate Change and Public Health

Climate variability and change alter social and economic dynamics and bring global inequalities [8]. This could result in economic migration (enhancing that from political turmoil), and slow-down access to natural and primary resources. Climate variability and change had impacts in historical times with respect to the development of many cultures. Changes have been observed in nutrient budgeting and nutrient re-cycling, and enhanced human pressure through population increase, virus and bacteria circulation all impacting public health. Based upon projected population increases, the total primary energy demand is expected to increase by ~60% during the first quarter of the Twenty-first century. Most of this energy comes from fossil sources and, unfortunately, only 1–2% is expected from renewable sources.

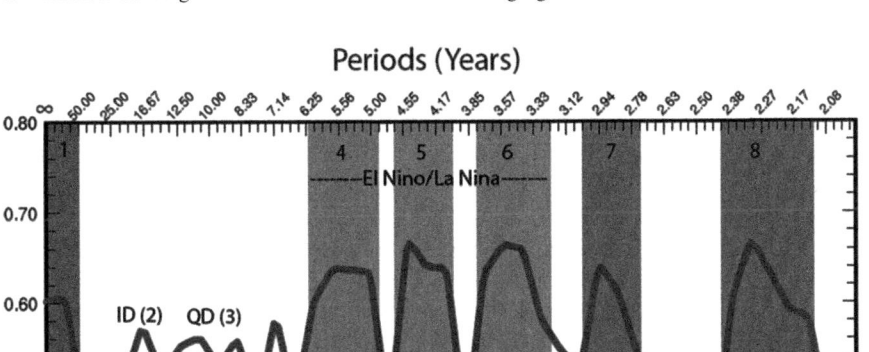

Fig. 19.1 During the twentieth century, a global statistical analysis in the frequency-domain of both sea-surface temperature and sea-level pressure, allowed identification of natural climate signals. Colored bands highlight those signals with their percentage of variance displayed on the ordinates, i.e., secular signal or penta-decadal (*blue band*, signal # 1), ENSO (*orange bands*, signals # 4, 5, and 6), the quasi-biennial signal (*green bands*, signals # 7, and 8). Signal # 2 ID or inter-decadal, and signal # 3 QD or quasi-decadal, are more local and found over the Pacific and Atlantic Oceans, respectively [10]. The anthropogenic climate change is to interact and modulate the above climate fluctuations

Such disequilibrium is already creating socio-economical chaos as well as local and regional vulnerability in terms of prices and supply/demand. All will have unforeseen impacts on the environment and consequently on public health, i.e., infectious diseases, respiratory and circulatory problems, pollution, allergens, impaired immune systems, and new and re-emerging diseases. Health issues will also be associated with poor water quality and induced malnutrition, which may require huge expenditures for poverty reduction.

Climate variability and change perturb important physical and biological systems to which human populations are biologically and culturally adjusted. Various environmental changes linked to natural and anthropogenic climate variability, loss of biodiversity and land-use changes will all have their own impacts on public health. Beneficial impacts such as decreases in cold-related deaths and reduced viability of mosquitoes in hot tropical regions are, however, anticipated to be outweighed by adverse impacts such heat-waves and epidemics/pandemics in new regions and globally. Direct influence of demographic factors (virus circulation) is conspicuous for infectious diseases transmitted from person-to-person. Most emerging (or re-emerging) infectious diseases are due partly to changes in "microbial traffics," that is, the introduction of new pathogenic agents from wildlife into new and unprepared populations, thus creating new risks. Dissemination and diffusion of diseases by vectors (and so-called 'reservoirs') into new habitats deserve specific attention. Processes may depend upon ecological and environmental factors even if the spreading of diseases is facilitated by population movements, crowding demographic effects, sanitation levels, and/or breakdowns in public health and information systems (HIS). Today, the recorded increase in occurrence of many infectious diseases already reflects the 'compounded effects' from climatic and environmental changes, demographic increases, and economical, social and technological changes.

19.1.3 Climate Change and Infectious Diseases

The challenge for assessing socio-economical impacts from infectious diseases (~75% of actual infectious diseases in humans are zoonoses) cannot be addressed without considering both abiotic and biotic environmental factors that affect the maintenance and transmission of diseases. The last quarter century has witnessed an explosion of environmentally related illnesses, i.e., diseases and disorders, with strong environmental forcing and adaptation or lack thereof. For infectious diseases, this includes increases in the prevalence, incidence and geographical distribution across wide taxonomic ranges, related to climate/environment changes and practical changes in land-use. The large spatio-temporal scales of these changes represent an important step in the understanding of diseases from the individual-centered traditional view of microbiology and medical epidemiology.

Direct health effects of climate variability and change include: changes in morbidity and mortality from heat-waves and thermal stress (such as in 2003 over southwest Europe and in 2007 over Italy and Greece); respiratory ailments associated with modified concentrations of aero-allergens (spores, moulds, fungus) and/or air pollutants; and health consequences from extreme weather events, including storms, cold waves, floods, storm surges, droughts and windstorms.

Indirect health effects result from perturbation of complex ecological systems, and include: alterations in the ecology, range, activity of vector-borne infectious diseases (i.e., Malaria, West Nile Virus from Africa to USA, Rift Valley Fever from Kenya to Senegal and Mauritania, Avian Flu, Chikungunya from the Indian Ocean to southern France and northern Italy, Dengue Fever from Central America to Florida, New Orleans, Texas, among others) [9]; alterations in the environment of water-borne diseases and pathogens (i.e., gastro-intestinal infections and *Vibrio* diseases including Cholera); alterations in the atmospheric boundary layer, and transmission of air-borne diseases (Meningococcal meningitis, respiratory ailments); alterations and regional changes in agricultural practices and food security (malnutrition, lack of fresh water....). Public health will also be affected by massive population movements on narrow coastal regions, and by regional conflicts over declining agricultural and water resources. Leishmaniasis is already endemic south of Europe and the Maghreb. Climate variability and change may extend the habitat of sandflies and phlebotome vectors northwards. The 'blue tongue' disease has been spreading northward over Western Europe. Increasing temperatures in some critical regions will directly influence the life stages of tick species responsible for the transmission of the Lyme disease.

19.1.4 Climate Variability and Change and Decision-Making Processes

Climate variability and change affect regional socio-economical cost/loss, reflecting the local balance/imbalance from temperature and soil moisture changes, use and abuse of fertilizers, and pest and pathogens activity. The decision-making models used will include at least:

1. Identification of "normal" impacts of disease (in lives and cost)
2. Definition of a "climate event" linked to a "health event" (epidemics, endemics, pandemics…)
3. Definition of "increased impacts" and losses (in lives and Euros)
4. Identification of effective methods to mitigate losses
5. Definition of costs (Euros) for implementation of the above and improve Health Information Systems (HIS)
6. Quantification of the savings (in lives and Euros) if a "health event" does not occur

Regional modelling studies consistently indicate that tropical and sub-tropical countries would be most affected, but varying and changing climate/environment at higher latitudes/altitudes must also be considered. Forecasting climate impacts on public health requires the development of scenario-based risk (i.e., hazards + vulnerability) assessments which must include potential consequences from complex demographic, social, political and economical disruptions. Integrated mathematical modelling must be used if one wants to estimate the future impacts of climate on health (see [7]). Such modelling requires both an understanding of and a modeling capability of each component in the chain of causation.

Nevertheless, uncertainties do remain and are due to future industrial-economic activities, interactions between and within natural systems, and differences in sensitivity of disease systems and vulnerability of human and animal populations. Non-linear uncertainties arise from the stochastic nature of the biophysical systems being modelled. Local anthropogenic deforestation may directly alter the distribution of vector-borne diseases while also causing a local increase in temperature. Differences in population vulnerability will also occur due to the heterogeneity of human culture and behaviour, including social relation).

19.1.5   Climate/Environmental Changes and Remote Sensing

Public health indicators and disease surveillance activities should be integrated with other in-situ observation systems such as Global Climate Observing System (GCOS), Global Ocean Observing System (GOOS), Global Terrestrial Observing System (GTOS), and Global Earth Observation System of Systems (GEOSS). Today, the use of satellites allows monitoring changes in environmental and climatic parameters at high resolution. The example and the detailed and integrated conceptual approach of *tele-epidemiology* for Rift Valley Fever (RVF) are given hereafter. Overall, GEOSS is to provide a continuum of observational spatio-temporal scales on both oceanic and terrestrial environmental structures.

19.2   Tele-Epidemiology

Infectious diseases remain a conspicuous challenge to public health today. In the context of climate variability and change and the rapidly increasing human population, some epidemics are emerging or re-emerging all over the world, such as RVF over West Africa, Dengue Fever over northern Argentina and southern USA, and Chikungunya over southern Europe.

19.2.1   The Conceptual Approach

Following the Johannesburg Summit of 2002, the new conceptual approach of tele-epidemiology has been put into action [5] in order to monitor and study the spread of human and animal infectious diseases which are closely tied to climate and environmental variability and changes. By combining satellite-originated data on vegetation (SPOT-image), meteorology (Meteosat, TRMM), and oceanography (Topex/Poseidon; ENVISAT, JASON) with hydrology data (distribution of lakes, water levels in rivers, ponds and reservoirs), with clinical data from humans and animals (clinical cases, serum use…) and entomological data, predictive mathematical models can be constructed.

This integrated and multidisciplinary approach includes:

1. Monitoring and assembling multidisciplinary in-situ datasets to extract and identify physical and biological mechanisms at stake
2. Remote-sensing/monitoring of climate and environment linking epidemics with "confounding factors" such as rainfall, vegetation, hydrology, population dynamics, and

3. Use of bio-mathematical models for epidemic dynamics, vectors aggressiveness and associated risks

An interactive health information system (HIS) on re-emergent and infectious diseases (RedGems, www.redgems.org) was thus born [6] The primary mission of the RedGems information site is to facilitate the development of Early Warning Systems (EWS) with a main objective of attempting to predict and mitigate public health impacts from epidemics, endemics and pandemics.

19.2.2 The RVF Case: The Adapt RVF Project

The various components of the above approach have been thoroughly tested with the emerging RVF in the Ferlo (Senegal). This successful approach has lead the Senegalese government to provide funding, and extend the approach initiated to all risk zones (i.e., hazards + vulnerability) in which populations and cattle are exposed [13].

The Ferlo region in Senegal became prone to RVF in the late 1980s with the appearance of infected vector/mosquitoes from the *Aedes vexans* and *Culex poicilipes* species [4,12,13]. The latter proliferate near temporary ponds and neighbouring humid vegetation. RVF epizootic outbreaks in livestock cause spontaneous abortions and perinatal mortality. So far, human-related disease symptoms are often limited to flu-like syndromes but can include more severe forms of encephalitis and hemorrhagic fevers. Socio-economic resources can be seriously affected. Human mortality cases have been recently reported in Mauritania.

The ultimate goal has been to use specific Geographical Information System (GIS) tools [11] and high resolution remote sensing (RS) images/data to detect the "beating" of the breeding ponds and evaluate RVF diffusion with areas at risk: the so-called Zone Potentially Occupied by Mosquitoes (ZPOM).

The integrated approach to determine the environmental risk levels of RVF is presented in Fig. 19.2. The upper left box in the figure identifies key entomological factors for *Aedes vexans* (flying-range, aggressiveness and embryogenesis), environmental factors (rainfall distribution, limnimetry and ponds dynamics) as well as pastoralist data such as the zones where animals are parked for the night. From the upper right box, the detection of lead environmental and climatic factors (mainly rainfall) favouring the mechanisms presented in the box, are highlighted. For example, localization and optimal pond conditions for the breeding and hatching of *Aedes vexans* can be modelled [1]). The central box refers to the zone potentially occupied by mosquitoes (ZPOM) derived from pond dynamics after a productive rainfall event including the flying ranges of *Aedes vexans*. The combination and integration of all the elements mentioned above lead to the notion of risks: hazards and vulnerability of hosts exposure. This original approach [2] bridges the physical and

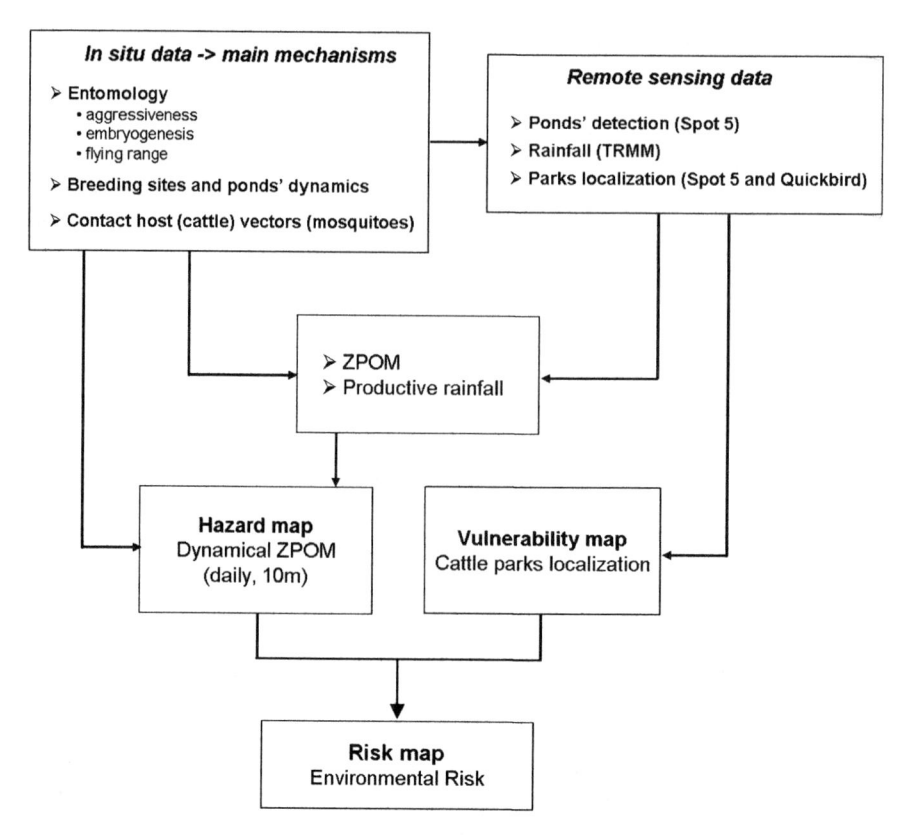

Fig. 19.2 Integrated conceptual approach. The basic components for the concept are presented in the top three boxes: in-situ data (*upper left*), remotely-sensed data (*upper right*), and ZPOM and 'productive' rainfall in terms of production of vectors/mosquitoes (*middle*). The bottom three boxes are for hazards (*bottom left*) and vulnerability (*bottom right*), both leading to the environmental risks (*very bottom*)

biological mechanisms, linking environmental conditions to the production of RVF vectors and the accompanying potential risks.

Possible hazards in the vicinity of fenced-in hosts are displayed in Fig. 19.3, where the Barkedji area is shown along with the mapped ZPOMs. Thus, parks and villages can easily be identified. For example out of 18 rainfall events obtained from TRMM for the 2003 rainy season, seven were considered as 'productive' with regard to *Aedes vexans* (based on entomological studies).

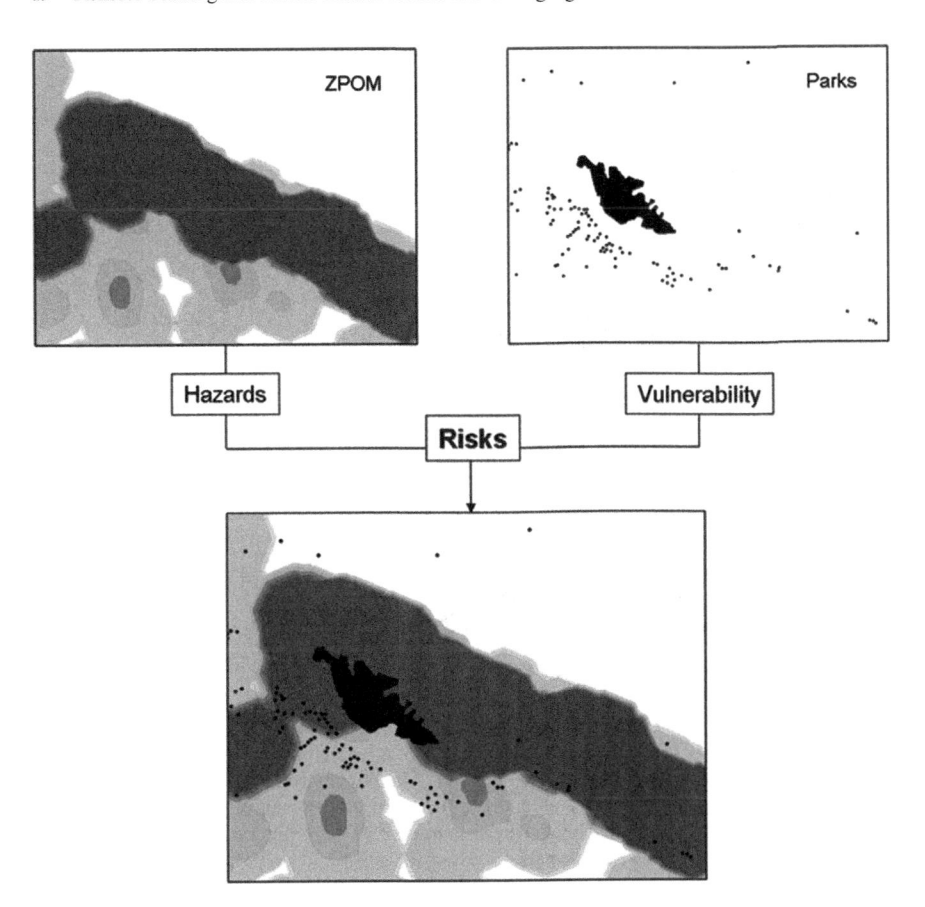

Fig. 19.3 Zone Potentially Occupied by Mosquitoes, or ZPOMs with ranked hazards from *yellow* (*low hazards*) to *red* (*high hazards*). ZPOM in the Barkedji area is obtained from the ponds distribution after a single rainfall event (*top left*). Localization of the Barkedji village and ruminants' fenced-in areas (vulnerability, from QuickBird) in *black* for the same area and period (*top right*). Potential risks i.e., = hazards + vulnerability are shown by super-imposing the two pictures (*bottom*)

19.3   Conclusion

Climate variability and change and environmental risks comprise mechanisms linking rainfall variability and trends, density of vectors/mosquitoes and their aggressiveness, and host vulnerability. The dynamical evolution of ZPOMs (Fig. 19.4) from ponds clustering has identified risks as a function of discrete and productive rainfall events. It dramatically displays the socio-economic problems to which populations and cattle of the region can be exposed. The socio-economic risks can thus

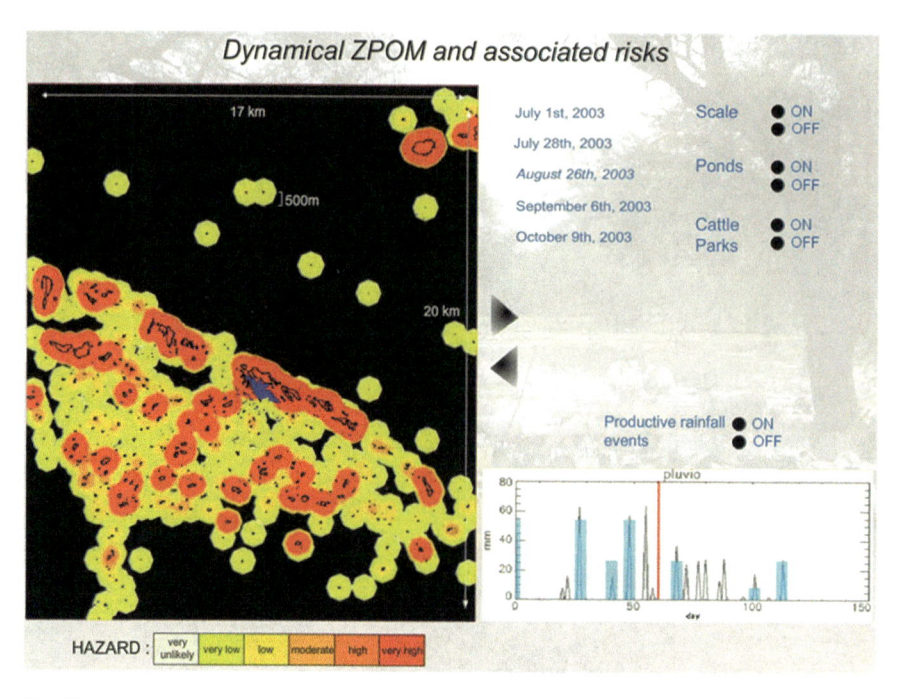

Fig. 19.4 Dynamic ZPOMs and ranked hazards. Dynamic ZPOMs with ranked hazards (from very unlikely and very low in *yellow*, to very high in *red*, *bottom scale*) and ponds distribution (in *blue*) during the 2003 rainy season. From the hyperlinked figure available in the on-line paper, by clicking on the two fat *black arrows*, animated ZPOMs from a 'productive' rainfall (highlighted in *blue*, at the *bottom right*) are displayed (*upper arrow* for *forward* motion, *lower arrow* for *backward* motion) along with the relative parks' locations (vulnerability). The starting date is June 28, 2003. ZPOMS for specific date can also be displayed. The vertical *red* marker is for accurate time positioning on a daily basis during the rainy season

be anticipated based on statistical evaluation of the seasonal rainfalls, which can be done a few months prior to the rainy season (based upon seasonal forecasts).

Mitigation of impacts can be accomplished though strategic displacement of the fenced-in areas during the course of the rainy season, vaccination, and destruction of vectors. Besides population movement, climatic and environmental changes are necessary conditions for the RVF virus to circulate and be transmitted. The conceptual approach presented here and which can be used in quasi real-time, is to be linked with biological modelling of virus transmission and circulation, as well as with classical epidemiological models. Ultimately, the fully integrated approach should help understanding the mechanisms leading to potential RVF epidemics and improve related EWS or "RVFews".

Recently, high-resolution TerraSar-X radar remote sensing used RVFews to become truly operational during the monsoon season [14]. Sequences of ZPOM maps (after a significant and 'productive' rainfall event) were sent for the first time to 'Centre de Suivi Ecologique' and 'Direction du Suivi Vétérinaire' de Dakar

(Senegal). The T³ Model from the last World Health Summit (Berlin, 2010) is being included in the *tele-epidemiolgy* approach. It includes the:

1. Transition phase: Coping with new challenges from new and re-emerging diseases
2. Translation phase: Innovating beyond benches and bedsides by using high res. Technology (including optical and radar remote sensing)
3. Transformation: Re-inventing public health politics, managerial and security issues, including new guidelines and TORs in a climate variability and change context, for multi-disciplinarity and EWS

The physical and biological mechanisms from other infectious diseases are to be developed by applying a similar methodology elsewhere (including higher-latitude regions) where climate and environment are also varying and changing rapidly. This is in the process of being implemented for Malaria epidemics over Burkina Faso (PaluClim project).

Acknowledgment Our research was part of recently funded AdaptFVR and PaluClim projects, facilitated by the "Gestion des Impacts du Changement Climatique (GICC)", a program supported by the French "Ministère de l'Ecologie, du Développement durable, des Transports et du Logement (MEDTL)" (http://www.gip-ecofor.org/gicc/). The authors would like also to thank Dr. Antonio Güell, Head of the Application and Valorization department at CNES, for his unconditional support. Tourre would like to thank Dr. Philippe Dandin, Director of the Climatology department at Météo-France and Dr. Mike Purdy recently appointed Columbia University's Executive Vice President for Research, for facilitating this applied research.

References

1. Bicout DF, Porphyre T, Ndione J-A, Sabatier P (2003) Modelling abundance of Aedes and Culex spp. in rain-fed ponds in Barkedji, Senegal. In: Proceedings of the 10th international symposium on veterinarian epidemiology and economics, Vina del Mar, Chile, 2003, p 84
2. CNES (2008) Method for tele-epidemiology (Méthode pour la télé-épidémiologie). Patent pending # PCT/FR2009/050735
3. Lacaux J-P, Tourre YM (2006) Impacts des changements climatiques. Le climat et sa variabilité ont-ils un impact sur la santé humaine ? Biofutur 270:22–24
4. Lacaux J-P, Tourre YM, Vignolles C, Ndione J-A, Lafaye M (2007) Classification of ponds from high spatial resolution remote sensing: application to rift valley fever epidemics in Senegal. J Remote Sens Environ 106:66–74
5. Lafaye M (2006) Nouvelles applications spatiales pour la santé: la télé-épidémiologie pour le suivi des fièvres aviaires. CNES Magazine, February 2006, pp 30–31
6. Marechal F, Ribeiro N, Lafaye M, Güell A (2008) Satellite imaging and vector-borne diseases: the approach of the French National Space Agency (CNES). Geospat Health 3(1):1–5
7. Martens P (2001) Climate change: vulnerability and sustainability. IPCC TAR report. www.grida.no/climate/ipcc_tar/wg2/539.htm. Accessed 12 Oct 2011
8. Plan Bleu (2008) Mediterranean basin: climate change and impacts during the 21st century. Report 2008, Le Plan Bleu, Sophia Antipolis, p 67
9. Takken W (2006) Environmental change and malaria risk: global and local implications. Springer, Dordrecht, p 150. ISBN-13: 978-1402039270

10. Tourre YM, White WB (2006) Global climate signals and equatorial SST variability in the Indian, Pacific and Atlantic oceans during the 20th century. Geophys Res Lett 33:L06716. doi:10.1029/2005GL025176

11. Tourre YM, Fontannaz D, Vignolles C, Ndione J-A, Lacaux J-P, Lafaye M (2007) GIS and high-resolution remote sensing improve early warning planning for mosquito-borne epidemics. Healthy GIS, GIS for Health and Human Services, ESRI, pp 1–4

12. Tourre YM, Lacaux J-P, Vignolles C, Ndione J-A, Lafaye M (2008) Mapping of zones potentially occupied by Aedes vexans and Culex poicilipes mosquitoes, the main vectors of Rift Valley fever in Senegal. Geospat Health 3(1):69–79

13. Vignolles C, Lacaux J-P, Tourre YM, Bigeard G, Ndione J-A, Lafaye M (2009) Rift Valley fever in a zone potentially occupied by Aedes vexans in Senegal: dynamics and risk mapping. Geospat Health 3(2):211–220

14. Vignolles C, Tourre YM, Mora O, Imanache L, Lafaye M (2010) TerraSAR-X high-resolution radar remote sensing: an operational warning system for Rift Valley fever risk. Geospat Health 5(2):23–31

Chapter 20
The Effect of Heat Stress on Daily Mortality in Tel Aviv, Israel

C. Peretz, A. Biggeri, P. Alpert, and M. Baccini

Abstract Weather-related morbidity and mortality have attracted renewed interest because of climate changes. During a multi-center project conducted within Europe, the apparent threshold temperature where the heat effect changes was found to be different for Mediterranean and north continental cities. In this paper, we study the V/J relationship between heat stress (Discomfort Index-DI) and mortality in Tel Aviv, a city within Asia, using daily data of mortality counts and meteorological variables for the period 1/1/2000–31/12/2004; using a Poisson regression and accounting for confounders. The relationship between the discomfort index DI (lag 0–3) and log mortality rates was J shaped for Tel Aviv. The DI threshold was found to be 29.3 (90% CrI = 28.0–30.7). Above this threshold, a 1 unit increase in DI was found to be associated with increased mortality of 3.72% (90% CrI = −0.23 to 8.72). NO_2 was also found to have a significant effect on mortality. As global warming continues, even though there exists a high awareness amongst the Israeli population of the negative health impacts of heat, there is still a vital need to develop local policies to mitigate heat-related deaths.

Keywords Heat stress • Mortality • Global warming • Mediterranean cities

C. Peretz (✉)
Department of Epidemiology, Faculty of Medicine, Tel Aviv University,
Tel Aviv, Israel
e-mail: cperetz@post.tau.ac.il

A. Biggeri • M. Baccini
Department of Statistics "G. Parenti", University of Florence, Florence, Italy

P. Alpert
Department of Geophysics, Faculty of Exact Sciences,
Tel Aviv University, Tel Aviv, Israel

H.J.S. Fernando et al. (eds.), *National Security and Human Health Implications of Climate Change*, NATO Science for Peace and Security Series C: Environmental Security, DOI 10.1007/978-94-007-2430-3_20, © Springer Science+Business Media B.V. 2012

20.1 Introduction

Weather-related morbidity and mortality have attracted renewed interest because of climate changes. Specifically, the short-term effects of temperature on mortality have recently been studied in Europe [1–3], Australia [4] the US [5, 6], East Asia [7], and other places [8], displaying a seasonal pattern with increased mortality in cold and hot temperatures (a V or J shape). Within Europe, in the multi-center project of PHEWE (Assessment and Prevention of acute Health Effects of Weather conditions in Europe), a threshold apparent temperature – a point where the heat effect changes, was found to be different between Mediterranean and north continental cities [1]. This difference is related to the fact that populations adapt to their local climate—physiologically, culturally and behaviourally.

In South-Eastern Mediterranean cities in West Asia, such as Tel Aviv, the climate-mortality association has not yet been investigated (except for Beirut in the years 1997–1999, [9]). Tel Aviv is the second largest city in Israel and is situated on the East Mediterranean coast, with warm to hot, dry summers and cool, wet winters. Compared to European Mediterranean cities, there are many more hot days throughout the year as well as episodes of resuspended wind-blown dust from the Sahara desert, mainly in spring. Air pollutants are potential confounders in the association between temperature and mortality [5]. In Tel Aviv, the main source of air pollution is heavy traffic and to a lesser extent power stations and industrial zones.

For the calculation of valid indices that define heat stress and zones of discomfort, many physiological and environmental factors are required. In Israel, we use the discomfort index (DI) as an index for human thermal comfort [10]. This index involves two environmental factors, temperature and relative humidity, similar to the index of apparent temperature (AT) which is common in studies on the short term effects of heat on health [11]. Since Tel Aviv differs from European Mediterranean cities in climate, culture and inhabitants' behaviour, especially with regard to a high use of air conditioners and a high awareness of water/fluid consumption, our aim was to study the V/J relationship between heat stress and mortality in Tel Aviv, while estimating a threshold heat-stress (DI) point (where the heat-stress effect changes) during the whole year in a time-series design.

20.2 Methods

20.2.1 Health Data

The data was collected for the Tel Aviv area for the period 1/1/2000–31/12/2004 in a time-series design [6].The Israel Ministry of Health provided daily mortality counts, referring to the city residences. Taking into account the results of previous studies and the biological plausibility of the health effects [12], the following causes of death were selected for all ages combined and specific age groups (0, 1–14, 15–44, 45–64, 65–74, 75+y): all causes (except external causes) ICD-9: 1–799; cardiac diseases ICD-9: 390–429; circulatory system diseases ICD-9: 440–459; and respiratory diseases ICD-9: 460–519. The gender was stamped in the data set.

The Israel Ministry of Environmental Protection provided data from monitoring stations located in the city for the entire study period. For this study, we focused on air temperature and relative humidity recorded every half hour. Mean daily averages of all stations were calculated for each variable. Quality control included a descriptive overview of the variables, detecting possible errors and extreme values, testing for homogeneity and correcting erroneous values where possible.

We focused on the effect of the discomfort index on mortality, according to the following formula that involves temperature (Temp) and relative humidity (RH) as both additive and multiplicative factors [10]:

$$DI = -0.394479 + 0.784533 \times Temp + 0.022226 \times RH + 0.0023765 \times Temp \times RH$$

This index has been used for more than four decades and is highly correlated with the effective temperature index and with the wet-bulb globe temperature (WBGT) heat stress index that was developed in the US Navy as part of a study on heat related injuries during military training. From a biometeorological perspective, this index is more logical than describing temperature and humidity separately. The common categorization of DI, based on studies of populations from different climate conditions and ethnicity is: <22 (no heat stress), 22–23.9 (mild), 24–27.9 (moderately heavy), and >28 (severe heat stress). The Israeli Defense Forces has adopted this categorization for guidelines for exercising in heat. Daily values of DI were computed based on daily average values of temperature (°C) and relative humidity.

The following pollutant half-hourly measurements were collected from six monitoring stations in Tel Aviv which are part of the air-quality network of the Ministry of Environmental Protection: CO (maximum 8-h moving average); O_3 (daily maximum, maximum 8-h moving average); NO_2 (daily maximum, daily average); SO_2 (daily average); TSP or Black Smoke (daily average); PM_{10} (daily average); and $PM_{2.5}$ (daily average). Monitor selection was based on local criteria, mainly on the completeness of measurements and representation of population exposure. A standardized procedure was used to fill in days with missing data [12]. Since different stations differ in monitored pollutants, we estimated a mean maximum/average daily value of all data available for each pollutant.

For this study, we used the maximum hourly value of nitrogen dioxide (NO_2) as an indicator of the overall daily air pollution level for the entire city.

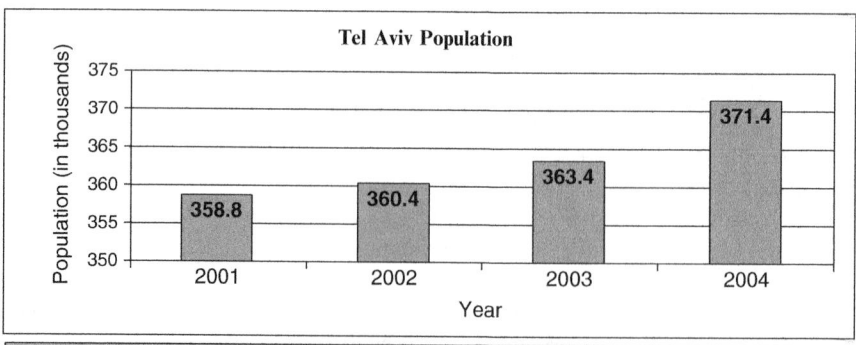

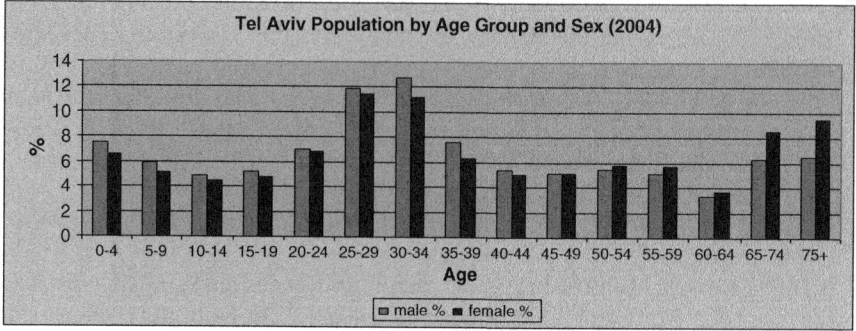

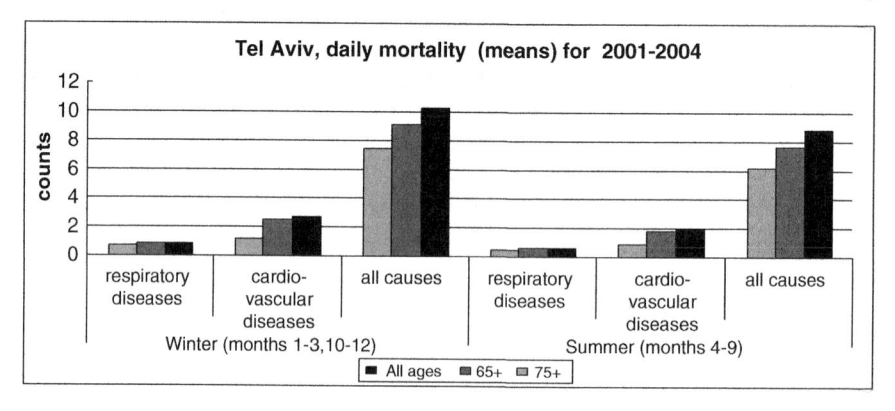

Fig. 20.1 Tel Aviv population (2001–2004)

20.2.2 Tel Aviv – General Characteristics

Characteristics of Tel Aviv are presented in Fig. 20.1. The area of the city is about 50.5 km^2 and the population in the years 2001–2004 grew from 358,800 to 371,400 inhabitants. On average, 8.8 inhabitants died per day during the summers and 10.3 died per day during the winters of 2001–2004. About 80% of households have air conditioners.

20.2.3 Statistical Modeling

A Poisson regression model was specified for the daily death count using a Bayesian approach [1]. We included in the model dummy variables for the day of week and calendar month and a linear term for the maximum hourly NO2 concentrations (lag 0–1). We modelled the relationship between DI (lag 0–3) and mortality by two linear terms constrained to joint in a point (threshold), using R software.

The model for the daily number of deaths (y_i) was the following:

$$y_i \sim Po(\mu_i)$$
$$log(\mu_i) = \alpha + month_i + wday_i + \beta \times poll_i + \gamma_0 \times T_i + \gamma_1 \times (T_i - threshold)_+$$
$$(T_i - threshold)_+ = \begin{cases} (T_i - threshold) & T_i > threshold \\ 0 & T_i \le threshold \end{cases}$$

The threshold, a value of DI which corresponds to a change in the effect estimate, was considered as an unknown parameter to be estimated. A normal prior distribution with large variance, centered in 24, was specified for the threshold. Non-informative prior distributions were specified on all regression coefficients. The joint posterior distribution for the model parameters was obtained using MCMC methods, with the software WinBugs 14. Posterior mean and 90% credibility intervals are provided for the parameters of interest.

20.3 Results

Table 20.1 presents summary statistics of meteorological data during the summer in Tel Aviv (months 4–9) in comparison to other Mediterranean European cities [18]. The mean AT temperature in Tel Aviv was found to be the highest among these cities and the relative humidity was also high. Figure 20.2 presents daily data of temperature, discomfort index and relative humidity as well as NO_2 levels during the study period (2001–2004) in Tel Aviv.

The relationship between the discomfort index DI (lag 0–3) and log mortality rates was J shaped for Tel Aviv (Fig 20.4). This indicated a linear excess risk to die for exposures to heat stress above a threshold. The DI threshold for Tel Aviv was found to be 29.3 (90% CrI = 28.0–30.7), a value considered to be a severe heat-stress (DI > 28). We report the heat effect as percent change in mortality associated with a 1 unit change in DI above/below the threshold. Above the threshold of 29.3, a 1 unit increase in DI was found to be associated with borderline significant increased mortality of 3.72% (90% CrI = −0.23 to 8.72). However, below the threshold a 1 unit decrease in DI was found to be associated with a non-significant

Table 20.1 Meteorological data in Tel Aviv and European Mediterranean cities, during the summer (months 4–9); mean, min-max

City	Study Period	Meteorological variables		
		AT[a]	Temperature[b]	Relative-humidity
Tel Aviv	2001–2004	32.4	25.5	71
		14.0–45.3	13.9–33.8	22–90
Valencia	1995–2000	29.5	22.3	66
		10.6–44.9	10.5–30	32–92
Athens	1992–1996	27.9	23.5	57
		7.9–41.6	7.6–34.3	23–89
Rome	1992–2000	26.1	20.5	72
		5.9–40.5	6.1–30.3	25–94
Milan	1990–2000	25.4	20.0	72
		2.7–40.8	2.5–29.4	26–100
Turin	1991–1999	23.4	18.5	74
		4.2–45.8	3.0–27.9	32–97
Barcelona	1992–2000	23.3	21.7	66
		6.5–36.9	8.6–34.2	29–99
Ljubljana	1992–1999	20.1	15.9	75
		−1.7 to 35.4	0.6–26.5	33–98

[a]Apparent temperature in °C, $AT = -2.653 + .994*temp + .0153*(dew)^2$
[b]In Tel Aviv it's a daily average AT while in the other cities it's a daily max AT

increase in mortality of 0.15% (90% CrI = −0.54 to 0.80), which can be considered to be 0% (Figs. 20.3 and 20.4). NO_2 was found to have a significant effect on mortality; an increase in 10 ppb was associated with an increase in mortality of 2.45% (90% CrI = 0.44–4.49).

20.4 Discussion

In this work we found a threshold heat-stress for Tel Aviv of DI = 29.3 (90% CrI = 28.0–30.7), a value 1.3 units higher than the lower value of severe heat-stress (DI = 28) [10]. In 2001–2004, 4% of the days were above this threshold and 15% were above DI = 28. This threshold value of DI = 29.3 for Tel Aviv is probably related to our specific health outcome, which was mortality rather than morbidity. Furthermore, we found that a 1-unit increase of DI above the threshold corresponded to a 3.7% increase in daily mortality, of borderline significance. A single day with a DI heat-stress exceeding the threshold noted above is sufficient to cause this increase in mortality, rather than requiring an extended heat wave. These results for Tel Aviv might not be relevant to other inland cities in Israel with different climatic conditions, especially with regard to humidity (*e.g.*, Beer Sheva and Jerusalem) or to rural

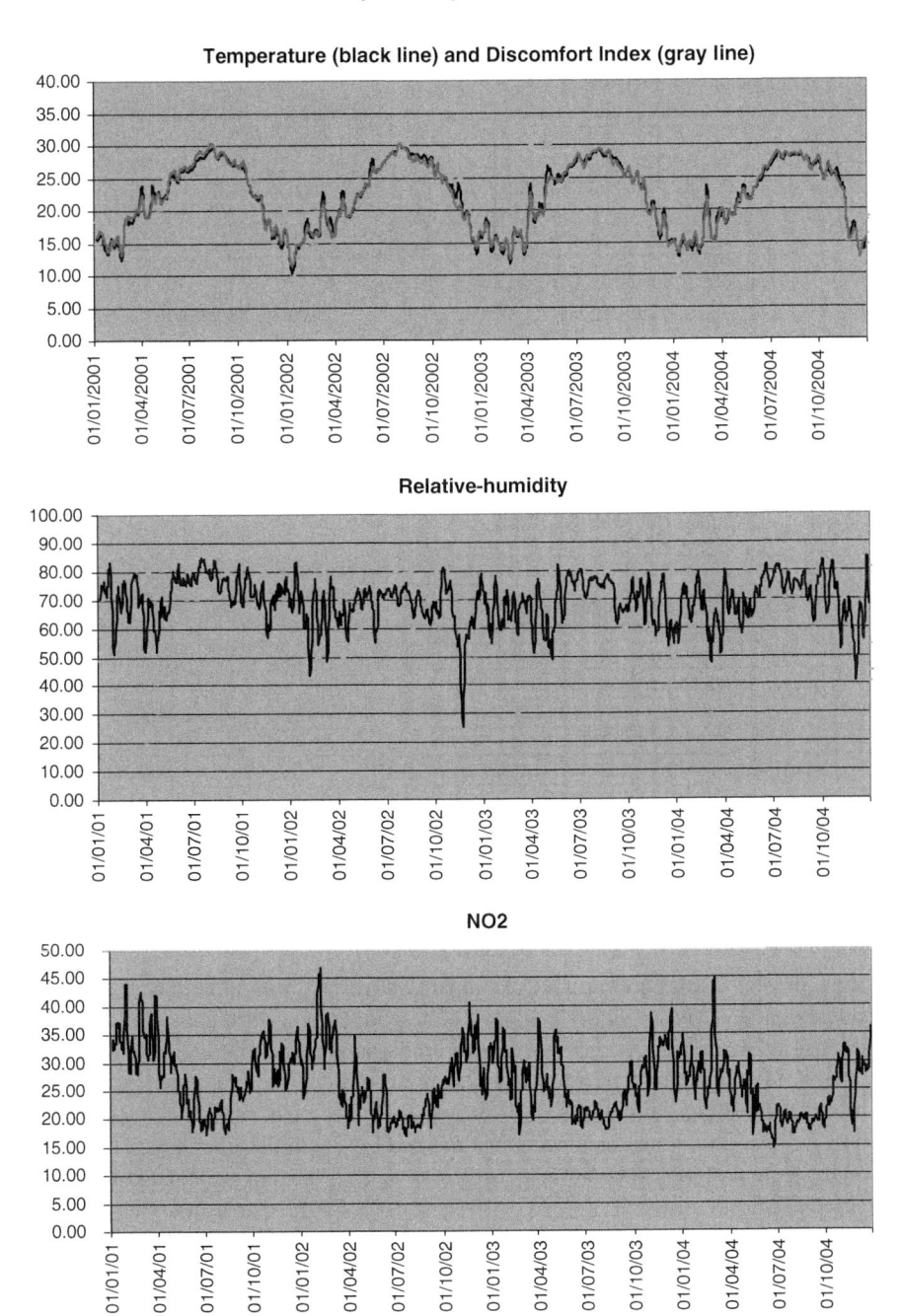

Fig. 20.2 Daily data of temperature (average in °C), discomfort index, relative humidity (average) and NO_2 levels (maximum in ppb), Tel Aviv 2001–2004; smoothed time series (7 day moving average)

a

The full posterior distribution

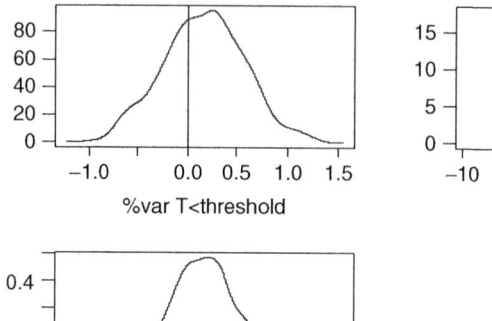

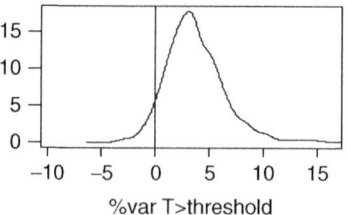

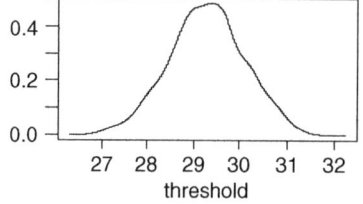

b

Posterior means and 90% credibility intervals

	Posterior mean	90% Credibility interval	
% variation under the threshold	0.15	−0.54	0.80
% variation over the threshold	3.72	−0.23	8.72
Threshold	29.28	27.96	30.65

Fig. 20.3 The posterior distribution of the percent variation under the threshold and over the threshold and the threshold (in DI)

areas. It is also difficult to directly compare our results with other epidemiological studies on heat and mortality because of methodological differences [11] such as: the use of heat-stress definition- either ambient average/max daily temperature [3, 4, 8, 9] or apparent temperature [1, 2, 6, 7]; the way of reporting on heat effects – above a threshold value [1, 3, 7–9] or not [2, 4, 6]; the approach of estimating a threshold value – based on a statistical method of maximum likelihood [1, 7, 8] or on inspection of the exposure-response curves [3, 9]; seasons included in the study, only the warm seasons [1, 2, 4, 6] or all seasons [3, 7–9]. After accounting for some methodological differences, our results are in agreement with those that have been obtained from European Mediterranean cities; that is, for a J shape relationship with a threshold of 29.4°C AT (29.3 DI) and quite a similar percent increase in mortality above the threshold, even though the higher average of AT temperature in Tel Aviv during the summer. However, the percent increase is less significant and this might be explained by the heterogeneity of Tel Aviv's population with regard to

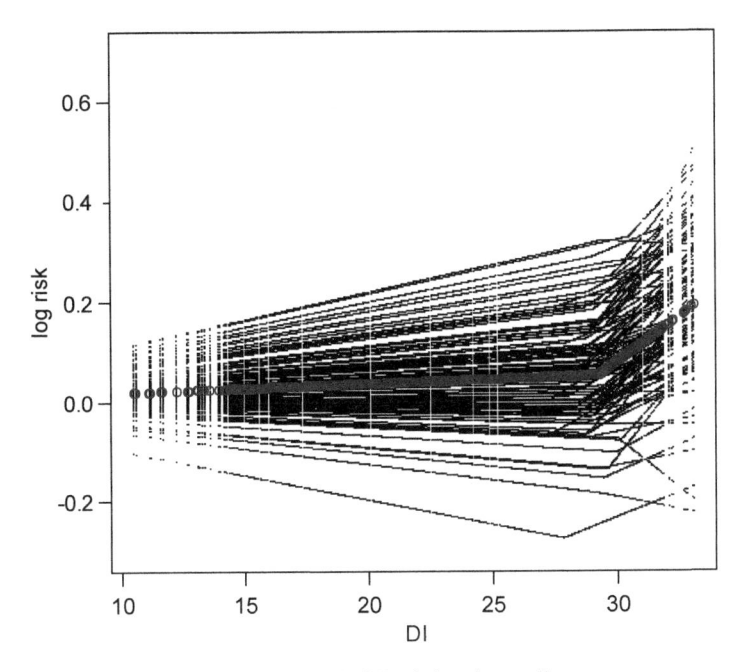

Fig. 20.4 Relationship between heat stress (in DI units) and mortality

socio-economic status, nationality (Jews and Arabs), immigrant status (newcomers from the former USSR or not), or by high use of domestic air conditioning in Israel, the air conditioning of public buildings due to mandatory building regulations [13], and behavioural adaption of the elderly that stay at home on hot days.

Interestingly, comparing Tel Aviv and Mediterranean European cities to Korean cities [11] with different levels of temperature and humidity, the mortality estimates above a similar threshold (Korea: 23.3–29.7°C of AT) resulted in a much higher effect in Korea – an increase of 6.73–16.3% (in six cities) in mortality per 1°C increase of AT compared to 3.7% (per 1 DI unit) and 3.1% (per 1°C AT) in Tel Aviv and Mediterranean European cities, respectively. Two recent American studies suggested that the effect estimates throughout California and other parts of the US are similar, even with different ranges of apparent temperatures. They both found an approximately 2% increase in mortality per a 10°F increase in apparent temperature [11], a smaller increase than reported above. It should be noted that these studies differ also in other methodological issues: in the study designs (time series or case-crossover); the lag-days that were considered (*e.g.*, 0, 0–4, 2-day average); the study time-period; the cause-specific mortality (*e.g.*, all cause or specific causes) and age-groups that were considered (*e.g.*, mortality of all ages or 65+); as well as in the statistical modelling.

The role of the following air-pollutants: O_3, PM_{10}, $PM_{2.5}$, CO and SO_2 as confounders and/or effect modifiers on the temperature-mortality association were found to be mixed; some investigators reported air pollutants as confounders or effect modifiers while others found no significant confounding or effect modification in their studies [11]. We found a significant effect of NO_2.

Our study has some major strengths. It is the first study of this kind in Israel. The exposure index was DI and not just temperature or relative humidity, and we adjusted for the possible confounding effects of air pollution. We suggest that further studies in Israel should focus on sub-districts and that they should examine heat-stress and different causes of death, define heat waves and examine their effect on mortality and consider max temperature during the day and max temperature during the night. To conclude, as global warming continues, the frequency, intensity and duration of heat stress days are likely to increase and even though there exists a high awareness amongst the Israeli population to the negative effect of heat on health, there is still a vital need for local policies to mitigate heat-related deaths.

Acknowledgements Many thanks to Alina Rosenberg who contributed to data preparation and to the Environment and Health Fund for financial support.

References

1. Baccini M, Biggeri A, Accetta G et al (2008) Heat effects on mortality in 15 European cities. Epidemiology 19(5):711–719
2. Almeida SP, Casimiro E, Calheiros J (2010) Effects of apparent temperature on daily mortality in Lisbon and Oporto. Portugal Environ Health 9:12
3. Iñiguez C, Ballester F, Ferrandiz J, Pérez-Hoyos S, Sáez M, López A, TEMPRO-EMECAS (2010) Relation between temperature and mortality in thirteen Spanish cities. Int J Environ Res Public Health 7(8):3196–3210
4. Hu W, Mengersen K, McMichael A, Tong S (2008) Temperature, air pollution and total mortality during summers in Sydney, 1994–2004. Int J Biometeorol 52(7):689–696
5. Basu R, Feng WY, Ostro BD (2008) Characterizing temperature and mortality in nine California counties. Epidemiology 19(1):138–145
6. Zanobetti A, Schwartz J (2008) Temperature and mortality in nine US cities. Epidemiology 19(4):563–570
7. Chung JY, Honda Y, Hong YC, Pan XC, Guo YL, Kim H (2009) Ambient temperature and mortality: an international study in four capital cities of East Asia. Sci Total Environ 408(2):390–396
8. McMichael AJ, Wilkinson P, Kovats RS et al (2008) International study of temperature, heat and urban mortality: the 'ISOTHURM' project. Int J Epidemiol 37(5):1121–1131
9. El-Zein A, Tewtel-Salem M, Nehme G (2004) A time-series analysis of mortality and air temperature in Greater Beirut. Sci Total Environ 330(1–3):71–80
10. Epstein Y, Moran DS (2006) Thermal comfort and the heat stress indices. Ind Health 44(3):388–398
11. Basu R (2009) High ambient temperature and mortality: a review of epidemiologic studies from 2001 to 2008. Environ Health 8:40

12. Michelozzi P, Kirchmayer U, Katsouyanni K, Biggeri A, McGregor G, Menne B, Kassomenos P, Anderson HR, Baccini M, Accetta G, Analytis A, Kosatsky T (2007) Assessment and prevention of acute health effects of weather conditions in Europe, the PHEWE project: background, objectives, design. Environ Health 6:12
13. Novikov I, Kalter-Leibovici O, Chetrit A, Stav N, Epstein Y (2011) Weather conditions and visits to the medical wing of emergency rooms in a metropolitan area during the warm season in Israel: a predictive model. Int J Biometeorol (ahead of print)

Chapter 21
West Nile Virus Eruptions in Summer 2010 – What Is the Possible Linkage with Climate Change?

Shlomit Paz

Abstract West Nile Virus (WNV) is a vector-borne pathogen of global importance. Many factors impact the transmission, epidemiology and geographic distribution of WNV. However, climate and especially warm conditions were found to be crucially important causes that instigated the outbreaks. New areas of the WNV transmission with the occurrence of human cases have been identified during summer 2010. According to the European Center for Disease Control (ECDC), infections by WNV have occurred in Greece, Romania, Hungary, Israel, Russia, and Italy. The precise reasons for the existence of the current outbreak of WNV infection in humans in Eurasia remain unclear. However, climatic factors are believed to have increased the abundance of mosquitoes and shortened the transmission cycle in the vectors, leading to increased human cases.

Mean monthly temperature and precipitation data show the extreme behavior of the air temperature as well as the rainfall patterns during summer 2010 in selected areas where WNV circulation occurred – Macedonia (Greece), Western Turkey, Southeastern Romania and Southwestern Russia. The results show that the warming tendency during the hot season over recent years continued in summer 2010. Moreover, the air temperature was extremely higher than normal in the selected study sites. This might have an impact on the risk for WNV outbreaks.

As for the precipitation, the picture is more complex. The increase in WNV cases could be related to the unusual increase in the rainfall amounts during that summer (Macedonia and SE Romania). Alternatively, WNV may increases after an extreme dry period (SW Russia), since standing water pools become richer in organic materials.

S. Paz (⊠)
Department of Geography and Environmental Studies, University of Haifa,
Mt. Carmel, Haifa, Israel
e-mail: shlomit@geo.haifa.ac.il

H.J.S. Fernando et al. (eds.), *National Security and Human Health Implications of Climate Change*, NATO Science for Peace and Security Series C: Environmental Security, DOI 10.1007/978-94-007-2430-3_21, © Springer Science+Business Media B.V. 2012

During the summer of 2010, Eurasia had to deal with exceptional heat-waves while a record of high numbers of extreme warm nights had been documented in parts of south-eastern Europe.

In summary, although the WNV transmission is multi-factorial, it seems that the increase in the summer temperature should be considered when evaluating the risk of WNV transmission.

Keywords West Nile virus • Vectorborne diseases • Temperature increase • Climate change

21.1   Introduction

One of the major concerns of climate change is the impact on human health. Global warming affects human health via pathways of varying complexity, scales and directness and different timing. Similarly, impacts (both positive and negative) vary geographically as a function of the physical and environmental conditions as well as the vulnerability of the local human population [27]. Climate is one of several factors that influence the distribution of vectorborne and zoonotic diseases (such as West Nile Virus, Malaria and Lyme disease). There is substantial concern that climate change will make certain environments more suitable for some of these diseases, worsening their significant global burden and potentially reintroducing them into geographic areas where they had been previously eradicated [14, 22].

For example, based on studies of vector and virus development, warming and increases in humidity are predicted to open up new zones for West Nile Virus (WNV) in North America [29, 30].

Indeed, since the temperatures are related to the life-cycle dynamics of both the vector species and the pathogenic organisms related to WNV, a considerable impact of global warming on the public is WNV altered transmission and geographical distribution.

21.2   WNV – Background

WNV is a member of the Japanese encephalitis serogroup, family Flaviviruses. The virus is transmitted between birds and transferred by carriers, especially by *Culex* mosquitoes [36]. The basic transmission cycle of WNV occurs in rural ecosystems, involving wild birds as the principal hosts and mosquitoes, largely bird-feeding species, as the primary vectors. Another cycle is in urban ecosystems (domestic birds and mosquitoes feeding on both birds and humans). The virus is amplified during periods of adult mosquito blood-meal feeding by continuous transmission between mosquito and susceptible bird species. Humans and horses are incidental and dead-end hosts for the viruses [2, 10, 13, 15, 19, 34].

West Nile Fever (WNF) in humans appears after an incubation period of 3–14 days, mostly at the end of the summer [1]. The virus was recognized as a cause of fatal human meningitis or encephalitis in elderly patients during an outbreak in Israel in 1957 [32]. WNV infection can be a serious disease with significant mortality in the elderly and those with underlying chronic disease and may have long-term health consequences [17].

WNV is endemic in Africa, in southwestern Asia, in eastern and southern Europe and in parts of the Mediterranean basin [6, 13]. The virus was first detected in the western hemisphere in New York City in 1999 [21, 33]. It has since moved rapidly westward and was detected in California in 2004 [16]. Today, WNV exists also in Central and South America and in the Caribbean [15], and is therefore a vector-borne pathogen of global importance.

21.3 The Linkage Between Climate Change and WNV Eruptions

Recent public health experience with the WNV outbreak in the United States reveals the complexity of such epidemics [22]. Many factors impact WNV transmission, epidemiology and geographic distribution. However, environmental temperatures were found to influence vector competence for arboviruses in general, and WNV in particular [4]. Indeed, warm conditions were detected to be crucially important causes that instigated the outbreaks [35] and summer temperature was found as one of the most important environmental variables modulating WNV activity in Europe [31]. High temperatures have been shown to speed up the replication development of the virus within the mosquito carriers, and this rapid amplification directly affects the likelihood of the mosquito reaching maturity and subsequently infecting other hosts [10, 28].

Warming of the mosquitoes' environment boosts their rates of reproduction and number of blood meals, prolongs their breeding season, and shortens the maturation period for the microbes they disperse [12, 18, 20, 28, 33]. It has been proven experimentally that high temperature during incubation affects the transmission of WNV in *Cx. pipines* mosquitoes and profoundly influences mosquito-to-vertebrate transmission rates [5, 6].

In a study on the linkage between extreme heat and the WNV outbreak in Israel, Paz [24] found that the minimum daily temperatures become the most important climatic factor that encourages the earlier appearance of the disease. Correlations were also detected for the relative humidity but temperature was found as more significant for increasing WNV. In a later study, Paz and Albersheim [25] presented the potential influence of extreme heat in the early spring on the vector population increase and on the disease's appearance weeks later.

The impact of precipitation patterns is complex: on the one hand, heavy rainfall during spring may enlarge the standing water resources at the beginning of the hot season. In contrast, in drought conditions, standing water pools become richer in the organic material that *Culex* needs to thrive. This may encourage birds to circulate around small water holes and thus increase the interactions with mosquitoes [3, 10, 11, 28].

21.4 WNV Eruption in Eurasia in Summer 2010 and Its Possible Linkage with the Extreme Meteorological Conditions

The WNV is endemic in eastern and southern Europe. Its transmission is ongoing in several EU Member States, as well as in other countries in the region. However, new areas of the virus transmission with the occurrence of human cases have been identified during summer 2010 [7]. Human infections by WNV have occurred in Greece, Romania, Hungary, Israel, Russia, and Italy [8].

The precise reasons for the existence of the current outbreak of WNV infection in humans in the region remain unclear. However, climatic factors are believed to have increased the abundance of mosquitoes and shortened the transmission cycle in the vectors, leading to increased human cases [9].

Based on mean monthly temperature and precipitation data from the Earth System Research Laboratory of NOAA, Paz [26] showed the extreme behavior of the air temperature as well as the rainfall pattern during summer 2010 in selected areas where WNV circulation occurred – Macedonia (Greece), Western Turkey, Southeastern Romania and Southwestern Russia (Figs. 21.1 and 21.2).

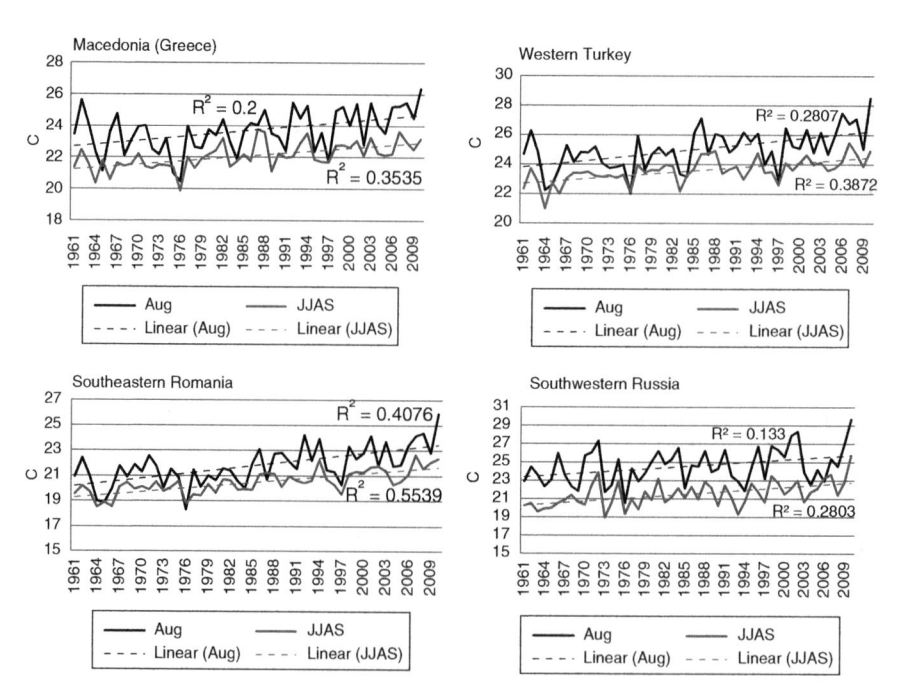

Fig. 21.1 Mean monthly air surface temperature (1961–2010) in selected areas where WNV circulation occurred during summer 2010 (Data source: Earth System Research Laboratory, NOAA). JJAS = mean monthly temperature for June, July, August and September

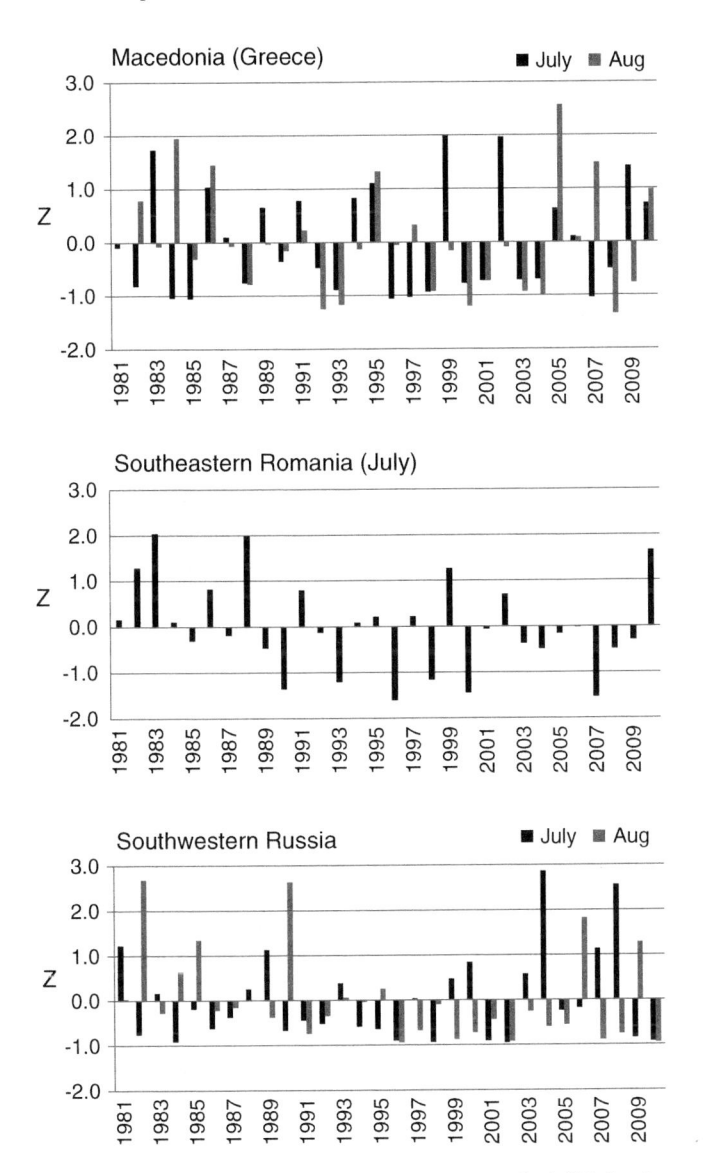

Fig. 21.2 Standardized monthly surface gauss precipitation rate (kg/m²/s) for the years 1981–2010, in selected areas where WNV circulation occurred during summer 2010 (Data source: Earth System Research Laboratory, NOAA)

21.5   Conclusions

The results show that the significant warming tendency during the hot season over recent years continued in summer 2010. Moreover, the air temperature was extremely higher than normal in the selected study sites (Fig. 21.1). This might have an impact on the risk for WNV outbreaks.

As for the precipitation, the picture is more complex (Fig. 21.2). The increase in WNV cases could be related to the unusual increase in the rainfall amounts during that summer (Macedonia and SE Romania). Indeed, meteorological data from 2010 for Central Macedonia (Greece) suggest that the temperatures were unusually high in July and August parallel with an unusual rainfall pattern [23]. Alternatively, WNV may increases after an extreme dry period (SW Russia), since standing water pools become richer in organic materials.

The year 2010 was of the top three warmest years since the beginning of instrumental climate records in 1850, as a part a continuing trend. Moreover, the decade of 2001–2010 was the warmest 10-year period. During the summer of 2010, Eurasia had to deal with exceptional heat-waves while a record of high numbers of extreme warm nights had been documented in parts of south-eastern Europe [37].

In summary, although the WNV transmission is multi-factorial, it seems that the increase in the summer temperature should be considered when evaluating the risk of WNV transmission [8]. Further research needs better resolution of the geographical distribution of the disease cases in humans as well as more meteorological data.

References

1. CDC, Division of vector-born infectious diseases (2006) West Nile virus: what you need to know. Available: http://www.cdc.gov/ncidod/dvbid/westnile/wnv_factsheet.htm. Accessed 9 Jan 2007
2. CDC, Division of vector-born infectious diseases (2009) West Nile virus – vertebrate ecology: transmission cycle. Available: http://www.cdc.gov/ncidod/dvbid/westnile/birds&mammals.htm. Accessed 10 July 2008
3. Chase JM, Knight TM (2003) Drought-induced mosquito outbreaks in wetlands. Ecol Lett 6:1017–1024
4. Cornel AJP, Jupp G, Blackburn NK (1993) Environmental temperature on the vector competence of Culex univittatus (Diptera: Culicidae) for West Nile virus. J Med Entomol 30:449–456
5. Dohm DJ, Turell MJ (2001) Effect of incubation at overwintering temperatures on the replication of WNV in New York Culex pipiens (Diptera: Culicidae). J Med Entomol 38:462–464
6. Dohm DJ, O'Guinn ML, Turell MJ (2002) Effect of environmental temperature on the ability of Culex pipiens (Diptera: Culicidae) to transmit WNV. J Med Entomol 39:221–225
7. ECDC (2010) West Nile virus transmission in Europe. Available: http://ecdc.europa.eu/en/activities/sciadvice/Lists/ECDC%20Reviews/ECDC_DispForm.aspx?List=512ff74f%2D77d4%2D4ad8%2Db6d6%2Dbf0f23083f30&ID=940&RootFolder=%2Fen%2Factivities%2Fsciadvice%2FLists%2FECDC%20Reviews. Accessed 4 Dec 2010
8. ECDC (2010) Meeting report: consultation on mosquito borne disease transmission risk in Europe. Paris, 26 Nov 2010
9. ECDC (2010) ECDC threat assessment, outbreak of West Nile virus infection in Greece, July–August 2010. Available: http://ecdc.europa.eu/en/healthtopics/Documents/1009_Threat%20assessment_West_Nile_Virus.pdf
10. Epstein PR (2001) WNV and the climate. J Urban Health 78:367–371
11. Epstein PR, Defilippo C (2001) WNV and drought. Glob Change Hum Health 2:105–107
12. Epstein PR (2005) Climate change and human health. New Engl J Med 353(14):1433–1436
13. Hubalek Z, Halouzka J (1999) WNF – a reemerging mosquito-borne viral disease in Europe. Emerg Infect Dis 5:643–650

14. IPCC (2007) Climate change: impacts, adaptation and vulnerability working group II contribution to the intergovernmental panel on climate change, fourth assessment report – summary for policymakers
15. Gibbs SEJ, Wimberly MC, Madden M, Masour J, Yabsley MJ, Stallknecht DE (2006) Factors affecting the geographic distribution of WNV in Georgia, USA: 2002–2004. Vector Borne Zoonotic Dis 6:73–82
16. Granwehr BP, Lillibridge K, Higgs S, Mason P, Aronson J, Campbell G, Barrett A (2004) WNV: where are we now? Lancet 4:547–556
17. Green MS, Weinberger M, Ben-Ezer J, Bin H, Mendelson E, Gandacu D, Kaufman Z, Dichtiar R, Sobel A, Cohen D, Chowers MY (2005) Long-term death rates, West Nile virus epidemic, Israel, 2000. Emerg Infect Dis 11:1754–1757
18. Marsh MM, Gross JM Jr (2001) Environmental geography – science, land use and earth systems. Wiley, New York, pp 202–203
19. Mclean R, Ubico SR, Docherty DE, Hansen WR, Sileo L, McNamara TS (2001) West Nile virus transmission and ecology in birds. Ann N Y Acad Sci 951:54–57
20. McMichael AJ (2003) Global climate change and health: an old story writ large. In: McMichael AJ, Campbell-Lendrum DH, Corvalán CF et al (eds) Climate change and human health – risks and responses. WHO, Geneva, pp 1–17
21. Nash D, Mostashari F, Fine A, Miller J, O'Leary D, Murray K et al (2001) The outbreak of WNV infection in the New York city area in 1999. New Engl J Med 344:807–814
22. NIEHS (2010) A human health perspective on climate change a report outlining the research needs on the human health effects of climate change April 22, 2010. Environmental Health Perspectives and the National Institute of Environmental Health Sciences, Research Triangle Park
23. Papa A, Perperidou P, Tzouli A, Castiletti C (2010) West Nile virus neutralising antibodies in humans in Greece. Vector Borne Zoonotic Dis 10(7):655–658
24. Paz S (2006) The West Nile virus outbreak in Israel (2000) from a new perspective: the regional impact of climate change. Int J Environ Heal Res 16(1):1–13
25. Paz S, Albersheim I (2008) Influence of warming tendency on *Culex pipiens* population abundance and on the probability of West Nile fever outbreaks (Israeli case: 2001–2005). Ecohealth 5(1):40–48
26. Paz S (2010) Climatic trends in Europe and in the Mediterranean region in relation to West Nile virus outbreaks. Presentation in the consultation on the mosquito borne disease transmission risk in Europe (ECDC and InVS), Paris, Nov 2010
27. Paz S, Xoplaki E, Gershunov A (2010) Scientific report of the workshop: impacts of Mediterranean climate change on human health. The European Science Foundation (ESF) – MedClivar, Jan 2010, 10 pp
28. Pats JA, Githeko AK, McCarty JP, Hussain S, Confalonieri U, de Wet N (2003) Climate change and infection disease. In: McMichael AJ, Campbell-Lendrum DH, Corvalán CF et al (eds) Climate change and human health – risks and responses. WHO, Geneva, pp 103–132
29. Reisen WK, Fang Y, Martinez VM (2006) Effects of temperature on the transmission of West Nile virus by *Culex tarsalis* (Diptera: Culicidae). J Med Entomol 43:309–317
30. Rosenthal J (2010) Climate change and the geographic distribution of infectious diseases. Ecohealth 6(4):489–495. doi:10.1007/s10393-010-0314-1, published online: 25 May, 2010
31. Savage HM, Ceianu C, Nicolescu G, Karabatsos N, Lanciotti R, Vladimirescu A et al (1999) Entomologic and avian investigation of an epidemic of WNV fever in Romania in 1996, with serologic and molecular characterization of a virus isolate from mosquitoes. Am J Trop Med Hyg 61:600–611
32. Spigland I, Jasinka-Klinberg W, Hofshi E, Goldblum N (1958) Clinical and laboratory observations in an outbreak of West Nile fever in Israel in 1957. Harefua 54:275–281
33. Tibbetts J (2007) Driven to extremes health effects of climate change. Environ Health Perspect 115(4):A196–A203
34. Tsai TF, Popovici F, Cernescu C, Campbell GL, Nedelcu N (1998) West Nile encephalitis epidemic in southeastern Romania. Lancet 352:767–771

35. Turell MJ, O'Guinn ML, Dohm DJ, Jones JW (2001) Vector competence of North American mosquitoes (Diptera: Culicidae) for WNV. J Med Entomol 38:130–134
36. Weinberger M, Pitlik SD, Gandacu D, Lang R, Nassar F, Ben-David D (2001) WNF outbreak, Israel, 2000: epidemiologic aspects. Emerg Infect Dis 7:686–691
37. WMO (2010) Press release no. 904: 2010 in the top three warmest years, 2001–2010 warmest 10-year period. Available: http://www.wmo.int/pages/mediacentre/press_releases/pr_904_en.html. Accessed 11 Apr 2011

Chapter 22
Global Water Security: Engineering the Future

Roger A. Falconer and Michael R. Norton MBE

Abstract The paper introduces some of the general challenges of global water security, particularly in poverty stricken regions such as Africa, and highlights the likely global impact of climate change, increasing pollution and population growth etc. on water resources, as outlined in recent studies. The nexus between water, food and energy is introduced, along with the concept of virtual water and the impact of the water footprint and the need for society, industry and governments to become more conscious of the water footprint, alongside the carbon footprint. Various practical solutions to enhancing security of supply are introduced and discussed, such as desalination and integrated water management in the form of 'Cloud to Coast', together with global actions needed. Finally, some water security challenges and opportunities for developed countries, such as the UK, are discussed, particularly with regard to the need to price water appropriately and the need to appreciate that the price of water should cover more than just the cost of delivery to the home. The paper concludes with the urgent need to raise the profile of global water security at all levels of society and through international bodies, for the benefit of humanity worldwide.

Keywords Climate change • Water resources • Water pollution • World population • Integrated water management • Eco-systems • Bio-diversity • Desalination • Water pricing

R.A. Falconer (⌧)
Director Hydro-environmental Research Centre, Cardiff School of Engineering, Cardiff University, The Parade, Cardiff, CF24 3AA, UK
e-mail: FalconerRA@cf.ac.uk

M.R. Norton MBE
Halcrow Group Ltd, Burderop Park, Swindon SN4 0QD, UK
e-mail: NortonMR@halcrow.com

H.J.S. Fernando et al. (eds.), *National Security and Human Health Implications of Climate Change*, NATO Science for Peace and Security Series C: Environmental Security, DOI 10.1007/978-94-007-2430-3_22, © Springer Science+Business Media B.V. 2012

22.1 Introduction

In recent years there has been growing international concern about the increasing crisis in Global Water Security, with security referring here to security of supply – both quantity and quality. As an example of this concern in April 2010 the Royal Academy of Engineering published a report entitled Global Water Security – An Engineering Perspective [1]. This report was produced by the UK Institution of Civil Engineers, the Royal Academy of Engineering and the Chartered Institution of Water and Environmental Management, through a Steering Group of 12 specialists working in the field. The Group took evidence from a wide range of international experts covering all aspects of water security. The main drivers for the report were concerns from numerous sources from within the UK Government, the professions and learned societies about the increasing challenges arising relating to water security and the implications for the UK, both nationally and internationally. Some of these challenges, threats and opportunities are introduced below.

The world consists of 1.4 billion km^3 of water, of which only 35 million km^3 are available in the form of freshwater, and of this resource only 105 thousand km^3 are accessible as a vital natural resource. This vital resource is required for a wide range of uses including: sustaining human life (i.e. consumption and sanitation), supporting for food production (with the addition of nutrients and sunlight), supporting energy production, sustaining industry, and maintaining our ecosystem, biodiversity and landscape.

There are 1.2 billion people living on this earth today with no access to safe drinking water; typically two million people die annually of diarrhoea – still one of the biggest causes of infant mortality on our planet today [2]. Traditional public health engineering still offers considerable challenges and rewarding career opportunities, second to none, to those young people aspiring to want to save lives or improve the quality of life of our fellow citizens living in developing countries and facing challenges of inadequate and/or poor quality water resources.

There are 2.4 billion people who do not have basic water sanitation and typically one million die annually of hepatitis A. Women in developing countries have to walk typically 3.7 miles to carry water for the family [3]; again engineers can, and do, make a huge contribution to the quality of life for these women. Floods often cause significant loss of life and destroy homes, with last year's floods in Pakistan leading to an estimated 21 million people being homeless. However, the disease associated with the after effects of such floods can often bring far more loss of life to communities and countries than the floods themselves [4]. It is estimated by the BMJ 2004 that at any one time more than half the hospital beds worldwide are occupied by people with water related diseases [5]. Other interesting – but startling – facts relating to water security include such information as: 'more people in the world have access to cell phones than have access to a toilet' [6]. Hence, the challenges of water security are immense and on a global scale.

22.2 Impacts of Climate Change and Population Growth

Water is at the heart of climate change and, along with the challenges cited above, global water security is expected to be exacerbated through shifting weather patterns, more intense hurricanes, typhoons, storms, floods, droughts, and the effects of glacial melt, snowmelt, evaporation, evapotranspiration and rising sea levels. Mean precipitation is expected to increase in the tropics and high latitudes, and decrease in the sub-tropics and mid-latitudes [1]. Water stress will deteriorate globally over the next century, with increased runoff in certain parts of the world (East Africa, India, China) and reductions in other regions (Mediterranean, North Africa) causing increasing problems. Average global temperatures are expected to rise by at least 2°C by the end of this century. If the temperature increases between 2°C and 5°C there will be major water resources problems globally, also resulting in significant sea water level rise and causing catastrophic coastal flooding in many parts of the world, such as Bangladesh. The implications of significant average global temperature rises above 2°C are highlighted in the Stern Report to the UK Government [7].

In 2009 the UK Government's Chief Scientific Advisor, Professor Sir John Beddington, raised the prospect of a "Perfect Storm" of global dimensions by 2030 with the impacts of global challenges such as climate change, food, energy and water security coming together to impact significantly on the lives of all people on earth. He noted that if we don't act now this will put at risk the wellbeing of many people on earth, especially the world's poor and most vulnerable. By 2030 the world's population is expected to increase from six to eight billion. Associated with this population growth we can expect the demand for food, energy and water to increase by 50% for food, 50% for energy and 30% for water (Fig. 22.1).

The water-food-energy nexus is crucial to our existence, with water being at the heart of everything; it is crucial for our energy supply, food, health, industry, trade etc. If we look at the water stress globally (defined as millions of litres of water available per person per year) from 1960 to 2010, we find that even in the southeast of England water supply is currently particularly stressed, and water stress is a growing major problem across right across the US.

Problems in water supply will relate not only to the 50% increase in human population over the next 30 years but also to urbanisation that is occurring all over the world, which is exacerbating this effect. In countries such as China, for example, people are moving into the major cities creating more mega-cities, while in the UK people are moving more and more to the southeast of England, which is not sustainable over the long term and we have to look at how we can make other parts of the UK more attractive for regional development in the future. Projects such as the Severn Barrage, previously considered as a project only for renewable energy, need to be considered also as creating a catalyst for re-distributing the population and encouraging people to live in parts of the country where the water stress is much lower [8].

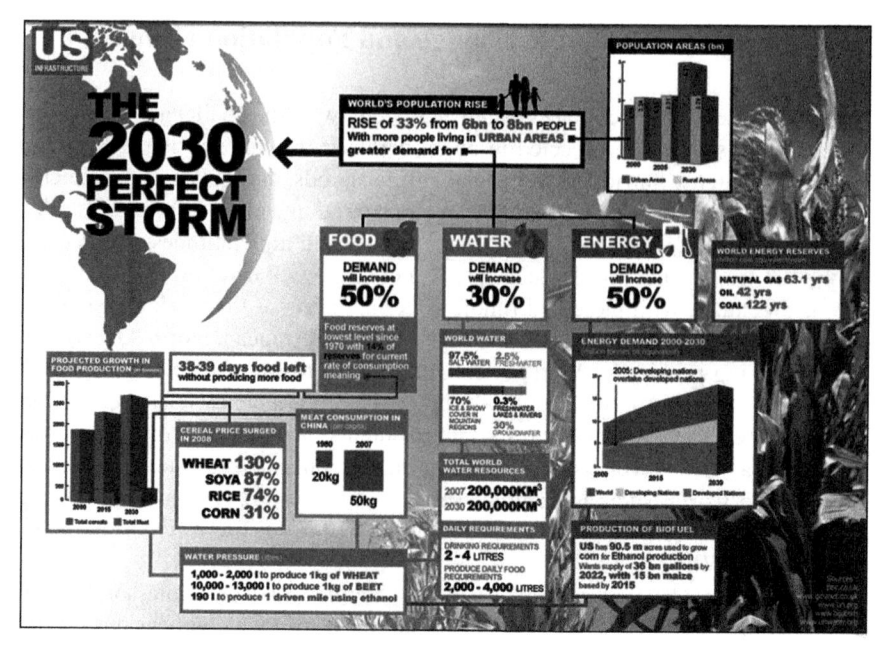

Fig. 22.1 The 'Perfect Storm' from a lecture by Sir John Beddington UK Government Chief Scientist (Source: US Infrastructure)

22.3   Water as the Vital Resource

As indicated above, increasing attention is now being paid to water as a vital resource in the context of the **water-food-energy nexus.** For example, water is required to grow food that sustains life; energy is required to treat and move water; energy production and industrial processes require fresh water; lifestyle changes result in increased energy and food consumption; increasing conflicts are arising between land for bio-fuels and land for traditional agriculture; etc. Impacting on all areas of the nexus we have: competition for finance; international trade flows; the environment and loss of biodiversity; climate change impacts on supply and demand; etc.

The percentage of water that is used globally for domestic, industrial and agricultural purposes is typically 8%, 23% and 69% respectively, with these figures varying from 13%, 54% and 33% across Europe, to 7%, 5% and 88% across Africa. Even though 13% of water abstractions in Europe are for drinking, the largest withdrawals are for irrigated agriculture and for food production. There is limited scope to increase the global area for crop irrigation, therefore there is a growing need for higher crop yields or improved irrigation to meet future food demand. Alternatively, we need to seek to grow more food in those parts of the world where there is sufficient land and rainfall. Furthermore, there is a growing need to produce food in such a way as to protect the natural resources it depends on, i.e. the soil, nutrients and water; particularly since we also depend upon these resources for other services, such as: drinking water, climate regulation, flood protection, filtering of pollutants etc.

22.4 Understanding the Problem

In the first instance we have to recognise the water cycle in providing freshwater in the form of rainfall, evaporation, condensation, precipitation, infiltration and run-off. In considering the 'Cloud to Coast' concept, being developed by Halcrow and Cardiff [9] where water transport solutions are being treated in an integrated manner, we note that for every 100 raindrops that fall on the land only 36 reach the ocean. Of the rest, most are held as soil moisture (referred to as 'green water') and which are used by our landscape, ecosystem and farmers. The role of green water is much undervalued and constitutes two thirds of our rainfall. That which enters lakes, rivers and aquifers (referred to as 'blue water') provides the water which we withdraw for our needs. It is this water that receives most attention. Finally there is grey water, which is the wastewater and surplus water which is returned to rivers etc. following consumption (Fig. 22.2).

Finally, there is the concept of virtual water. This is the water which has been used to grow food or produce goods, for domestic purposes or export, depending on the local or national economy. It may have green, blue and grey water components. Virtual water leads into the concept of the water footprint, which is analogous to that of the carbon footprint. To produce 1 kg of wheat requires 1,300 l of water [10], whereas to produce 1 kg of beef requires 15,000 l of water, i.e. over ten times as much water. Looking at other commodities, it takes 140 l of virtual (or embedded)

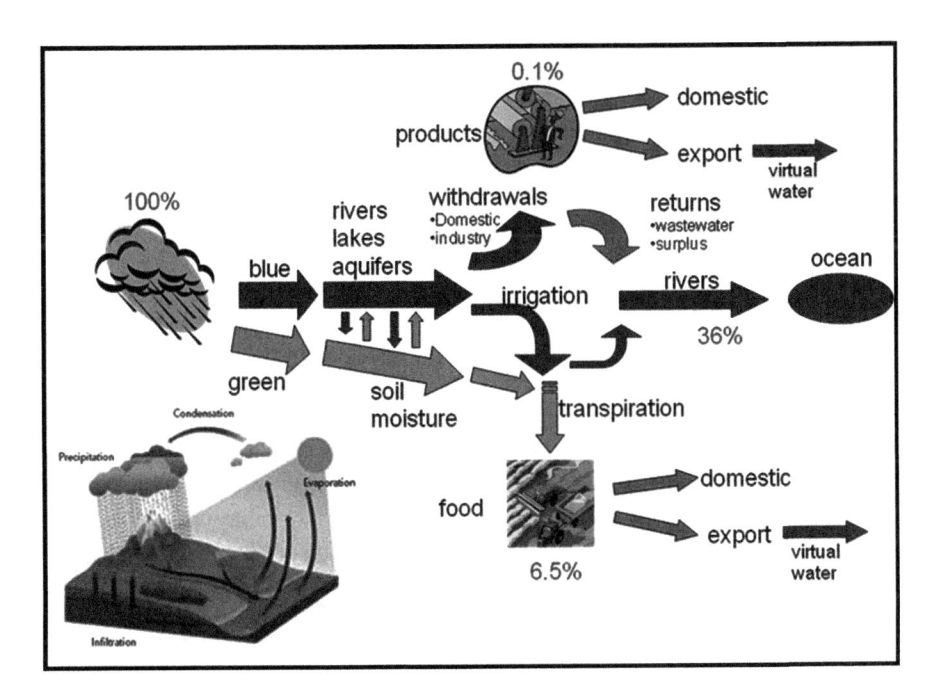

Fig. 22.2 The water cycle: including *green* water, *blue* water, *grey* water. (Source [1])

water, nearly a bath full (150 l), to produce one cup of coffee, and that water is often used in another country – such as Brazil – when the coffee is drunk in a country in Europe. One pair of cotton jeans requires 73 baths full of virtual water, which are attributable mainly to the cotton production, and that water is likely to be used in countries such as Egypt, where there are already serious water shortages.

The water footprint can be expressed in many ways; by person, nation, industry or product. Applied to nations, the concept permits assessment of external and internal footprints, often closely linked to trade. Taking the UK as an example, the average footprint of a person is estimated to be 4.3 m^3 per day, of which only 150 l, i.e. 1/30th, is that which is used in homes supplied by water companies for domestic use. The remainder is the virtual water embedded in the food eaten, the beverages drunk, the clothes worn, the cars driven, etc. Currently water footprints focus almost exclusively on volume and this approach is an excellent tool for raising awareness, but it does not necessarily represent the impacts of water use. To achieve this there is also a need to consider water stress and quality.

The embedded water footprint of the 25 European Union countries bears most heavily on India and Pakistan, which are the primary sources of cotton supply to the EU. The drying up of the Aral Sea is one example which can be partly attributed to cotton production, though this is not the only cause of the drying up of this water body. The point to appreciate, however, is that the demand for embedded water products in one country can have very serious impacts elsewhere in the world, such as Egypt, for example. International trade has the potential to save water globally if a water intensive commodity is traded from an area where it is produced with high productivity to an area with lower productivity. However, there is a continuing lack of correlation between countries hydrologically best suited to growing food or crops such as cotton and those that actually do. Furthermore, as economies of countries such as China and India continue to grow, then changing food diets and the increasing demand for clothes and material goods such as cars etc., will place an added stress on global water security, partly through the increasing imports of virtual water.

22.5   Developing Potential Solutions

Desalination is one possible solution in large coastal cities, but this process is still relatively expensive and imposes a large carbon footprint through large energy demands. Research studies being undertaken within our Hydro-environmental Research Centre at Cardiff University have also found that salinity levels can accumulate along the coastal region and this cannot be sustainable in the long term. Computational fluid dynamics simulation studies along the Arabian coast of the Persian Gulf have predicted increasing salinity levels along the coast due to the rapid growth in desalination plants in the region and this must have long term impacts for the hydro-ecology of this highly stressed water body. Field data and ecological observations confirm these numerical model predictions [11].

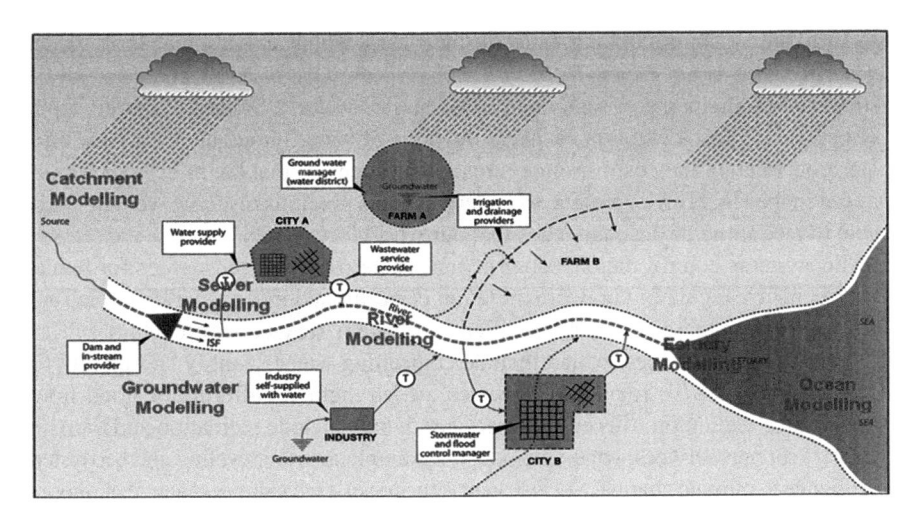

Fig. 22.3 C2C cloud to coast: integrated water management solutions

Conservation and water re-use is often a short term solution to a longer term problem. Storage involves water transfer and better integrated water management, with a much more holistic approach to river basin management being required than used hitherto. To increase global water security, improved water quality in river basins and estuarine and coastal waters is required, along with a reduction in global water pollution. It also goes without saying that global population growth needs controlling. Integrated water management requires a Cloud to Coast (C2C) approach that treats the water cycle with an integrated systems solution, bringing together the professionals who currently specialise in modelling various components of the system. They include hydraulic engineers, hydrologists, biologists and the like, with the distribution from the cloud to the coast, through the catchment, groundwater, sewers, rivers, estuaries, needing to be treated as one. The Hydro-environmental Research Centre at Cardiff University is currently refining su1ch an integrated approach with the Halcrow Group Ltd., wherein complex numerical hydro-environmental modelling tools are linked through open MI to provide integrated solutions with mass and momentum conservation at the boundary interface etc. (Fig. 22.3) [12].

Turning to the value of water, this poses the question: Is water a human right or is it an economic good? Economic theory informs us that it is easier to encourage funding if the true economic value of water is realised. Without it we get a price-cost differential and long-term sustainability becomes unlikely. However, to what extent is water a human right and, if so, whose responsibility is it to deliver it and meet the costs? True water pricing and trading is rare, but Australia and Chile have introduced it in their water scarce regions and they maintain that it has resulted in lower water consumption and significant increases in agricultural productivity. In the UK the average cost of water per cubic metre is £3 ($4.8), paid to the private water companies. This provides the consumer with approximately 1 week of water for drinking, washing, cooking, toilet flushing, car washing and in some places garden watering. This is not expensive in comparison with what else one could buy in the street for £3, including: a sandwich, 2 l of bottled water, a Starbucks coffee, a pint of beer, etc. These comparisons place the price of water into context and one must question whether the cost of water is really so expensive that the price could not be raised in the UK? If we continue to undervalue this precious resource we will not be able to face some of the challenges that our world faces today.

Ecosystems control the character of renewable freshwater resources for human well-being by regulating how precipitation is partitioned into evaporative, recharge and runoff processes. These so-called ecosystem services are categorised as: (i) provisioning services, which include controlling water quantity and quality for consumptive use; (ii) regulatory services, which include buffering of flood flows and climate regulation; (iii) cultural services, which include recreation and tourism; and (iv) support services, which include, for example, nutrient cycling and ecosystem resilience to climate change. Society generally doesn't measure or manage economic values exchanged other than through markets. The invisibility of nature's flow into the economy is a significant reason behind ecosystem degradation.

Forests and wetlands play roles in determining the level of local and regional rainfall, the ability of the land to absorb and retain water, and its quality when used. Avoiding greenhouse gas emissions by conserving forests is estimated to be worth $3.7 trillion.

In returning to the challenge of achieving Global Water Security, we believe there are five tests that need to be satisfied to achieve this goal including,

- Affordable drinking water supplies for all, to promote public health
- Sustainable sources of water for industry and its supply chain, to promote economic health
- Integrated management of water resources (including quantity and quality), for all users
- Policy and trade reforms, which encourage sustainable water resources development and which discourage conflict, and
- Mobilisation of substantial volumes of public and private funding, via transparent and fair regulatory regimes in order to fairly price water.

22.6   Conclusions

In conclusion, there are undeniable and understandable international concerns about the increasing threat of Global Water Security, in terms of the quantity and quality of water supply. These concerns need to be addressed at the international level by concerted actions, with these actions focusing on – but not exclusively – the following recommendations:

- An increasing awareness of the water-food-energy nexus and the inter-dependency of each of these resources on one another,

- The need for the water footprint and the concept of virtual water to be better understood and more widely used by: industry, governments and the public,
- Better technologies and practises are needed for more efficient agriculture, without detrimental effects on the aquatic ecosystem,
- New sustainable sources of water are needed from desalination plants (together with more energy efficiency), recycling and water harvesting,
- Water needs to be more fairly priced on the global trading markets, particularly with regard to pricing the cost of virtual water,
- Intergovernmental bodies, such as the World Trade Organisation, need to elevate issues of Global Water Security up their own and governmental agenda, and
- The Public must be more engaged to become more involved in the issues and challenges of Global Water Security.

Ultimately, water is one of the most precious resources on earth and man cannot live on this planet without this most precious of resources; for our very existence it needs to be available in quantity and of a sufficiently high quality.

References

1. Global Water Security – an engineering perspective (2010) Available at: http://www.raeng.org.uk/news/publications/list/reports/Global_Water_Security_report.pdf
2. UNICEF WHO (2009) Report – Diarrhoea: why children are still dying and what can be done. Available at: http://whqlibdoc.who.int/publications/2009/9789241598415_eng.pdf
3. National Geographic (2010) Water: our thirsty world. A special issue. Official Journal of the National Geographic Society, April, 217(4)
4. Kay D, Falconer RA (2008) Hydro-epidemiology: the emergence of a research agenda. Environ Fluid Mech 8(5–6):451–459
5. Human Development Report – Beyond scarcity: power, poverty and the global water crises (2006) Available at: http://hdr.undp.org/en/media/HDR06-complete.pdf
6. Water Facts (2011) Available at: http://water.org/learn-about-the-water-crisis/facts/. Accessed 1 Apr 2010
7. Stern N (2006) Stern review of the economics of climate change. HM Treasury, London. Archived from the original on 31 Jan 2010. Available at: http://www.webcitation.org/5nCeyEYJr. Retrieved 31 Jan 2010
8. Falconer RA (2011) Global water security: an introduction. Sci Parliament J Parliament Sci Comm 68(1):34–36
9. Halcrow C2C: cloud to coast solutions (2010) is available at: http://www.halcrow.com/Areas-of-expertise/Water/Water-security/Cloud2Coast-C2C-solutions/
10. Hoekstra AY, Chapagain AK, Aldaya MM, Mekonnen MM (2011) The water footprint assessment manual. Earthscan Publishing, London, p 201
11. Al-Osairi Y, Falconer RA, Imberger J (2010) Three-dimensional modelling hydro-environmental modelling of the Arabian (or Persian) Gulf. In: Proceedings of the 9th international conference on hydroinformatics, Vol 1. Chemical Industry Press, Tianjin, 2010, pp 132–138
12. Lin B, Wicks JM, Falconer RA, Adams K (2006) Integrating 1-D and 2-D hydrodynamic models for flood simulation. Proc Inst Civil Eng Water Manage 159(WM1):19–25

Chapter 23
Assessing Local Water Conflicts: Understanding the Links Between Water, Marginalisation and Climate Change

Lukas Ruettinger

Abstract The discourse and academic work around water conflicts is often focused on international water conflicts. As a consequence, although local water conflicts are common and affect the everyday life of many communities around the world, they are frequently overlooked. Analytical tools and trainings on local water conflicts are thus scarce. The Water, Crisis and Climate Change Assessment Framework (WACCAF) helps to close this gap. It guides users through an analysis of the different factors that play a role in local water conflicts. The goal of this tool is to better understand the conflict potential of competition around water resources, in order to prevent a water crisis from escalating into a conflict. It can also help to understand an existing water conflict and identify ways to solve it. The WACCAF specifically focuses on how the interaction between marginalisation and unequal water access and availability can create conflict (potential). These findings are then placed in wider social and historical contexts by looking at past conflicts and general marginalisation patterns in society. The analysis is completed by understanding the factors that decrease the potential for conflict, in particular cooperation and conflict resolution mechanisms.

Keywords Water conflict • Conflict management • Conflict analysis • Cooperation • Natural resources • Climate change • Marginalization

23.1 Introduction

In 1999 a conflict between two Yemeni villages over a well close to the city of Ta'iz escalated, which needed the intervention of army. Six people died and 60 left injured. In 2000, after privatising the water supply, protests erupted over deficient

L. Ruettinger (✉)
Adelphi, Caspar-Theyss-Strasse 14a, 14193 Berlin, Germany
e-mail: ruettinger@adelphi.de

H.J.S. Fernando et al. (eds.), *National Security and Human Health Implications of Climate Change*, NATO Science for Peace and Security Series C: Environmental Security, DOI 10.1007/978-94-007-2430-3_23, © Springer Science+Business Media B.V. 2012

water supply and high prices in the Bolivian city of Cochabamba. The police and military intervened, leading to violent clashes that injured 100 people and left one person dead. In 2004, after over 10,000 farmers protested against a dam on the Dadu River in the Sichuan Province of China, riot police was deployed to quell the protest, leaving at least one person dead. In 2008, during a conflict between two pastoralist clans over water boreholes in arid Northern Kenya, four people died.[1]

Local water conflicts like these are common and affect the everyday life of many communities around the world. However, much of the discourse and academic work on water conflicts focuses on international and transboundary water conflicts (for example [34, 35]). The same is true for analytical tools and trainings on water conflicts. To help close this gap adelphi has developed the *Water, Crisis and Climate Change Assessment Framework* (WACCAF) as part of the European Union's Initiative for Peacebuilding. The WACCAF provides a framework that guides the user through an analysis looking at the different factors that play a role in local water conflicts. It is based on the state of the art in the field as well as adelphi's own research. The goal of this tool is to better understand the conflict potential of competition around water resources in order to prevent a water crisis from escalating into a conflict. It can also help to understand an existing water conflict and identify ways to solve it. As such, the WACCAF can support actors working in the water sector or in crisis prevention.

This article gives an overview of the WACCAF and an introduction into local water conflicts. It starts by explaining the approach and basic assumptions of the framework. The second part illustrates the main factors playing a role in water conflicts by first focusing on the resource itself and then putting the conflict in its broader social context. Finally, factors decreasing the conflict potential – like conflict resolution mechanisms and cooperation – are introduced and explained.

23.2 Approach and Basic Assumptions

Choosing the word *crisis* instead of *conflict* as part of the name of the WACCAF points to the first important understanding the framework is based on: the importance of the context and the situation in creating conflict potential. A crisis can be understood as an unstable condition or complex and difficult situation that can change abruptly. Taking a water crisis as the starting point, the WACCAF tries to identify and understand the factors that can turn this crisis into a conflict – in other words to help understand how the potential for conflict is created. A conflict does not necessarily need to be violent: water conflicts can manifest themselves in multiple ways, from verbal exchanges and confrontations entailing damages to infrastructure or riots, to violent escalations [1, 6].

---

[1] Data from the Pacific Institute for Studies in Development, Environment, and Security database on Water and Conflict (Water Brief).

These conflicts can arise between different social groups. From the resource perspective, they can be divided into two main groups: first, water users groups, like agricultural, domestic or industrial users [3]. Examples are farmers, fishermen, paper mills and hydroelectric power plants but also individual households using water to cook, wash or for recreational activities like fishing. The second main group are natural resource management groups, referred to hereafter as management groups. They control the availability of and/or access to water resources. Management groups can either manage the resource itself, like informal local irrigation communities and village councils or public and private water service providers or they can manage ecosystems that are important for water access and availability and thus govern water indirectly, like government authorities managing wetlands or protected nature reserves. *Water access* refers to the users' need to *access* a water resource at certain points, like a river bank or water tap, in order to be able to use it. *Water availability* on the other hand refers to the water that can actually be used. It does not only mean that there is enough water (quantity) but also that it is not polluted (quality).

The main assumption the WACCAF rests upon is that competition between these groups over water access and availability can turn into conflict. This is especially pertinent if competition increases, for example through increasing demand or water scarcity. However, there is no automatism or simple causal chain of events. The likelihood that competition turns into conflict is dependent on a variety of factors and their interaction. Yet some of these factors and interactions are more important than others. Research on local water conflicts suggests that if *unequal water access and/or availability* affect *marginalised groups*, it can exacerbate already existing tensions and make communities more prone to conflict [1, 2, 4, 5, 7]. An important factor here is perception: one group has to realise that another group is impacting their water access and/or availability and it has to perceive this situation as unjust.

This is where marginalisation becomes a part of the equation. Marginalisation is the exclusion of a group from economic, social and political means of promoting one's self-determination [8] and it often goes hand in hand with unequal water access or availability. For example, the *socio-economic marginalisation* of certain groups is often connected to a lower socio-economic status. This lower socio-economic status can lead to unequal water access and/or availability, for instance urban poor who cannot afford clean water or poor farmers who do not have the financial and technical means to dig wells. Socio-economic marginalisation in turn is often connected to *political marginalisation,* as marginalised groups are often excluded from decision-making processes and do not have the same voting rights or means to express their interests in the political system. Normally, marginalisation of a certain group is not restricted to the water sector. In fact, if a group is experiencing unequal water access or availability it is more likely that this group is marginalised in general [10]. This also means that inequality in the water sector is much more likely to be perceived as unjust, since it feeds into already existing grievances, thus raising tensions. However, inequality, grievances and marginalisation do not lead only to increased tensions; they often also bring about stronger group identities both in the marginalised as well as in the more powerful group. These group identities are a powerful mobilisation resource and strategy, which can later be used to escalate a conflict, especially when it turns violent [9].

23.3 Linking Marginalisation and Unequal Water Access and Availability

Based on these assumptions the WACCAF helps to assess how marginalisation and unequal water access and availability interact. Water access and availability are dependent on number of variables. The WACCAF structures them in three sets of factors: (1) the role of water management and infrastructure, (2) the environment and human impact, and (3) the role of climate change. To assess the potential for a local water conflict, these three sets of factors are the starting point. The goal is to understand the specific links between marginalisation and unequal water access and/or availability in each of the three sets of factors. This does not mean that interactions between the different categories should be disregarded; it just serves as a structure that helps to break down the analysis into smaller tasks.

23.3.1 The Role of Water Management and Infrastructure

The first and most obvious entry point for understanding unequal water access and availability is the management of the resource itself. When trying to understand water management, it is important to analyse the institutions that set rules, allocate water, collect tariffs and set up infrastructure. A clear sign of trouble is when these water management institutions mirror the socio-economic or political marginalisation of certain groups. This might start with the government deciding to favour more lucrative and powerful economic sectors or groups as part of their development policies. Common examples are preferred water access for large-scale agriculture, industries or urban developments. This often leads to negative impacts on the water access and availability of small scale farmers or poorer population groups.

Corruption can aggravate this marginalisation by giving more powerful groups and resource-rich the possibility to influence water management institutions. They can ignore rules and regulations allowing them to deplete and pollute water stocks, exploit and destroy valuable ecosystems and steal money that was meant to build and maintain public water infrastructure. However, poor governance that leads to unequal water access and availability can also be due to insufficient financial, technical and managerial capacities. For example, agricultural water supply is often not taken into account in urban planning or infrastructure is set up in ways that later lead to conflicts [11]. Also, the lack of water management institutions can easily lead to problems, through overuse of the resource.

An especially conflict-prone situation is when water management changes: a well publicised example of this is the privatisation of the water supply in the Bolivian city of Cochabamba. Discontent over rising prices, especially among the urban poor, met unresolved grievances built by years of marginalisation and conflict. This lead to protests and violent clashes between protesters, law enforcement and the military, leaving more than a hundred people injured and one person dead [12].

23.3.2 Environment and Human Impact

Another way of looking at water access and availability is by analysing the impacts humans have on the environment. This can happen directly through pollution and overuse, but also indirectly via the destruction of ecosystems that provide important functions in the water cycle – such as forests or wetlands, which serve as water storage and filters [13]. This can become problematic when one group impacts or controls an ecosystem or water resource in a way that restricts the water access and availability of another group. In this regard the control over land and land ownership are especially important, because they often give one group the possibility to impact or control a water resource or ecosystem. Examples of impacts include individual land owners who decide to cut down a forest, or governments restricting access to wetlands by declaring them a nature reserve. A good starting point is to analyse how certain governance institutions and policies influence these dynamics. Again, formal and informal governance institutions, and their policies, can mirror the marginalisation of certain groups. This can be very obvious, as in the case of land tenure systems that disadvantage certain groups [14]. It can also be more indirect, such as subsidies for only certain farmers, which allow them to dig deeper wells; this in turn lowers the water table leading to less water for poor subsistence farmers who do not have the capital, nor received the subsidies, to dig deeper wells [1, 2].

Governance problems like corruption, lacking capacities and failures in implementing environmental legislation often create or aggravate pollution, overuse and ecosystem degradation [10]. For example, in Peru, the government policies to expand the extractive industry sector combined with deficient environmental regulation have lead mining companies to use poor environmental practices, many resulting in water pollution. This has created conflicts with local communities that have in the past escalated in the killing of environmental activists and farmers [15].

23.3.3 The Role of Climate Change

The last set of factors that is important to understand in regard to water access and availability is climate change. From a security perspective climate change is often understood as a threat multiplier. This means that climate change increases the conflict potential by putting additional stress on a crisis situation [16, 17]. With regard to water, global warming is predicted to accelerate the process by which water is transferred within our climate system. This will most likely cause substantial changes in precipitation patterns and intensity, impacting water users that depend on rain as a water source, for instance farmers using rain-fed agriculture. Also, extreme weather events like droughts, floods and storms will increase. At the same time, rising temperatures will accelerate the melting of ice and snow cover impacting the timing and discharge rate of river flows [3, 18–20]. These effects can interact, as in the case of India where climate change is altering groundwater recharge rates by accelerating the melting of Himalayan glaciers combined with erratic precipitation falls [21, 22].

Another major impact of climate change is the expected rise of sea levels caused by the melting of glaciers, ice caps and ice sheets coupled with the thermal expansion of the sea due to higher water temperatures. This will lead to floods along coastal areas as well as salt water intrusions into coastal aquifers, thus affecting water quality [3, 18–20].

The consequences of climate change are often exacerbated by other human impacts or environmental factors. In many places people settle too close to coastal areas or flood plains, thus increasing their vulnerability to the effects of climate change. Additionally, ecosystems that provide protection from extreme weather events – such as mangroves and coral reefs that provide protection against floods – as well as ecosystems that regulate the local climate, like forests, may be destroyed [23]. These dynamics can be connected to socio-economic marginalisation, since poor population groups often have no alternative to settling in marginal areas or to using ecosystems in an unsustainable way, in order to provide for their livelihoods. Also, poor population groups might not have the financial and technical capacities to develop resilience or adapt to climate change [25]. For example, subsistence farmers normally depend on agriculture as their sole source of income. If this income source breaks away because of changes in water availability, they will likely not be able to adapt, due to insufficient education or financial capital. Lacking access to financial and technical capacities for climate change adaption can also be a sign for political marginalisation. While adaption measures can help to reduce the vulnerability of poor and marginalised communities [24], they may also lead to new conflicts, for example, if new dams are built to improve water supply or if demand management puts restrictions on certain water users.

23.4 The Broader Context

Understanding the factors influencing water availability and access, and the role marginalisation plays, is an important step. However, it is not enough for understanding the potential for conflict. The social and historical contexts beyond water availability and access are just as important. As mentioned before marginalisation is rarely limited to water availability and access. Most of the time, it is connected to broader marginalisation and past injustices might have already led to conflicts around other issues. In general, the potential for water conflict is much higher if there is already a history of conflict or if there are ongoing conflicts. Therefore, it is important to understand how broader marginalisation and past conflicts play into the emergence of disputes over water. Often, there are 'hidden conflicts' underlying the water conflict [26]. Examples are civil wars, independence movements or conflicts and tensions along ethnic lines. These past or ongoing conflicts have most likely led to polarization and strong group identities, especially if violence was involved [27]. Often political leaders play a decisive role in these conflicts. They can help create or aggravate cleavages between groups by using them to mobilize their constituency.

Besides conflicts between user groups, conflicts in the region or neighbourhood can also increase the potential for conflict. For example, neighbouring conflicts can create refugee flows, leading to increased competition over resources in the receiving country. Not only refugees cross borders, however, and often small arms become more easily available providing the means to turn a conflict violent; even the conflict itself might "spill over" the border [28], for example if rebels use neighbouring countries as a base or retreat area [29, 30].

The analysis of the broader context thus links the local water conflict to bigger conflict structures and factors beyond the water sector. Without such an analysis any assessment of a water conflict will be incomplete.

23.5 From Conflict to Cooperation

However, it is not enough to simply look at the factors that contribute to an increase of conflict potential. It is just as important to understand the factors that decrease the conflict potential or manage conflicts in a non-violent manner. Conflicts as such are a normal part of society. They can be a powerful and important catalytic force of change and of righting wrongs such as the marginalisation of social groups. Conflict resolution mechanisms, if they are legitimate, inclusive, representative, and transparent can help to manage a conflict in a non-violent manner. However, if perceived as partisan, illegitimate, corrupt or unrepresentative, they will most likely create new grievances contributing to the potential for conflict.

Besides conflict resolution mechanisms, another important mechanism to preventing conflict is cooperation between (potential) conflict parties. This includes business and trade but also cultural associations and joint resource management institutions. Over the long term cooperation opens communication channels, builds trust and creates relationships. These channels and relationships can later be used to manage disagreements [31, 32]. This also works the other way around: cooperation in the field of environment and water can reduce the potential for other conflicts and thus be a tool for peacebuilding and conflict prevention beyond the water sector.

A crucial question in this regard is how to get conflicting parties to sit down and start cooperating if they do not want to. Cooperation only works if all parties have an interest in cooperating. Thus a thorough understanding of the different interests and positions can provide a starting point. This understanding can be used to create an interest in cooperating, either by providing benefits and incentives to cooperate or by punishing non-cooperation. It is very important that any attempt to prevent or manage a conflict is done in a way that does not create any unintended and aggravating consequences. In the world of development cooperation this concept is called 'do no harm' [33].

23.6 Connecting the Dots

After (1) assessing the role of marginalisation and the factors that lead to unequal water access and/or availability; (2) putting this analysis in its broader social context by looking at broader marginalisation and other conflict structures; and (3) identifying factors that can decrease the conflict potential, all these dots have to be connected into a comprehensive assessment of the water conflict (existing or potential). While the WACCAF guides through the assessment of a (potential) water conflict by pointing out the different factors that are important to look at and illustrating common conflict dynamics, this part is up to the user of the WACCAF. Every conflict is unique and highly dependent upon the context, thus making it impossible to provide a fixed blueprint for an assessment. Thus the WACCAF should not be understood as a straightjacket for the analysis but more as a guide that provides the main dots; connecting them is the task of the framework's user.

There are many ways the final assessment can help to solve local water conflicts or help to prevent a water crisis from escalating into conflict: first, it can help raise awareness among conflicting parties and decision makers or help develop more conflict-sensitive water infrastructure and development projects; second, the WACCAF itself can also be used to build capacities by using it as part of a training course on local water conflicts; lastly, it can be used for a broader academic research on water conflicts, to further our understanding of the dynamics that turn a water crisis into a conflict.

References

1. Houdret A (2008) Scarce water, plenty of conflicts? Local water conflicts and the role of development cooperation, Institut für Entwicklung und Frieden, Duisburg. http://inef.uni-due.de/page/documents/PolicyBrief03_en.pdf. Accessed 4 June 2010
2. Houdret A, Kramer A, Carius A (2010) The water security nexus: challenges and opportunities for development cooperation, Deutsche Gesellschaft für Technische Zusammenarbeit (GTZ), Eschborn
3. Gleick PH et al (2009) The world's water 2008–2009. Island Press, Washington, DC
4. Gehrig J, Rogers MM (2009) Water and conflict: incorporating peacebuilding into water development. Catholic Relief Services, Baltimore
5. Richards A (2002) Coping with water scarcity: the governance challenge. Institute on Global Conflict and Cooperation, San Diego
6. Yasmi Y, Schanz H, Salim A (2006) Manifestation of conflict escalation in natural resource management. Environ Sci Policy 9(6):538–546
7. Lecoutere E, D'Exelle B, Van Campenhout B (2010) Who engages in water scarcity conflicts? A field experiment with irrigators in semi-arid Africa. MICROCON, Brighton. Available at: http://www.microconflict.eu/publications/RWP31_EL_BD_BVC.pdf. Accessed 25 Jan 2011
8. Burton M, Kagan C (2005) Chapter 14 Marginalization. In: Nelson G, Prilleltensky I (ed) Community psychology: in pursuit of liberation and well-being. Palgrave Macmillan, London, 293–328
9. UNDP (2004) Human development report 2004: cultural liberty in today's diverse world. Oxford University Press, New York

10. UNDP (2006) Human development report 2006: beyond scarcity – power, poverty and the global water crises. United Nations Development Programme, New York
11. Harris K (2008) Water and conflict: making water delivery conflict-sensitive in Uganda, Saferworld, CECORE, REDROC, YODEO. Available at: http://www.saferworld.org.uk/publications.php/355/water_and_conflict. Accessed 5 June 2010
12. Beltrán EP (2003) Water, privatization and conflict: women from the Cochabamba valley. Heinrich-Böll-Stiftung, Washington, DC
13. Millennium Ecosystem Assessment (2005) Ecosystems and human well-being: synthesis, 1st edn. Island Press, Washington DC
14. Amman HM, Duraiappah AK (2004) Land tenure and conflict resolution: a game theoretic approach in the Narok district in Kenya. Environ Dev Econ 9(03):383–407
15. Bebbington A, Williams M (2008) Water and mining conflicts in Peru. Mt Res Dev 28(3/4):190–195. doi:10.1659/mrd.1039, Aug–Nov 2008
16. Carius A (2009) Climate change and security in Africa challenges and international policy context, paper commissioned by the United Nations Office of the Special Adviser on Africa (OSAA), Expert group meeting "Natural resources, climate change and conflict in Africa: protecting Africa's natural resource base in support of durable peace and sustainable development", 17–18 Dec 2009, conference room E, United Nations, New York
17. Brown O, Crawford A (2009) Climate change and security in Africa a study for the Nordic-African foreign ministers meeting
18. Bates B et al (2008) Climate change and water. IPCC Secretariat, Geneva
19. Ludwig F et al (2009) Climate change adaptation in the water sector. Earthscan, London and Sterling
20. Kabat P, Bates B (2003) Climate changes the water rules – how water managers can cope with today's climate variability and tomorrow's climate change. WaterandClimate, Netherlands
21. Zemp et al. (2008) Global glacier changes: facts and figures, World Glacier Monitoring Service (WGMS), Zurich, Switzerland. http://www.grid.unep.ch/glaciers/pdfs/cover.pdf. Accessed 6 Mar 2010
22. Prakash A, Sama R (2006) Social undercurrents in a water-scarce village. http://conflicts.indiawaterportal.org/sites/conflicts.indiawaterportal.org/files/conflicts_vadali.pdf. Accessed 6 Mar 2010
23. Trumper K et al (2009) The natural fix?: the role of ecosystems in climate mitigation, a UNEP rapid response assessment, United Nations, Cambridge
24. UNFCCC (2007) Climate change: impacts, vulnerabilities and adaptation in developing counties. United Nations Framework Convention on Climate Change, Bonn
25. Raleigh C (2010) Political marginalisation, climate change, and conflict in African Sahel states. Int Stud Rev 12(1):69–86
26. Means K et al (2002) Community-based forest resource conflict management. A training package. Food and Agriculture Organization of the United Nations, Rome
27. Enns D (2007) Identity and victimhood. Questions for conflict management practice, Berlin, Germany: Berghof occasional paper no. 28. http://www.berghof-conflictresearch.org/documents/publications/boc28e.pdf. Accessed 6 Mar 2010
28. Geiss R (2009) Armed violence in Fragile states: low-intensity conflicts, spillover conflicts, and sporadic law enforcement operations by third parties, ICRC
29. Smith D (2004) Trends and causes of armed conflict, Berghof Conflict Research. http://berghof-handbook.net/documents/publications/smith_handbook.pdf. Accessed 6 Mar 2010
30. Fearnely L, Chiwandamira L (2006) Understanding armed conflict and peacebuilding in Africa. IDASA, Pretoria
31. Mathieu P, Benali A, Aubriot O (2001) Dynamiques institutionnelles et conflit autour des droits d'eau dans un système d'irrigation traditionnel au Maroc. Revue Tiers Monde 42(166):353–374
32. FOEME (2005) Good water neighbors: a model for community development programs in regions of conflict. Developing cross border community partnerships to overcome conflict and advance human security. Friends of the Earth Middle East, Amman/Bethlehem/Tel Aviv

33. OECD (2001) The DAC guidelines helping prevent violent conflict. Available at: http://www. oecd.org/dataoecd/15/54/1886146.pdf. Accessed 6 Mar 2010
34. Wolf AT, Yoffe SB, Mark G (2003) International waters: identifying basins at risk. Water Pol 5(1):29–60
35. Wolf AT, Kramer A, Dabelko G, Carius A (2005) Managing water conflict and cooperation. In: The Worldwatch Institute (ed) State of the world 2005: redefining global security. WW Norton & Company, New York & London

Chapter 24
Climate Change Impacts on River Catchment Hydrology Using Dynamic Downscaling of Global Climate Models

Ch. Skoulikaris and J. Ganoulis

Abstract Impacts from climate change may affect vital factors of environmental and human security related to water uses, such as domestic water supply, hydropower and industrial production, agricultural irrigation and ecosystems needs. This is particularly important in regions with arid and semi-arid climates like the Mediterranean and the South Eastern Europe. In this presentation the coupling of different climate change models with a distributed hydrological model was developed in order to explore the impact of climate change on water resources at the river basin level. Firstly, the coupled atmosphere-ocean global climate model ECHAM5/MPIOM, developed by the Max Planck Institute for Meteorology in Hamburg, Germany, was used to provide boundary conditions of the regional climate model CLM. Simulation results at 6 hourly intervals were provided for Europe, using a spatial grid resolution of 20×20 km. Secondly, the spatially distributed hydrological model MODSUR-NEIGE (MODélisation des transferts de SURface en présence de NEIGE, in French), developed by the School of Mines, Paris, France, was used for dynamically downscaling boundary conditions provided by CLM over a spatially variable grid ranging between 250 m and 2 km in size. In this way temperature, precipitation and evapotranspiration distributions were adapted to local conditions, such as the river watershed relief, local geology and land uses. The methodology is illustrated for the Mesta/Nestos river basin, which is shared between Bulgaria and Greece and is part of the worldwide UNESCO-HELP initiative. The upstream northern part of the basin (Mesta) is in Bulgaria, and the downstream part (Nestos)

Ch. Skoulikaris • J. Ganoulis (✉)
Department of Civil Engineering, Aristotle University of Thessaloniki,
54124, Thessaloniki, Greece
e-mail: iganouli@civil.auth.gr

H.J.S. Fernando et al. (eds.), *National Security and Human Health Implications of Climate Change*, NATO Science for Peace and Security Series C: Environmental Security, DOI 10.1007/978-94-007-2430-3_24, © Springer Science+Business Media B.V. 2012

is in Greece. Impacts from climate change have resulted in a significant reduction of river flow with serious consequences for hydropower production and agricultural activities mainly near the Nestos delta.

Keywords Climate change • Modelling • Dynamic downscaling • Impacts • Hydrology

24.1 Introduction

Scale effects are crucial for estimating impacts from climate change on the river basin hydrology. At the river catchment scale, important parameters are the catchment's more or less complex orography, hydrography, geology, soil characteristics and land use. In order to take into account the spatial distribution of these parameters, hydrological models should use grid resolution ranging usually from 100 m to 1 km.

Climate models simulate changes of climate variables, such as temperature, air moisture, precipitation and solar radiation at horizontal grids of a larger size, usually ranging from 200 to 500 km. Such large scale resolution of climate characteristics is based on Global Circulation Models (GCMs), which consider the movement of the whole terrestrial atmosphere and result in global or synoptic climate features with a very rough approximation of future variations in hydrological characteristics at the river basin scale.

Recent progress in climate modelling simulations report results on grids of sizes ranging from 1 to 50 km [7, 9]. These models, called Regional Climate Models (RCMs) or Local Climate Models (LCMs) result from dynamic downscaling of the global climate variables and could be used in order to provide the boundary conditions for distributed hydrological models at the river basin scale.

In this presentation the physically-based hydrological distributed model MODSUR-NEIGE is coupled with the European local climate model CLM in order to simulate impacts from climate change on hydrological characteristics at the river catchment scale. The simulation results were obtained for the transboundary river basin of Mesta/Nestos, shared between Bulgaria and Greece.

As shown in Fig. 24.1, impacts from climate change on hydrological characteristics may generally be obtained in three steps:

1. Concentrations in the atmosphere of green house gases emissions can be estimated for the coming decades by use of chemistry models. This study used the scenarios of the concentration of future emissions adopted by the International Panel of Climate Change (IPCC),
2. Local Climate Models (LCMs) simulations are obtained by dynamic downscaling of Global Circulation Models (GCMs),
3. Results of LCMs are introduced as boundary conditions to a distributed hydrological model in order to obtain climate change impacts on the river's hydrology.

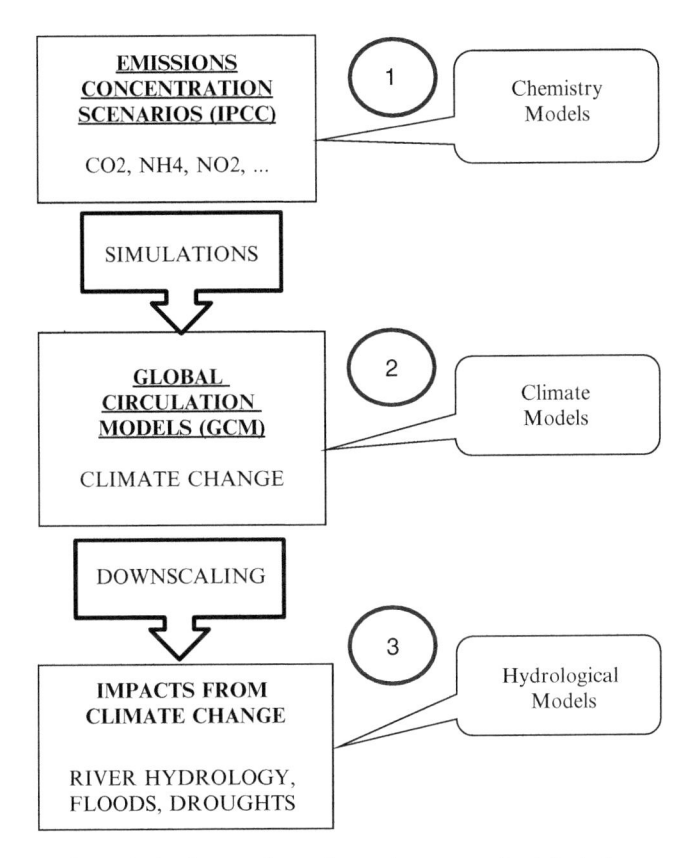

Fig. 24.1 Steps for analysing impacts from climate change on a river's hydrology

24.2 Climate Modelling Downscaling

Dynamical downscaling is obtained by introducing, as boundary conditions of specific Regional Climate Models RCMs ("forcing"), the results produced by Global Circulation Models (GCMs) [9]. In our area of study, namely Greece and Bulgaria, the CLM regional climate model was used, which was developed by dynamically downscaling GCMs at the European scale. CLM stands for the Climate version of the "Local Model" (CLM). It is a non hydrostatic climate model and has been used in Europe for simulations on time scales up to centuries and spatial resolutions between 1 and 50 km. The boundary conditions of the CLM are provided at 6 hourly intervals by the simulation results of the coupled atmosphere-ocean global climate model ECHAM5/MPIOM, developed by the Max Planck Institute for Meteorology in Hamburg, Germany.

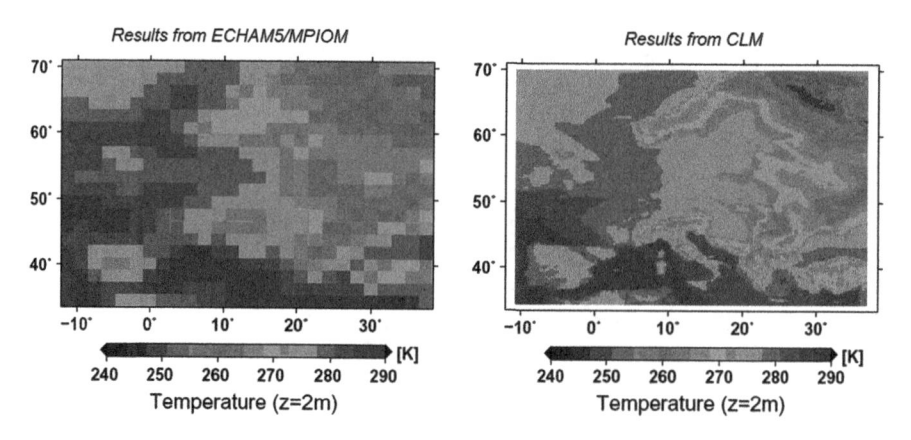

Fig. 24.2 Temperatures computed at a height of 2 m asl over Europe on 01-07-1984 at 06:00 h

The ECHAM5 model [4, 6] is an updated version of the ECHAM4 spectral model and uses a T21 Gaussian grid with a spatial resolution equivalent to 5.6° longitude × 5.6° latitude and 19 atmospheric layers (L19). It is coupled with the MPIOM ocean-sea ice component [7]. MPIOM is a simplified equation model (C-Grid, z- coordinates, free surface) using hydrostatic and Boussinesq fluid hypotheses. The horizontal resolution of MPIOM varies gradually between a minimum of 12 km close to Greenland and 150 km in the tropical Pacific. The ECHAM5/MPIOM model has been adopted by IPCC as one of the models used for the simulation of the SRES scenarios of the Fourth Assessment Report (AR4).

In Fig. 24.2 a comparison is shown between the temperature fields generated by ECHAM5/MPIOM and CLM for the same time and date. The CLM grid resolution is 0.165° (data stream 2), 0.2° (data stream 3), which is approximately 20 × 20 km. The output clearly indicates that CLM provides more spatially refined results than the ECHAM5/MPIOM model. The average of the differences between CLM and ECHAM5/MPIOM values reveal a bias of about –2.5 K and about 1 mm/day, i.e. the CLM simulates lower temperature and more precipitation than the GCM.

24.3 Hydrological Downscaling

Results of simulations of CLM under climate change conditions using IPCC scenarios of greenhouse gases emissions were used as forcing conditions for running the MODSUR-NEIGE distributed hydrological model (Fig. 24.3).

MODSUR is a physically based distributed hydrological model, developed at the Ecole Nationale Supérieure des Mines de Paris [2, 8] in order to simulate the spatial and temporal evolution of a river and the water table flows. Since the amount of snow in the Rila and Pirin mountains of Bulgaria at the head of the basin is

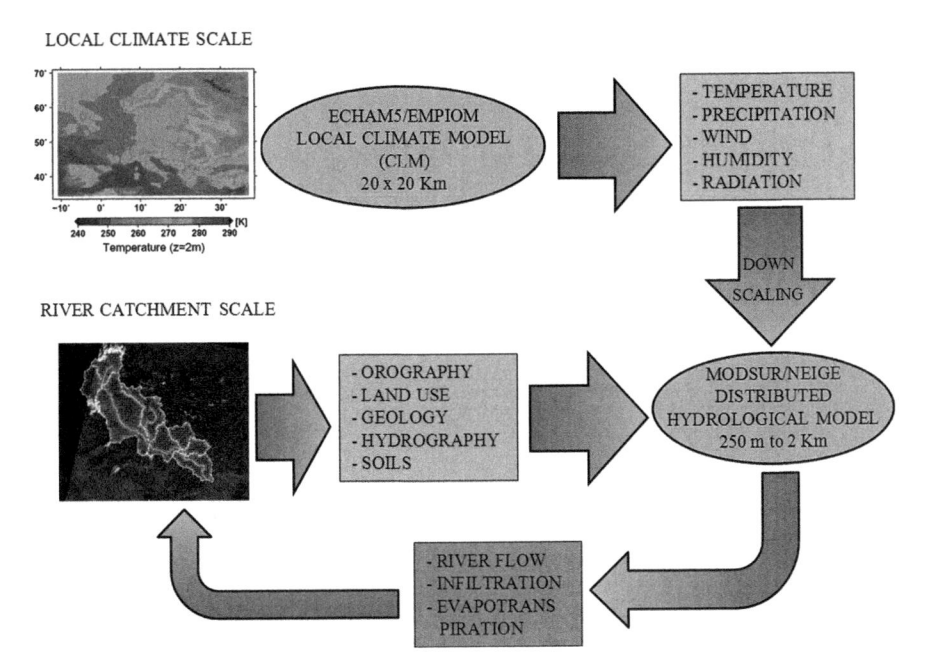

Fig. 24.3 Coupling the MODSUR/NEIGE hydrological model to the local climate model LCM

important for the Mesta/Nestos river flow, a more advanced version of the model, known as MODSUR-NEIGE, was used in order to also account for the snow cover regime.

The model is based on a dense spatial grid (Fig. 24.3) made of variably sized square cells. Characteristics of the surface domain (runoff directions, altitude, soil and land-use) are attached to each cell. The grid topology is based on the so-called four neighbours rule (four-connectivity). Each cell may only be connected to cells of the same dimension, or cells which are four times larger or four times smaller [10]. The surface water is transferred through the runoff network or networks to the catchment outlet. The ensemble of connected cells builds a runoff network, which directs the flow down to the catchment outlet.

24.4 Results of Simulation

For several years the subject of climate change and its influence in the area of southwest Bulgaria [1] and the Mesta in particular [3] have been investigated on the various aspects of mountain forest ecology and water resources.

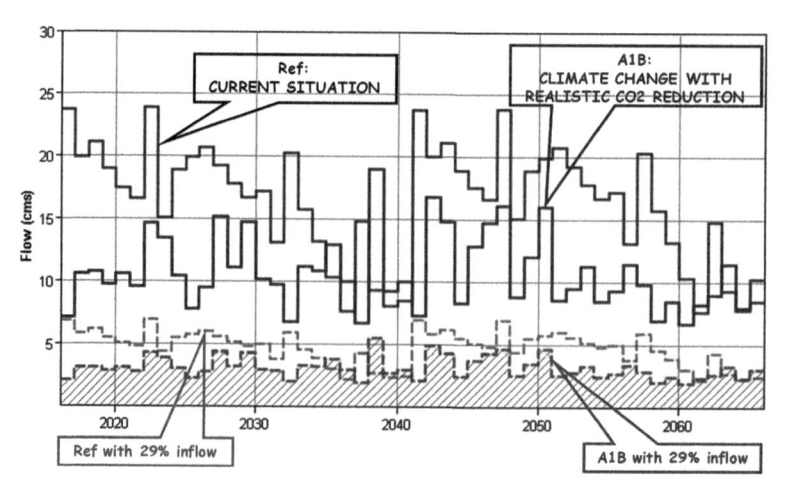

Fig. 24.4 Water inflows from Bulgaria to Greece for the Ref and A1B climate scenarios. 29% inflow from Bulgaria refers to the bilateral agreement between the two countries

Our study of climate change impacts on the water regime of the Mesta/Nestos river basin was conducted with the use of the IPCC/A1B scenario [5] for the period between January 2016 and December 2065. The A1B scenario is based on the realistic assumption of adequate reduction of CO2 emissions if in the future fossil fuel and green energy technologies are used equitably. In order to compare the data produced by the coupling of the climate change and hydrologic models, it was decided to create a "reference" flow regime, namely Ref, by duplicating and projecting in the future a stable past flow regime (January 1970 to December 1995), which would be representative of the average conditions of flow during the contemporary period. Additionally, different scenarios based on the water inflows from the upstream part of the basin in Bulgaria to the Greek part downstream were developed in order to explore future water uses in the Greek part of the basin. It should be noted that according to the bilateral treaty of 1995, known as the "Agreement between the Government of the Hellenic Republic and the Government of the Republic of Bulgaria for the Waters of River Nestos", which was ratified in Sofia on December 22nd, 1995 (Law 2402/1996, OG 98/A/4-6-1996), it is stipulated that Greece should receive 29% of the runoff from Bulgaria. However, in reality Greece currently receives 85–90% of the Mesta waters [10].

The simulation procedure demonstrated that the available water volumes in the Ref climate scenario are more plentiful than those in the A1B scenario; see Fig. 24.4. The average flow is equal to 16.32 m³/s and to 10.72 m³/s in the Ref and A1B scenario, respectively. On the other hand, the A1B scenario has more extreme flows, both in terms of timing and frequency, with a maximum flow of 81.20 m³/s in the year 2022. In the Ref scenario the same value is equal to 60.90 m³/s and is predicted to occur in 2019. As for the case where Greece receives 29% of the total upstream runoff, phenomena of extended and extensive water scarcity are obvious since the annual average river flow at the border is less than 5 m³/s in both scenarios.

24.5 Conclusions

Using a dynamic downscaling technique, temperature, precipitation and evapotranspiration characteristics were estimated at the river catchment scale for the next 50 years. The global climate model ECHAM5/MPIOM, which uses 6 hourly intervals and a grid resolution typically ranging between 200 and 500 km, was nested with the CLM regional climate model of 20×20 km spatial resolution. Both models were developed by the Max Planck Institute for Meteorology, Germany and adopted by IPCC. The results obtained from the CLM were introduced into the distributed hydrological model MODSUR-NEIGE in order to simulate daily river flows with a spatial grid resolution ranging from 250 m to 2 km. The results of simulating different climate change scenarios for the Mesta/Nestos river basin show that future time series of average annual river flow decrease, which means difficulties in ensuring future socio-economic security in the region mainly in terms of producing sufficient hydro-electrical energy and meeting the irrigation needs in the Nestos river delta region.

References

1. Alexandrov V, Genev M (2003) Climate variability and change impact on water resources in Bulgaria. Eur Water 1/2:25–30
2. Golaz-Cavazzi C (1995) Exploitation d'un modèle numérique de terrain pour l'aide à la mise en place d'un modèle hydrologique distribué. DEA UPMC, 71pp
3. Grunewald K, Scheithauer J, Monget J-M, Brown B (2008) Characterization of contemporary local climate change in the mountains of southwest of Bulgaria. Climatic Change 95(3–4): 535–549
4. Roeckner E et al (2003) The atmospheric general circulation model ECHAM 5. Part I: model description. MPI-Report 349
5. IPCC (2001) In: Houghton JT, Ding Y, Griggs DJ, Noguer M, van der Linden PJ, Dai X, Maskell K, Johnson CA (eds), Climate change 2001: the scientific basis: contribution of working group I to the third assessment report of the intergovernmental panel on climate change. Cambridge University Press, Cambridge and New York, 881 pp
6. Jungclaus JH, Botzet M, Haak H, Keenlyside N, Luo J-J, Latif M, Marotzke J, Mikolajewicz U, Roeckner E (2006) Ocean circulation and tropical variability in the coupled model ECHAM5/MPI-OM. J Clim 19(16):3952–3972
7. Kotlarski S, Block A, Böhm U, Jacob D, Keuler K, Knoche R, Rechid D, Walter A (2005) Regional climate model simulations as input for hydrological applications: evaluation of uncertainties. Adv Geosci 5:119–125
8. Ledoux E, Girard G, de Marsily G, Deschenes J (1989) Spatially distributed modeling: conceptual approach, coupling surface water and ground water. In: Morel-Seytoux HJ (ed) Unsaturated flow hydrologic modeling – theory and practice, NATO ASI Series S 275. Kluwer, Boston, pp 435–454
9. Mearns LO, Giorgi F, Whetton P, Pabon D, Hulme M, Lal M (2003) Guidelines for use of climate scenarios developed from regional climate model experiments. DDC of IPCC TGCIA. http://www.ipcc-data.org/guidelines/dgm_no1_v1_10-2003.pdf
10. Skoulikaris Ch (2008) Mathematical modelling applied to the sustainable management of water resources projects at a river basin scale: the case of the Mesta-Nestos. Joint PhD thesis, Aristotle University of Thessaloniki, Greece and Mines ParisTech, Paris

Chapter 25
Implications of Climate Change for Marginal and Inland Seas

Peter O. Zavialov, Andrey G. Zatsepin, Peter N. Makkaveev, Alexander Kazmin, Vyacheslav V. Kremenetskiy, and Vladimir B. Piotuh

Abstract As confined water bodies, the semi-enclosed and inland seas are particularly vulnerable to the global change. The climate change trends observed in the marginal seas reflect a spectrum of interactions between the ocean, atmosphere and the continents, and this is why their responses are generally more complex than those characteristic for the open ocean. In this article, we start with the ongoing climate change processes in the deep ocean versus the semi-enclosed and marginal seas, and then discuss case studies based on the recent data collected from marginal or inland seas, namely, the Black Sea, the Kara Sea, and the Aral Sea. We show, in particular, that over much of the last few decades, the sea surface temperature changes in the Black Sea had the opposite sign compared to those in the world ocean. The sea temperature within the uppermost oxygenized layer of the Black Sea exhibited significant correlation with North Atlantic Oscillation. Furthermore, we show that in the Kara Sea, which is poorly covered by observational data, pH is likely to have been growing during the last two decades, in contrast with the global acidification trends, possibly, in connection with the long-term variability of fluvial runoffs from Yenisey and Ob rivers modulated by wind forcing. The Aral Sea represents an extreme response of a large inland water body to climate change and anthropogenic impacts. The Aral Sea level has dropped over 26 m since 1960, and its volume decreased by a factor of 10. The desiccation was primarily caused by anthropogenic diversions of water from the tributary rivers, but about 30% of the level drop was associated with climate change at the regional scale. We discuss the ongoing changes in the ionic salt composition of the Aral Sea water accompanying the desiccation.

Keywords Climate change • Marginal and inland seas • Land-sea-air interactions

P.O. Zavialov (✉) • A.G. Zatsepin • P.N. Makkaveev • A. Kazmin • V.V. Kremenetskiy
• V.B. Piotuh
P.P. Shirshov Institute of Oceanology, Russian Academy of Sciences, Moscow, Russia
e-mail: peter@ocean.ru

H.J.S. Fernando et al. (eds.), *National Security and Human Health Implications of Climate Change*, NATO Science for Peace and Security Series C: Environmental Security, DOI 10.1007/978-94-007-2430-3_25, © Springer Science+Business Media B.V. 2012

25.1 Introduction

The effects of climate change in the ocean are well established now. Their principal manifestations include the growth of temperature and net heat content in the upper layer of the ocean, the sea level rise (associated with the thermal expansion of water and ice melting), rather complex changes in salinity pattern and meridional overturning circulation, positive trends in wave height and steepness at most locations, and pronounced ocean acidification [4]. For the period from 1955 to 2003, the increase of the heat content in the upper 700 m layer is estimated as $10.9 \cdot 10^{22}$ J, which corresponds to the trend of 0.14 Wm^{-2} (Fig. 25.1). The observed changes in salinity show similar patterns in different ocean basins. The subtropical waters have became saltier and the subpolar surface and intermediate waters have freshened in the Atlantic and Pacific Oceans during the period from the 1960s to the 1990s [4]. The global sea level rose at about 1.7 mm $year^{-1}$ during the twentieth century, with the maximum trends documented for the northwestern Indian ocean and eastern Pacific. The sea level is projected to rise during the twenty-first century at a greater rate, most likely 2–4 mm $year^{-1}$. The increased uptake of the atmospheric CO_2 by the ocean has resulted in the ocean acidification worldwide. To date, the global average surface pH has decreased by about 0.12 over the instrumental record period, and is projected to further decrease by 0.15–0.35 at the global scale until 2100 (Fig. 25.2).

However, the climate change processes in the marginal and inland seas (hereinafter MIS) are much less well-understood and quantified. Do the ongoing trends and future projections mentioned above for the ocean also hold for the MIS? In this paper, we address three illustrative case studies. First, we discuss the sea surface

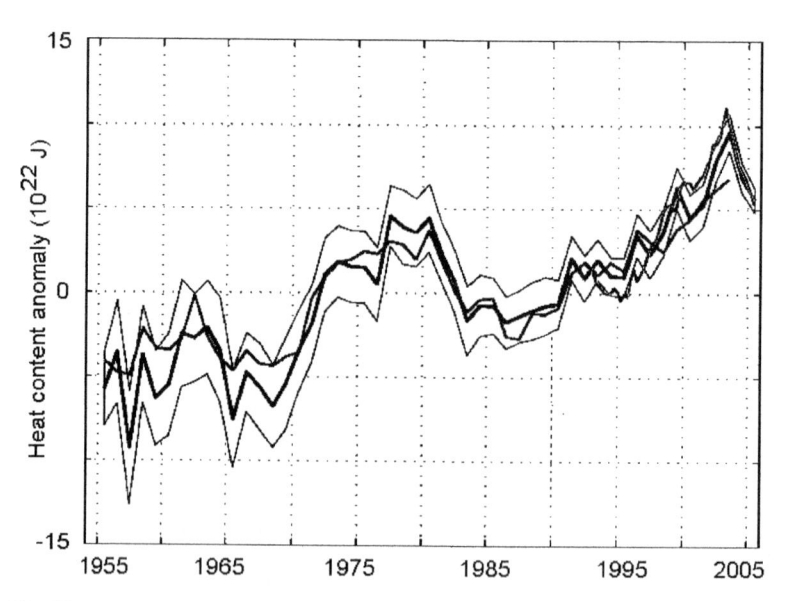

Fig. 25.1 Time series of global annual ocean heat content (10^{22} J) for the 0–700 m layer (Modified from [4])

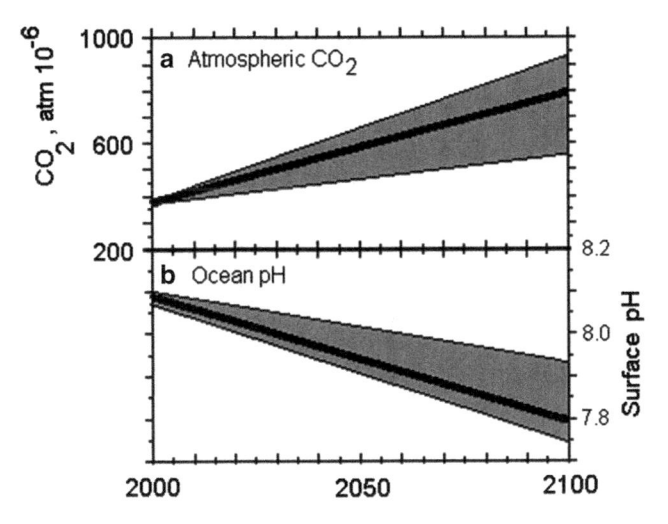

Fig. 25.2 Projected changes in the atmospheric GO2 and global ocean surface pH for different scenarios of climate change until 2100 (Drawn based on the data given in [4])

temperature (SST) variability in the Black Sea, based on numerous monitoring cruises of the last two decades, as well as the available historical data. We then address the changes of pH in the Kara Sea, a marginal sea in the Russian Arctic strongly exposed to continental discharges, as observed in two research expeditions of Shirshov Institute separated by an interval of 14 years. Finally, based on a number of field campaigns in the area, we discuss in some detail the desiccation of the Aral Sea, a large marine-type inland water body whose water budget turned deficit because of both anthropogenic impacts and climate change.

25.2 Data

The data on the Black Sea used in this study were obtained in several cruises of R/V *Akvanavt* of the Shirshov Institute in 1997–2007 (Fig. 25.3). The total number of hydrographic stations occupied during this period was 727. We also use historical data sets such as MCSST (weekly mean multichannel SST at ~18 km resolution based on AVHRR night-time measurements for the period 1982–2002), NCEP/NCAR reanalysis data (monthly mean SST, air temperature and surface wind components on 1 degree grid for the period 1950–2005), and GISST (monthly SST on 1 degree grid for the period 1950–1994).

The data on the Kara Sea were collected during two research cruises to the area, one in September of 1993 (R/V *Mendeleev*), and the other one in September of 2007 (R/V *Keldysh*). The areas sampled in the both expeditions were located in the south-western part of the sea off the Yamal Peninsula and north of it to Novaya Zemlya, and the eastern region adjacent to the Ob River mouth and the area north of it up to St. Anna trench.

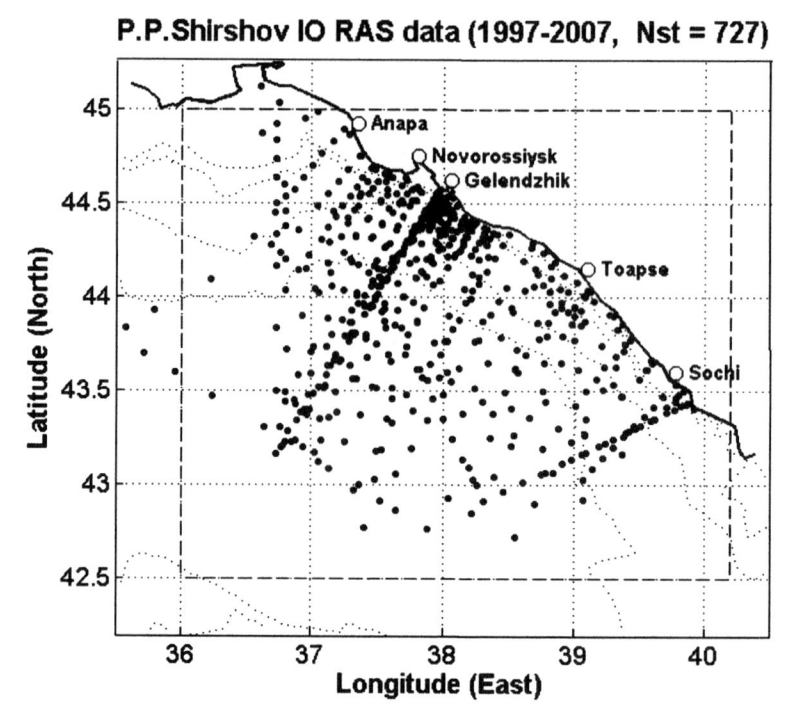

Fig. 25.3 Hydrographic stations occupied by the Shirshov Institute in 1997–2007 and used in this study

The observations in the Aral Sea were conducted in 12 field surveys during the period from 2002 to 2010. In each of the campaigns, hydrographic (CTD profiling), chemical, and biological data were obtained from motor boats, covering the western and the eastern basins of the Sea. Analyses of the collected water samples allowed following the changes of the ionic composition accompanying the lake desiccation.

25.3 Results and Discussion

25.3.1 Case Study 1: SST in the Black Sea

The variability of the Black Sea SST and associated physical processes on the scales from synoptic to seasonal is well understood (e.g., a comprehensive review can be found in [5]). However, the variability at the interannual to interdecadal scales is less well known. A reconstruction for the mean winter SST since 1950, based on our synthetic historical data set, is shown in Fig. 25.4. It can be seen that during much of the last few decades, the SST was decreasing. The decrease commenced in the early 1960s and continued until the mid 1990s, at a rate of about 0.05°C year^{-1}. Then there was an abrupt increase in 1994–1995, followed by another negative

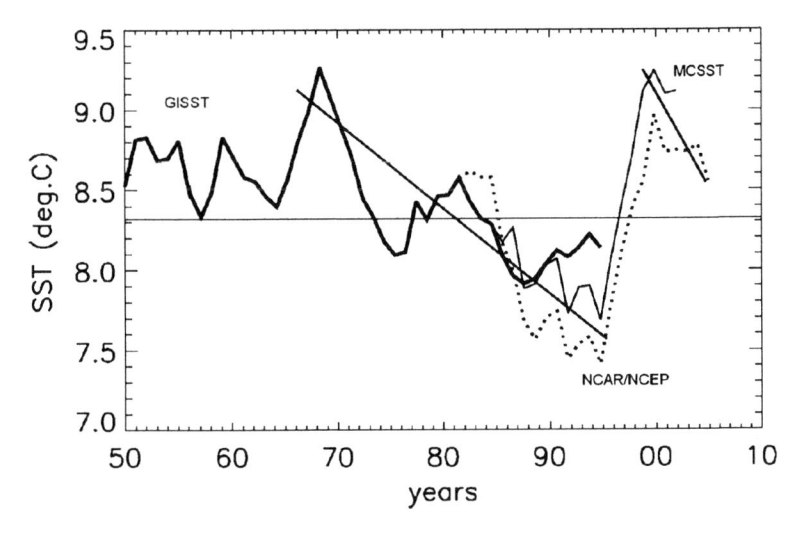

Fig. 25.4 Trends of mean winter SST in the Black Sea for 1950–2010. The straight lines highlight the periods with trends towards cooling

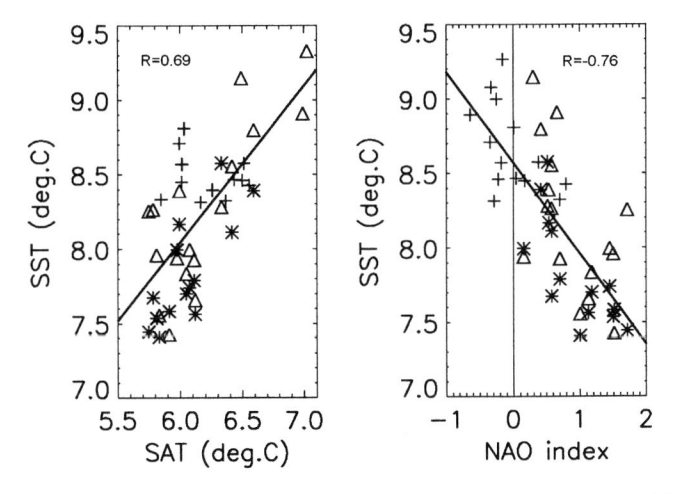

Fig. 25.5 *Right panel*: correlation between the winter SST in the Black Sea and NAO index *Left panel*: correlation between the winter SST in the Black Sea and surface air temperature

trend period since the beginning of the 2000s. This pattern is different from the SST behaviour documented for the North Atlantic.

Hence, in contrast to the warming trends of the SST typical for the Northern Atlantic, the pattern of long-term SST variability in the Black Sea is better described as an intermittent periods of SST increase/decrease with the duration of approximately 6–10 years and fast (1–2 years) transitions between them. The SST is strongly correlated with the North Atlantic Oscillation (NAO) index (Fig. 25.5, right panel). It has been long known the wind variability over the Black Sea exhibits correlations with NAO. This wind variability includes two dominant wind regimes,

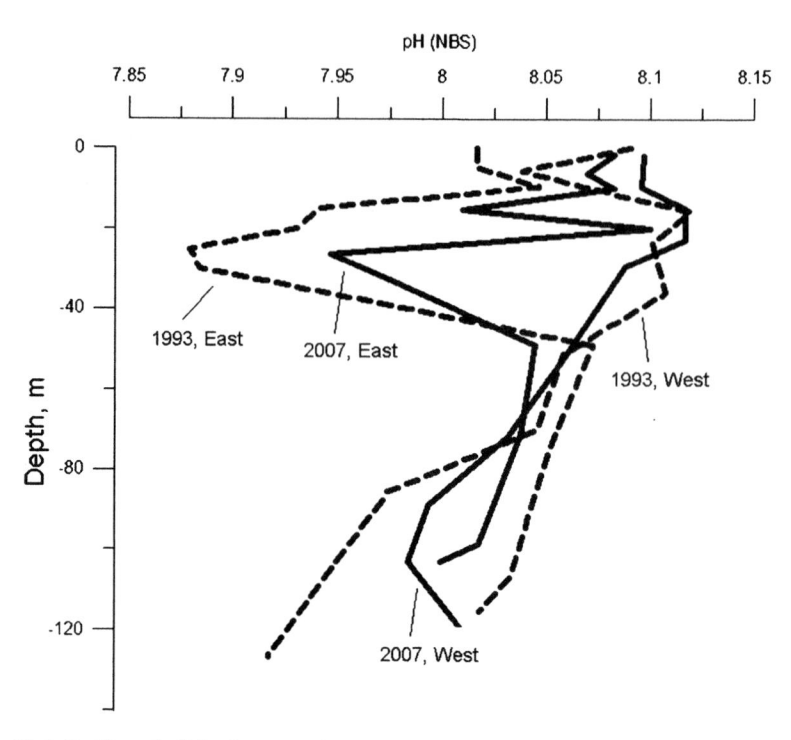

Fig. 25.6 Profiles of pH in the upper layer of the Kara Sea (western part – Yamal Peninsula region, eastern part – Ob region)

namely, the regimes characterized by prevailing N or SW winds. An important new finding is that the switching between two regimes occurs quite abruptly at values of NAO index slightly below zero. At weak to moderate positive NAO index, the SW wind regime prevails and the wind components are correlated with the NAO index. The long-term variability of winter SST in the Black Sea is significantly correlated with the variability of surface air temperature (SAT, Fig. 25.5, left panel). In turn, SAT is highly correlated with the meridional component of the surface wind stress. Hence, strengthening/weakening of SW winds, or weakening/strengthening of N wind, cause SAT and SST to increase/decrease.

25.3.2   Case Study 2: pH in the Kara Sea

Based on the data collected in the Kara Sea during the cruises of September, 1993 (R/V *Mendeleev*), and September, 2007 (R/V *Keldysh*), we evaluate the changes of pH in the southwestern and the southeastern parts of the Sea. The western sampling area is located north of Yamal Peninsula, and the eastern one is adjacent to the Ob River mouth.

It can be seen in the pH profiles (Fig. 25.6) that the pH values have actually increased by about 0.1 in the uppermost 50 m layer western area, while remaining

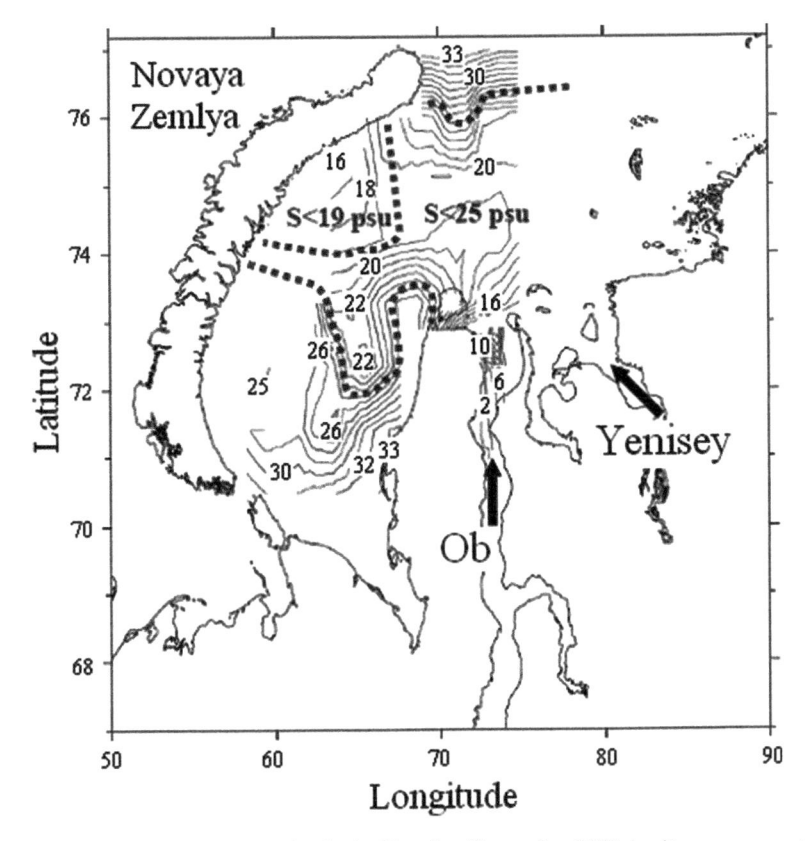

Fig. 25.7 Surface salinity distribution in the Kara Sea (September 2007, in situ measurements in a cruise of R/V *Keldysh*)

essentially unchanged in the eastern area. Such an increase is apparently inconsistent with the global ocean acidification pattern. This discrepancy can be hypothetically attributed to the effect of river discharges. The Kara Sea is exposed to the fluvial outflow from Ob and Yenisey rivers at the rate of 1,300 km^3 year^{-1}, on the long-term average. In 2007, an unknown fraction of the runoff veered eastward (Fig. 25.7), but the remaining volume formed a vast region of freshwater influence (ROFI) across the Sea, and a minimum salinity area adjacent to Novaya Zemlya. The eastern sampling area was within this ROFI, and the western one outside it. A similar situation occurred during the cruise of 1993. It is known, however, that Ob and Yenisey runoffs demonstrated energetic variability and significant climatic trends during the 1990s and 2000s, including negative trend for late summer and autumn [6]. This may have resulted in corresponding variability of pH within the ROFI in the Sea.

This example, as well as other available data suggests that unlike in the open ocean, in the MIS surrounding Russia, the long-term variability of pH followed a complex pattern rather than reflected by any distinct climatic trend signal.

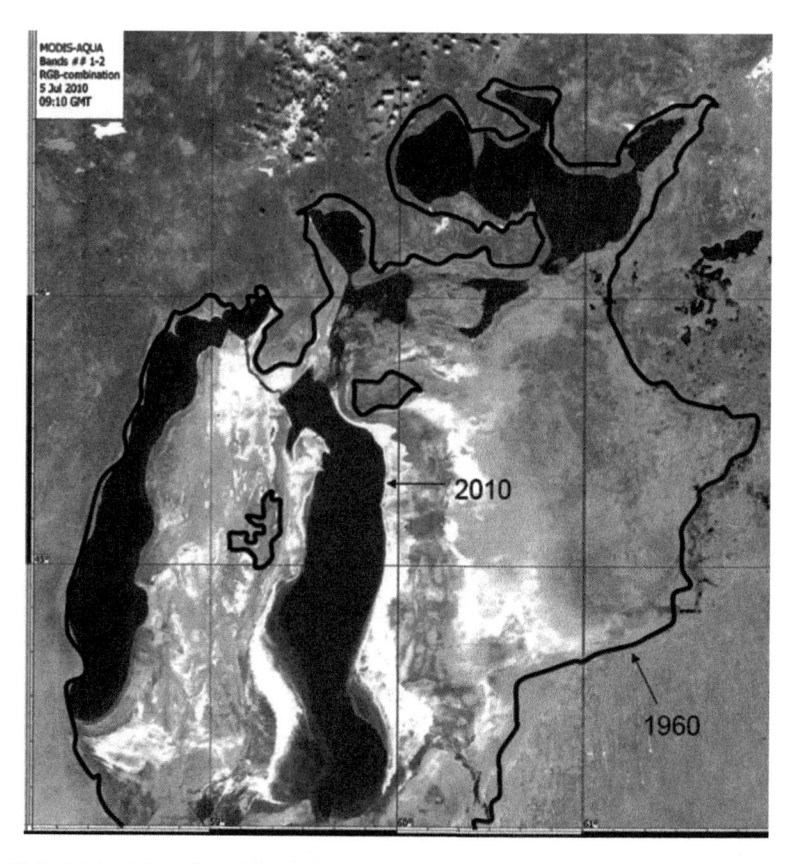

Fig. 25.8 Original shore line of the Aral Sea (as of 1960) superimposed on a recent satellite image of the lake (MODIS-Aqua scanner, July 5, 2010). The northernmost water body is the small sea, the western and the eastern basins of the large sea are seen in the southwestern part of the figure, interconnected through the narrow strait

25.3.3   Case Study 3: Salt Composition of the Aral Sea

In 1960, the Aral Sea level started to drop continuously (Fig. 25.8). It is generally believed that the shallowing followed an increase of water diversions from the tributary rivers for agriculture. Models, however, suggest that the desiccation was triggered, at least partly, by natural climate variability (e.g., [3]). With its today's volume of only about 90 km^3, the lake has lost over 90% of its water (Fig. 25.9). The shrinking of the Aral Sea has been accompanied by a continuous salinity build-up. The most recent salinity figures available on the date of this writing (data obtained in September 2010) are 121 g/kg, constituting a notable increase by a factor of 13 since 1960.

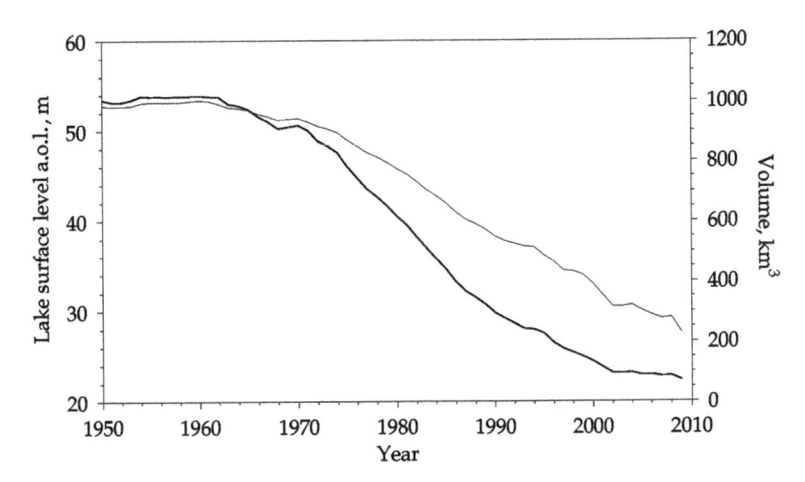

Fig. 25.9 The Aral sea level drop (*fine curve*) and volume shrinking (*bold curve*) for 1950–2010

Table 25.1 Relative contents of the principal ions in the Aral Sea water in 1952 and 2008

Ion	Cl$^-$	SO$_4^{2-}$	HCO$_3^-$	Na$^+$	Mg^{2+}	K$^+$	Ca^{2+}
Content (%) 1952	34.5	31.1	1.5	21.9	5.2	1.2	4.6
Content (%) 2008	43.3	22.6	0.6	24.8	6.7	1.5	0.5

The desiccation resulted in significant alterations in not only physical state of the lake, but also its chemical regime. As a consequence of the stratification, the bottom portion of the column turned anoxic and contained hydrogen sulfide. The H$_2$S contamination was first observed in 2002 [7] and persisted since then, at concentrations ranging from 5 to 80 mg/l, although intermittent ventilation episodes associated with winter convection were also documented [8]. The depth of the upper limit of the sulfide-containing layer varied from 15 to 35 m.

During the desiccation, the ionic salt composition of the Aral Sea has been subject to significant changes because of the chemical precipitation accompanying the salinity build-up [3]. The total masses of minerals sedimented from the onset of the desiccation through 2008 are estimated as follows: gypsum 2.3 billion tons, mirabilite 1.9 billion tons, halite 0.4 billion tons, carbonates of calcium and magnesium 0.2 billion tons [8].

The large scale precipitation of the salts has led to corresponding changes in the salt composition of the remaining water mass of the lake, so that the today's figures differ considerably from those known for the pre-desiccation period (e.g. [2]). The relative concentrations of the principal ions before and after the desiccation are shown in Table 25.1.

It is evident that the relative contents of the ions changed significantly during the desiccation. The most pronounced changes occurred with Ca^{2+} whose relative concentration decreased nine-fold, from 4.6% to only 0.5%. The sulfate ion, also consumed in gypsum precipitation, decreased in relative content as well, from 31% to 23%. In consequence, the mass ratio SO_4^{2-}/Cl^- decreased from 0.90 to 0.52, i.e., by 42%. The profound changes of the physical and chemical conditions of the Aral Sea led to severe consequences for the biological systems of the lake (e.g., [1]).

25.4   Conclusion

The climate change processes in the marginal seas and inland water bodies are understood and quantified to much lesser extent than those in the open ocean. The marginal and inland seas develop complex responses to the climate change, which are, generally, quite different from those typically manifested in the ocean. In such seas, the interannual and decadal variability associated with the terrestrial and atmospheric influences tends to prevail over the long-term global trends generated by the radiation forcing. Therefore, at least in some of the MIS, the temperature actually exhibits cooling trend rather than warming. In many cases, the changing continental freshwater discharges modulate the thermohaline structure and vertical stratification of MIS, as well as the hydrochemical regime and even the ionic composition of their waters. As a consequence, at least some of the MIS are characterized by a long-term increase of pH, as opposed to the ocean acidification tendency linked to an increase of CO_2 uptake elsewhere in the oceans. The inland water bodies are particularly sensitive with respect to the global change, mirroring climate change impacts as well as the anthropogenic pressures.

Acknowledgments   The works partly presented in this article were supported by the Russian Foundation for Basic Research, Russian Academy of Sciences, and CLIMSEAS Project within EU FP7.

References

1. Arashkevich EG, Sapozhnikov PV, Soloviov KA, Kudyshkin TV, Zavialov PO (2009) *Artemia parthenogenetica* (Branchiopoda: Anostraca) from the Large Aral sea: abundance, distribution, population structure and cyst production. J Marine Syst 76:359–366. doi:10.1016/j.jmarsys.2008.03.015
2. Blinov LK (1956) Hydrochemistry of the Aral sea. Gidrometeoizdat, Leningrad, 152 pp [in Russian]
3. Bortnik VN, Chistyaeva SP (eds) (1990) Hydrometeorology and hydrochemistry of the seas of the USSR, vol VII, Aral Sea. Gidrometeoizdat, Leningrad [in Russian]
4. IPCC (2007) The physical science basis. Contribution of working group I to the fourth assessment report of the intergovernmental panel on climate change. In: Solomon S, Qin D, Manning M, Chen Z, Marquis M, Averyt KB, Tignor M, Miller HL (eds). Cambridge University Press, Cambridge/New York

5. Oguz T, Tugrul S, Kideys AE, Ediger V, Kubilay N (2005) Physical and biogeochemical characteristics of the Black Sea. The Sea 14:1331–1369, Chapter 33
6. Yang D, Ye B, Shiklomanov A (2004) Discharge characteristics and changes over Ob river watershed in Siberia. J Hydrometeorol 5:595–610
7. Zavialov PO (2005) Physical oceanography of the dying Aral Sea. Springer Praxis Books, Chichester, 154 pp
8. Zavialov PO, Ni AA (2009) Chemistry of the Large Aral Sea. In: Kostianoy AG, Kosarev AN (eds) Aral Sea environment. Hdb Env Chem. Springer, Berlin/Heidelberg, 16 pp, doi:10.1007/698_2009_3

Chapter 26
Orographic Precipitation Simulated by a Super-High Resolution Global Climate Model over the Middle East

Pinhas Alpert, Fengjun Jin, and Haim Shafir

Abstract A super-high resolution (20 km) global climate model data and Climate Research Unit (CRU) data were employed to investigate the seasonal precipitation regime over the Middle East, and the main research focus is on the orographic rainfall effects over a large part of Turkey by using these two different datasets.

Results show that the 20 km regional precipitation over high mountains behaves differently in the 20 km resolution as compared to the CRU data for the time period of 1979–2002. The orographic precipitation over Turkey simulated by the 20 km GCM shows that, the amount of seasonal precipitation has significant relation with the altitude, which is not as pronounced in the CRU data. The area mean precipitation from the 20 km GCM is higher than that of CRU both for the wet and the dry seasons, with the mean value of about 25% and 39% higher, respectively. Results suggest that the higher resolution model is essential, especially in capturing the orographic precipitation over high altitudes.

Keywords Orographic precipitation • Climate model • Mediterranean • Middle East

26.1 Introduction

Orographic precipitation is a central part of the interaction between the land surface and the atmosphere. Not only it is important for natural ecosystems and for the management of human water resources but it also has significant ramifications for other physical components of the Earth system. For example, on short timescales,

P. Alpert (✉) • F. Jin • H. Shafir
Department of Geophysics and Planetary Sciences, Porter School of Environmental Studies,
Tel-Aviv University, Tel Aviv, Israel
e-mail: pinhas@post.tau.ac.il

H.J.S. Fernando et al. (eds.), *National Security and Human Health Implications of Climate Change*, NATO Science for Peace and Security Series C: Environmental Security, DOI 10.1007/978-94-007-2430-3_26, © Springer Science+Business Media B.V. 2012

natural hazards such as flash floods, landslides, and avalanches are impacted by precipitation intensity in mountainous regions (e.g., [5–7]). The physical mechanisms involved in orographic precipitation comprise a rich set of interactions encompassing fluid dynamics, thermodynamics, and micro-scale cloud processes, as well as being dependent on the larger-scale patterns of the atmospheric general circulation [4, 12]. During the past, many studies on this subject over different mountain areas, have been published. A comprehensive discussion on this subject can be found at Roe, [12]. Alpert et al. [2] provide a review on the prediction of meso-gamma scale orographic precipitation.

The orographic effect on the rainfall over the Middle East (ME) has been investigated during the past [3, 4]. This note focuses on the significant difference between the fine and the coarse resolution in climate runs in the accurate simulation of the orographic precipitation over the ME with implications to all high mountains rainfall.

26.2   Data

A super-high resolution 20 km grid GCM data developed at the Meteorological Research Institute (MRI) of the Japan Meteorological Agency (JMA), was used here. It is a climate-model version of the operational numerical weather prediction model used in the JMA. The simulations were performed at a triangular truncation 959 with a linear Gaussian grid (TL959) in the horizontal. The transform grid uses $1,920 \times 960$ grid cells, corresponding to a grid size of about 20 km. The model has 60 layers in the vertical with the model top at 0.1 hPa. A detailed description of the model is given in Mizuta et al. [11]. The two runs of the 20 km GCM cover the time periods 1979–2007 for current/control and 2075–2099 for the future. Due to the research focus, only the control run of the 20 km GCM was employed here. In addition, the global time series dataset based on rain gauge measurements (land only) from the climate research unit (CRU, in brevity; Mitchell and Jones, [10]) was employed here for comparison. The CRU grid horizontal resolution is 0.5×0.5 degree, and the time period available is 1901–2002. In order to make these two sets of data to be comparable, the 1979–2002 time periods for both data sets was selected. The study area covers a central high-mountain part of Turkey; four different sizes of domains were selected, for which the largest domain is about 80,000 km^2 (Fig. 26.2a).

26.3   Results and Discussion

26.3.1   The Whole Mediterranean

Figure 26.1 shows that the wet season mean precipitation from the 20 km GCM is quite consistent with the observations based CRU data. Over the Mediterranean (Med in brief) region, the latitudinal gradient is the predominant feature, with a

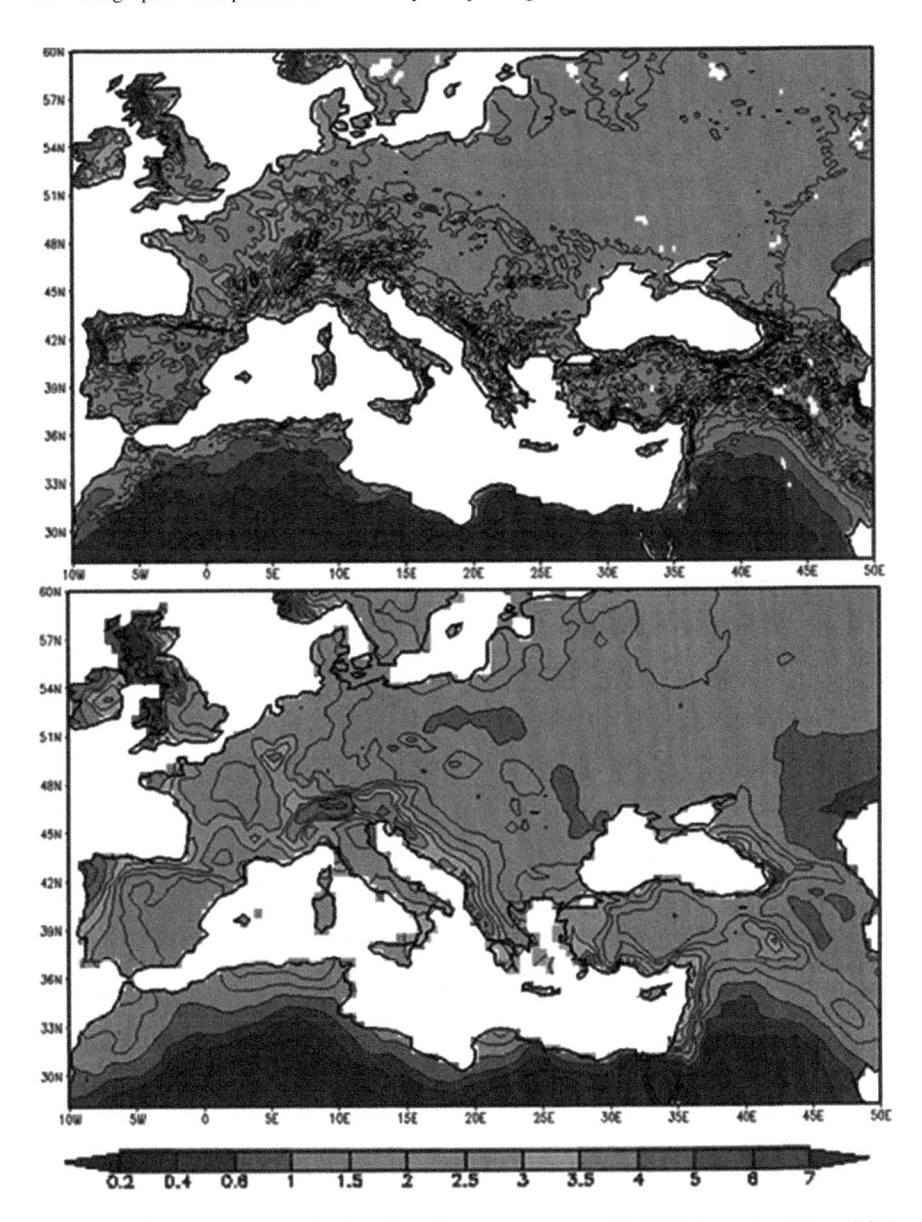

Fig. 26.1 Mean wet season (October–March) precipitation (1979–2002) from the 20 km GCM (*upper*) and CRU (*bottom*). Unit: mm/day

much drier area located at the south of the Med coastline, and a wetter area over the northern coastline of the Med basin. However, a closer examination shows that the amount of the precipitation peaks simulated by the 20 km model are larger than that from the CRU. For instance, this can be well noticed over the Fertile Crescent,

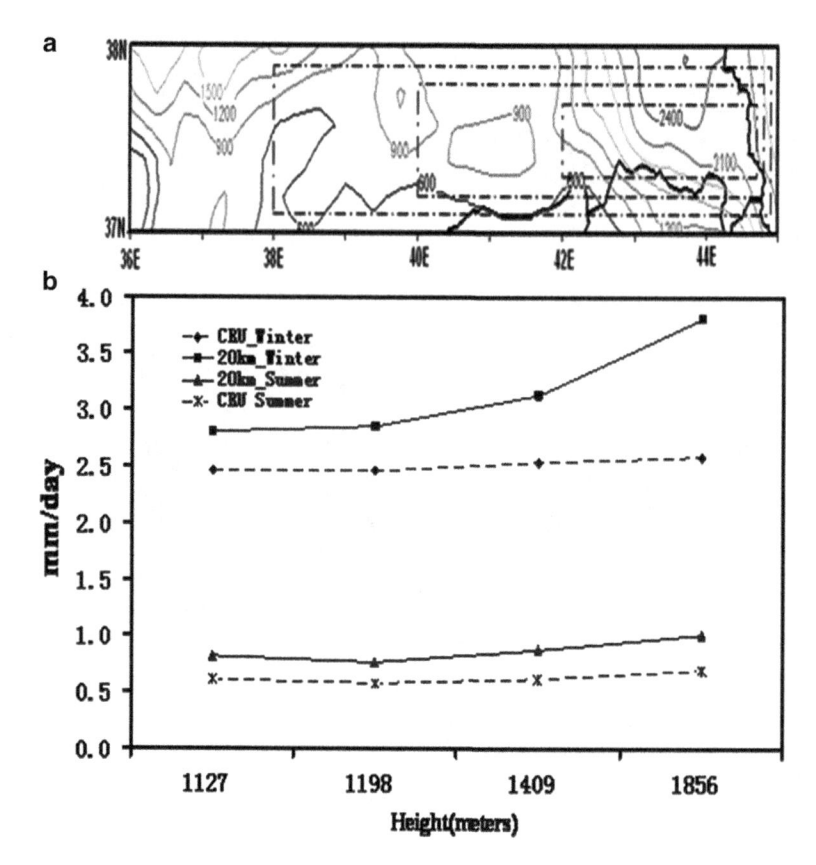

Fig. 26.2 (a) The altitude map for the selected four domains over Turkey. Unit: meters. (b) Area mean precipitation changes with the mean altitude for the selected four domains both for wet season (October–March; denoted winter in the figure) and dry season (May–September; denoted summer in the figure) for the 20 km control run as well as CRU (1979–2002). Unit: mm/day. The average altitudes of the four domains vary from 1,127 to 1,198, 1,409 and 1,856 m as indicated by the X-axis

the south coast line of Black Sea and over the Alpine mountain region. However, Kitoh et al., [9] showed that the simulated precipitation from 20 km GCM is surprisingly close to that based on the daily raingauge data over the ME [13].

26.3.2   High-Altitudes Orographic Precipitation over Turkey

Figure 26.2 shows the differences between the 20 km GCM and the CRU data in simulating the orographic rainfall, both for the wet season (October–April) and dry season (May–September), over the high mountainous section of Turkey. Figure 26.2b indicates that, over the same domain, the area mean precipitation from the 20 km

GCM is higher than that of CRU both for the wet and the dry seasons, with the mean value higher by about 25% and 39%, respectively. Of-course, in absolute values the differences are much larger in the wet season.

Especially for the highest altitude domain of 1,856 m average altitude, these differences reach the largest values of about 48% and 46% for the wet and dry seasons. Notice that the absolute average differences are of about 1.3 and 0.25 mm/day over that inner domain which is as large as 3×0.5 deg (an approximate area of 15,000 km^2).

One major result is that the wet season area mean precipitation is increasing with the altitude as can be clearly seen in the 20 km GCM, but it is not as significant in the CRU data. Also, Jin et al. [8] showed that the outstanding performance of 20 km GCM in simulating the precipitation through the E. Med south-to-north cross-section, suggesting that the high resolution model indeed has an essential role in capturing the orographic rainfall effect.

26.4   Summary

The super-high resolution 20 km GCM data shows its capability in capturing the seasonal precipitation over the Middle East region quite well. The seasonal precipitation simulated by the 20 km GCM is also quite close to that of CRU. However, the effect of orographic precipitation over the high-mountainous region of Turkey with the 20 km GCM shows that, the amount of seasonal precipitation has significant relation with the altitude, which is not as pronounced in the CRU data based on observations. One probable reason for that may be the severe lack in rainfall observations over high mountains as discussed by e.g. Alpert [1]. Due to the orographic effect, the area mean precipitation from the 20 km GCM is higher than that of CRU both for the wet and the dry seasons, with the mean value of about 25% and 39% higher, respectively. Results suggest that the higher resolution model is essential, especially in capturing the orographic precipitation.

References

1. Alpert P (1986) Mesoscale indexing of the distribution of precipitation over high mountains. J Clim Appl Meteorol 25:532–545
2. Alpert P, Shafir H, Cotton WR (1994) Prediction of Meso-γ scale orographic precipitation. Trends Hydrol 1:403–441
3. Alpert P, Shafir H (1989) A physical model to complement rainfall normals over complex terrain. J Hydrol 110:51–62
4. Alpert P, Shafir H (1991) On the role of the wind vector interaction with high-resolution topography in orographic rainfall modelling. Q J R Meteorol Soc 117:421–426
5. Caine N (1980) The rainfall intensity: duration controls on shallow landslides and debris flows. Geogr Ann Ser A 62:23–27

6. Caracena F, Maddox RA, Hoxit LR, Chappell CF (1979) Mesoscale analysis of the Big Thompson storm. Mon Weather Rev 107:1–17

7. Conway H, Raymond CF (1993) Snow stability during rain. J Glaciol 39:635–642

8. Jin F, Kitoh A, Alpert P (2009a) The atmospheric moisture budget over the eastern Mediterranean based on a high-resolution global model – past and future. Int J Climatol (in revision)

9. Kitoh A, Yatagai A, Alpert P (2008) First super-high-resolution model projection that the ancient Fertile Crescent will disappear in this century. Hydrol Res Lett 2:1–4. doi:10.3178 HRL.2.1

10. Mitchell TD, Jones PD (2005) An improved method of constructing a database of monthly climate observations and associated high-resolution grids. Int J Climato 25:693–712. doi:10.1002/joc.1181

11. Mizuta R, Oouchi K, Yoshimura H, Noda A, Katayama K, Yukimoto S, Hosaka M, Kusunoki S, Kawai H, Nakagawa M (2006) 20-km-mesh global climate simulations using JMA-GSM model mean climate states. J Meteorol Soc Jpn 84:165–185

12. Roe GH (2005) Orographic precipitation. Annu Rev Earth Pl Sci 33:645–671

13. Yatagai A, Alpert P, Xie P (2008) Development of a daily gridded precipitation data set for the Middle East. Adv Geosci 12:1–6

Chapter 27
How Effective Could 'Landscape Management' Tool Address Mitigation of Cultural and Natural Threats on Coastal Wetlands System?

Adnan Kaplan

Abstract This paper identifies landscape management as a tool for mitigating natural and cultural challenges of climate change, human and environmental security in the context of Gediz Delta (Management Plan) while scrutinizing the management plan for the ecological and socio-economic sustainability of the delta. Thus landscape management acts as a mechanism for strengthening the efficacy of the plan in the years ahead.

To this end, implementation of the management plan should accommodate thoroughly landscape management in legal, administrative and technical frameworks for the benefits of a self-sustainable coastal wetlands' system across the delta.

Keywords Landscape management • Coastal wetlands system • Wetland management plan • Gediz Delta • Adaptive management • Climate change • Conservation planning

27.1 Introduction

The concept of Environmental Security falls into the broader notion of human security. Human security is viewed as a paradigm shift from the traditional notion of security related to military and political issues to environmental, public health and quality of life issues. Water is one of the key assets central to human security [1]. By occupying zones of transition between terrestrial and marine ecosystems, coastal wetlands,

A. Kaplan (✉)
Department of Landscape Architecture, Faculty of Agriculture, Ege University,
35040, Bornova, İzmir, Türkiye (Turkey)
e-mail: adnankaplan@gmail.com; adnan.kaplan@ege.ed.tr

H.J.S. Fernando et al. (eds.), *National Security and Human Health Implications of Climate Change*, NATO Science for Peace and Security Series C: Environmental Security, DOI 10.1007/978-94-007-2430-3_27, © Springer Science+Business Media B.V. 2012

including swamps, salt marshes, mangroves, intertidal mudflats, seagrass beds and shallow subtidal habitats, are the interface of the coastal landscape [7]. They are exposed to a wide variety of natural disasters and man-made threats with an effect ranging between basin and local scales.

Decreasing fresh water and the destructive impact of sea towards inland areas have been the results of climate change that bring in adverse effects on physical structure and ecological processes of coastal wetlands. Natural disasters, chronic natural hazards and climate change are geophysically interacting phenomena, as are social, economic and ecological vulnerability and resilience. In many cases, enhancing resilience or decreasing vulnerability to acute and chronic hazards concurrently builds adaptive capacity to climatic and hydrologic changes as well as mitigates future climate change. The steadily increasing stresses from climate change, coupled with increased probabilities of more frequent, acute stresses, renders climate change a latent danger that could frustrate isolated attempts to enhance social well-being and national security [3].

Conflating environmental and human security, climate change and coastal wetlands into the realm of landscape management, this paper identifies this management type as a tool to mitigate natural and cultural challenges in the context of Gediz Delta Management Plan while screening the plan for the sustainability of the delta. Thus landscape management acts as a mechanism for strengthening the efficacy of the plan in the years ahead.

27.1.1 Natural and Cultural Threats and Landscape Management

Environmental management must start from sound knowledge of the functioning of abiotic and biotic elements of the wetland ecosystem, and of their relations with processes upstream and downstream in the catchment. In addition, socio-economic systems of wetland users need to be analyzed, especially those of local communities. In order to attain sustainable use, a participative approach must be applied which engages wetland users in the process of establishing and carrying out management measures [10]. Management process of any protected area involves designating conservation decisions, and subsequently planning, managing and monitoring [12]. Management of coastal wetlands at landscape level necessitates adoption of some scientific and concrete measures into planning and management pursuits that is able to conserve or enhance natural processes and ecological stability [4].

Landscape management as a tool of more highly integrated level of ecosystem management integrates the entire ecological, technological and social system of a landscape. Landscapes have to be evaluated constantly for their natural and socio-economic parameters as well as for their non-material values. Such a landscape-ecological approach addresses the interdependency of wetland functioning and other processes in the catchment through minimizing the cumulative adverse effects

Table 27.1 The nature of landscape management as a tool

Features	Explanation
Scale	Region-landscape-habitat
Content of work	Policy – planning – management – conservation – restoration
Example	Gediz Delta, comprised of coastal wetlands (system) at regional scale
Document	Gediz Delta management plan
Rationale	– Understanding the functioning of landscape structure and change of any wetland ecosystem within the regional hinterland
	– A participative approach accounting socio-economic interests of users (local communities)
	– Conserving and enhancing natural processes and ecological stability
Objectives	Sustainability, biodiversity and wise use of wetlands
Phases	– Identification of critical factors (natural and cultural threats)
	– Developing and implementing a set of management actions
	– Monitoring changes in wetland(s)
	– Adjusting current management practices appropriately
Adaptiveness	Scenario-based planning incorporated into active adaptive management to address natural and cultural challenges.
Mitigation	Adaptation and mitigation strategies and policies designed to minimize the effects of natural and cultural threats

Fig. 27.1 Vulnerable southern part of Gediz Delta (http//www.izmir.bel.tr, 13 May 2011)

of wetland use and alteration on society as a whole [10]. This enables the landscape management a particular value in managing habitats in and around wetlands' system at multiple scales (Table 27.1).

27.1.2 Case Study: Gediz Delta

Gediz Delta is situated on the western coast of Türkiye (Turkey), and it constitutes a large north-west tract −40.000 ha in size- of İzmir metropolitan city (Fig. 27.1).

About 20.000 ha of the delta embody a wide variety of coastal wetlands including estuary, river, lakes, mudflats, salt marshes, reed beds, lagoons, saltpans, shallows, hills and farmland, and a major part of it was designated as the Ramsar Site in 1998. Gediz River stretches out some intricate meanders westward to Aegean Sea passing by some major cities and many rural settlements. Gediz Delta is hereby a confluence of Gediz River with a length of over 400 km and Ege (Aegean) Sea, involving brackish, salt and freshwaters. The content and biodiversity in major segments of almost these wetlands change dramatically depending upon seasonal flux of either freshwater or sea and river flows.

The delta was comprised of 15 landscape types constituting coastal wetlands system. Most parts of landscapes in the region have been under the severe impacts of natural and cultural threats, and largely conflicting land uses in varied degrees [4]. Urban and industrial expansion and natural phenomena such as coastal erosion and drought have long resulted in a changing landscape (fragmentation and destruction of landscape units, scenic and physical attributes of landscapes, survival of creatures etc.), dwindling water resources (withdrawals of ground waters and cut off of water cycle, lack of fresh water and salinity, water pollution and erosion, destruction of natural coastal lines, flooding, destructive effect of sea waves, stormwater surge, land subsidence, …) and changing earth surface (soil erosion and pollution, destruction of natural coastal lines, …), which have posed challenges across the delta [5]. Coastal marshes and lagoons in the western and southern part of the delta are now under destructive impact of sea waves since the entity of fresh water and accumulation of sediment have long not supported these highly vulnerable landscapes (Fig. 27.2). On the other hand, basic changes in climate pattern within the last 50 years have resulted in drought that decreased the supply of fresh water while increasing susceptibility of the delta to erosion and sea waves.

27.1.3   Gediz Delta Management Plan

The management plan for Gediz Delta was prepared to set forth a balance between utilization and protection of existing delta ecosystems. Based on the new guidance of Ramsar Convention along with the Regulation of National Wetlands, the plan was officially prepared. Participatory approach with relevant stakeholders was distinguished in planning process. The plan was aimed to be multi-functional and multi-dimensional at social, economic and ecological aspects. Thus, it provides a set of management actions to be applied through public and private initiatives under the Ministry of Environment and Forestry. The guidance was addressed to protect and enhance a multitude of wetland ecosystems [4].

Based on sustainable management of coastal wetlands and other landscape types within the delta, the ultimate goal of the management plan is twofold, to conserve ecological balance, biodiversity and landscape structure of the delta, and to improve socio-economic wealth of local communities. Besides ecological stability of the

Fig. 27.2 Coastal marshes have been exposed to the severe impacts of urban development and sea water alike

delta, socio-economic structure of the delta has been analyzed to manage administrative outlook, living conditions and land use pattern of the region. In order to account these all, the plan introduces 14 basic directives with their own particular actions [9].

The coverage of the socio-economic analysis involved human activities, land uses and their impacts on coastal wetlands. This work also revealed the effects of discharge of industrial plants, pesticides and solid wastes over human security [11]. Taken these all together, Table 27.2 showcased these effects and the management actions against them.

27.1.4 Engagement of Landscape Management with the Delta Management Plan

Natural challenges and cultural interventions together with land uses have long resulted in some particular socio-ecological problems on the delta. Among them are fragmentation of landscapes, destruction of habitats, decreasing underwater table, disruption of coastal line, aggravation of soil and water salinization, land and soil erosion and the exploitation of freshwater [4]. In addition to causing direct habitat loss, urbanization impacts the structure and function of coastal wetlands through

Table 27.2 Gediz Delta management plan and its engagement with climate change-, human and environmental security based threats

Natural and cultural threats and/or stresses	Actions in the management plan
– Insufficient amount of fresh water is unable to serve to the total requirements of wetlands, agricultural fields and other land uses such as housing, industry etc., – Unproductive water use regime in agriculture	– Water-sensitive irrigation techniques have been extensively implemented, and using recycled water of the urban water treatment plant is needed, – Supplying fresh water for reed beds, coastal marshes through water channels, river beds and newly created reservoirs or ponds, – Drip irrigation in closed water systems for Gediz river basin
Pollution of Gediz river	Requiring a broad perspective of the basin management upon which the interested administrative and legal bodies should be committed to carrying out their assignments.
– Depositing of sludge wastes on the delta instead of making economic and ecological use of it, – Sludge causes negative effects (source of contamination) to human security	New treatment technologies to recycle water and other wastes for agriculture and wetlands should be implemented, and precautions against the penetration of these wastes into the delta should be put into practice.
Urban sprawl on the delta has gradually stripped off some agricultural lands.	Sustainable and multiple agricultural practices- including agriculture, animal husbandry and fishery – should be underway to discourage all sorts of urban interventions and to demonstrate their own economic support to local communities.
Uncoordinated implementation of management plan	Monitoring phase of the management plan calls for a coordinated working flow among interested public bodies for the next 5 years.
Drought and sea waves have triggered the lack of fresh water, sea level rise, higher frequency of storms, disturbance in hydrological cycles, changing wave characteristics, salt intrusion.	Ecological restoration program would be scheduled to rehabilitate ecosystems and habitats for landscape and species conservation.
Urban encroachment, highways, residential and industrial expansion	The management plan designates ecologically significant zones and delimits urban development.

effect on the hydrological and sedimentation regimes and the dynamics of nutrients and pollutants -the major drivers of wetland dynamics [7]. Vulnerability reduction and resilience enrichment requires a three-pronged approach: distributing decision-making authority to local and regional administrations; enhancing protections against environmental degradation; and diversifying, transferring or pooling risks across time, space and institutions [3].

With these all in mind, we aimed to inquire the nature of landscape management as a tool and its engagement with Gediz Delta Management Plan respectively at the panel entitled 'Landscape Management: Gediz Delta Management Plan'. It was organized by the Chamber of Landscape Architects İzmir Division on May 13, 2010. And the speakers focused on preparation and implementation phases of the plan, and on criticizing it technically and administratively with a view of landscape management. Thereafter academic and professional insights in the panel book (in press) would serve to the second screening period of the plan (from 2012 on). Climate change and its effect over human and environmental security have greatly taken up the agenda. Climate change effects such as fresh water and food shortages as well as multiple negative impacts on the environment have been a result of unsustainable land use, greenhouse gas emissions and management practices for many decades. In addition, it is underlined that water and land pollution across Gediz River cannot be resolved within the margin of the delta-scale management plan; it needs a basin-wide approach particularly. Such an approach should be linked with the management plan to mitigate this huge regional challenge (i.e. top down approach). So the management plan in the next decades should definitely address some academic and practical outputs of the panel as follows;

– Establishment of a single, but well-grounded authority (i.e. landscape management directorate) under the Ministry of Environment and Forestry, which fully be responsible for managing such sort of plans throughout the country,
– Ensuring a much more participative management process in both legal and administrative frameworks,
– Recognizing the ecologically delicate nature of coastal wetlands system as both barriers and interface between marine and terrestrial systems while delivering mitigation strategies against unsustainable land use and climate change effects,
– As stated in the management plan, ecological restoration of coastal wetlands should take place to compensate or recover the sea water impacts that physically change coastal landscapes,
– Monitoring efficient management of the plan officially,
– Recognizing that the ecological and economic cost of recovery of flooding, pollution and drought is much more than the budget for mitigation strategies that could, to some extent, decrease the growing impacts of natural and cultural threats.

27.2 General Discussion

Landscape management planning is long-term oriented and, therefore, will never receive priority over short term issues. It is often considered as interference in politics. Thus, the best examples of modern landscape planning in the developing world might be the planning and management of protected areas such as parks that, in many cases, activates processes and institutions that will also be instrumental in introducing landscape management into 'everyday' landscapes.

The enclosure of climate change aspects into the theory, modeling and application of spatial planning has recently moved to the forefront of landscape and planning research. Understanding the impacts of climate change on ecosystems, landscapes and land uses is an essential basis for well-grounded decisions on adaptation and mitigation strategies and politics at a local and regional scale. Spatial planning is expected to provide the instrumental framework for the implementation of these strategies and measures. Nevertheless, there is a major lack of methods to scale climate change effects down to regional and local scale and to project or estimate direct and indirect effects such as loss of biodiversity, flood risks, sea level rise, soil erosion, landslides, droughts, heat waves, permafrost decline, snow coverage decline, forest fires, increasing greenhouse gas emissions from soils, as well as economic impacts on tourism, water, agriculture and forest production [8].

Projected future climate change will undoubtedly result in even more dramatic shifts in the state of many ecosystems. These shifts will provide one of the largest challenges to natural resource managers and conservation planners. Many adaptation strategies that have been proposed for managing natural systems in a changing climate are reviewed. Most of the recommended approaches are general principles and many are tools that managers are already using. What is new is a turning toward a more agile management perspective. To address climate change, managers will need to act over different spatial and temporal scales. The focus of restoration will need to shift from historic species assemblages to potential future ecosystem services. Active adaptive management based on potential future climate impact scenarios will need to be a part of everyday operations. Climate change will force managers and planners to evaluate multiple potential scenarios of change for a given system and then to develop alternative management goals and strategies for those scenarios. Given the critical role that adaptive management is likely to play in addressing climate change, one of the most important research needs involves gaining a better understanding of how to implement adaptive management. What is the best way to explore multiple climate-change scenarios in an adaptive management setting? What will need to be monitored? How often will monitoring need to be done? There has long been a call for increasing the amount of adaptive management that is actually implemented. Implementing adaptive management in a changing climate will both allow us to learn more about the ecological effects of climate change and to provide flexible management approaches [6].

Achieving a whole landscape planning and management arrangement requires an integrated approach that links all currently stand-alone planning and management systems that have a major bearing on the state and health of the coastal zone. This should be underpinned by legislative reform and lead to concordance between local authority statutory planning schemes, the various plans and policies of government agencies exercising managerial responsibility within the coastal zone and the emergent regional state government arrangements [2].

27.3 Conclusion

Some local sectors such as agriculture, animal husbandry and fishery (are likely to) increase economic revenues of local communities. The survival of these activities is highly dependent on biodiversity and ecological stability of the delta. This proves that local economic welfare is contingent with the presence of coastal wetlands system, excluding some large-scale projects boosting more residential development, industry and highways. That is the underlying factor why local communities are in favor of protecting coastal wetlands system and rural landscapes.

Managers already have many of the tools necessary to address climate change. The vast majority of these tools are those recommended for protecting biodiversity and managing natural resources in general. What is needed in the face of climate change is a new perspective. This new perspective will require expanding the spatial and temporal scale of management and planning [6].

Landscape management in the delta should involve collecting ecological, social and economic inventories to produce a comprehensive database, and legally identify landscape monitoring (auditing) and restoration works.

To sum up, landscape management as an effective tool should be officially recognized in legal, administrative and technical frameworks. And implementation of the management plan should address thoroughly the landscape-based ecosystem approach for the benefits of a self-sustainable coastal wetlands' system as well as of the local communities across the delta and beyond. Since conservation and management of (inter)nationally significant landscapes stipulate preparation of official management plans and abide by the related legal framework and environmental legislations, these should be accounted as a viable opportunity in assembling climate change and sustainable planning/management studies to addressing the mitigation of natural and cultural pressures and/or threats. So these sort of plans and hereby 'landscape management' tool must in the coming years prioritize multiple-effects of climate change over cultural and natural landscapes more rigorously worldwide.

References

1. Bobylev N (2009) Multiple criteria decision making and environmental security. In: Illangasekare TH, Mahutova K, Barich JJ (eds) Proceedings of the NATO advanced research workshop on decision support for natural disasters and intentional threats to water security, 22–25 Apr 2007, Springer/Kluwer, Dubrovnik, pp 213–228
2. Dale PER, Dale MB, Dowe DL, Knight JM, Lemckert CJ, Choy DCL, Sheaves MJ, Sporne I (2010) A conceptual model for integrating physical geography research and coastal wetland management, with an Australian example. Prog Phys Geogr 34(5):605–624
3. Hultman NE, Bozmoski AS (2006) The changing face of normal disaster: risk, resilience and natural security in a changing climate. J Int Aff 59(2):25–41
4. Kaplan A, Hepcan Ş (2009) An examination of ecological risk assessment at landscape scale and the management plan. In: Illangasekare TH, Mahutova K, Barich JJ (eds) Proceedings of

the NATO advanced research workshop on decision support for natural disasters and intentional threats to water security, 22–25 Apr 2007, Springer/Kluwer, Dubrovnik, pp 239–250

5. Kaplan A, Hepcan Ş, Kurucu Y, Bolca M, Esitlili T, Tırıl A, Akgün A (2009) Ecological risk analysis of Gediz Delta based on remote sensing and geographic information system. Ege University Scientific Study Project Report, Project no. 2005-zrf-014, İzmir

6. Lawler JJ (2009) Climate change adaptation strategies for resource management and conservation planning. Ann N Y Acad Sci 1162:79–98

7. Lee SY, Dunn RJK, Young RA, Connolly RM, Dale PER, Dehayr R, Lemckert CJ, McKinnon S, Powell B, Teasdale PR, Welsh DT (2006) Impact of urbanization on coastal wetland structure and function. Austral Ecol 31:149–163

8. Meyer BC, Rannow S, Loibl W (2010) Editorial: climate change and spatial planning. Landscape Urban Plan 98:139–140

9. Ministry of Environment and Forestry (2007) The management plan for Gediz Delta. Turkish Republic of Ministry of Environment and Forestry, General Directorate of Nature Protection and National Parks, Ankara

10. Nath B, Hens L, Compton PA (eds) (1999) Instruments for environmental management. Routledge, London

11. Sönmez İÖ, Onmuş O (2006) Gediz Deltası Yönetim Planı Sosyo-Ekonomik Analiz Raporu (Report on Socio-economic Analysis of Gediz Delta Management Plan). İzmir KuÐ Cenneti Koruma ve Geliştirme Birliği ve Ege Doğa Derneği, Yayınlanmamış Rapor, İzmir

12. Yücel M (2009) Korunan alanlar ve yönetimi (Protected areas and their management). In: Akay A, Özen MD (eds) Peyzaj Yönetimi (Landscape management). Türkiye ve Orta Doğu Enstitüsü, Ankara, pp 81–102

Chapter 28
Statistical Eco-Indexes for Estimation of Changes in Ecological State of Natural Waters Due to Anthropogenic Impact and Climate Change

Iryna Kh. Bashmakova and Alexander Smirnov

Abstract Natural waters are characterised by numerous hydrochemical and hydrobiological parameters that strongly vary in space and time. To highlight the basic features of ecological state of a water body, and to trace its temporal evolution, we integrate information on different parameters measured in different units into statistical integral eco-indexes. The method offers the possibility to reliably assess water quality in water bodies or parts thereof. In this paper, the method involved is presented in terms of sanitary-microbiological and hydrochemical indexes by the example of the River Biferno (Molise, Italy). Eco-indexes are proved to be efficient and reliable instruments for tracing the state of the water ecosystem as dependent on anthropogenic and natural impacts, including those caused by climatic factors. They can be used, in particular, for analysing data from long-term observations – to detect historical trends and to give statistical forecasts.

Keywords Human health • Maximum Allowable Concentration (MAC) • Microbiological and toxicological eco-indexes • Water quality

I.Kh. Bashmakova (✉)
Division of Atmospheric Sciences, University of Helsinki, Helsinki, Finland

Marine Hydrological Institute of the National Academy of Sciences, Sevastopol, Ukraine

Department of Radio Physics, N.I. Lobachevski State University of Nizhniy Novgorod
e-mail: iryna.bashmakova@helsinki.fi

A. Smirnov
Arctic and Antarctic Research Institute, St. Petersburg, Russia

H.J.S. Fernando et al. (eds.), *National Security and Human Health Implications of Climate Change*, NATO Science for Peace and Security Series C: Environmental Security, DOI 10.1007/978-94-007-2430-3_28, © Springer Science+Business Media B.V. 2012

28.1   Introduction

Statistical methods based on appropriate choice of hydrochemical, toxicological and sensitive organisms-bioindicators are widely used for assessing the level of pollution, probabilities of extreme pollution events, and risks for human health [5, 12]. Among various protocols offered for environmental monitoring by regulatory organizations (e.g., EU Bathing Water Quality Directive [7–9, 14, 15]), the most important ones are: concentration of organic matter (including those of anthropogenic origin), toxicants (chlorine-organic pesticides and heavy metal compounds), epidemiologically dangerous intestinal bacteria and the rates of their accumulation in water, living organisms and sediments [6].

Eco-indices presented in this paper are used for the following purposes: to assess the ecological state of water bodies of various types; to detect (at early stages) negative effects of various pollutant on water ecosystem; and to quantify ecological risks and damages caused by anthropogenic impacts, in particular by widening spectrum of pollutants or climatic and other natural factors. For example, even a minor increase in water temperature, due to reduction of atmospheric precipitation over land together with growth of air temperature, causes adjustment of water ecosystems from optimal functioning, which entails changes in bio-chemical processes, hydrobiont productivity, concentrations of organic matter and toxic substances, species composition and abundance of bacterioplankton. Eventually this could result in intensified development of pathogenic microflora [13].

Such changes can be quickly and reliably detected using appropriate eco-indexes, sensitive even to minor disturbances in aquatic ecosystem. Therefore, probability distributions of environmentally hazardous factors, quantified through appropriate eco-indexes, can be used as environmental risk indicators. Then the environmental damage can be estimated in terms of differences between actual and retrospective values of relevant eco-indexes. Within this approach one may expect essential correlations between our eco-indexes and known indexes of climate change, such as Drought Indexes (PDSI, BMI, SPI, etc.).

28.2   Definition of Eco-Indexes

28.2.1   The Maximal Allowable Concentration (MAC) Approach

Let us assume that a water body, or its particular part, is characterised by N parameters $P_j, j = 1, 2, \ldots, N$, that could be of different nature (hydrochemical or hydrobiological) and measured in different units. First, all parameters are made dimensionless using as scales their maximum allowable concentrations (MAC):

$$D_j = \frac{P_j}{(\text{MAC})_j}. \tag{28.1}$$

Then we determine an integral, multi-component index:

$$E^{AUL} = \frac{1}{N} \sum_{j=1}^{N} \frac{P_j}{(MAC)_j}. \tag{28.2}$$

Such MAC-based indexes characterise deviations of the factual state of a water body (or its part in question) from the acceptable standard [1, 3, 4].

28.2.2 The Basic-State Approach

To extend the above integral eco-indexing approach to a wider range of parameters (beyond those characterised by MACs), in particular, for comparison of different parts of a water body accounting for parameters of very different nature (including physical parameters, such as temperature or turbidity), we must use alternative measures instead of MACs. Let us consider a water body covered with M sampling sites: $s = 1,2,...,M$. Each site, s, provides measurements of N parameters: $P_{s,j}, j = 1,2,...,$ N. First, we determine for each parameter P_j the water-body mean value $\overline{P}_j$ and the squared standard deviation σ_j^2:

$$\overline{P}_j = \frac{1}{M} \sum_{s=1}^{s=M} P_{sj}, \quad \sigma_j^2 = \frac{1}{M} \sum_{s=1}^{s=M} \left(P_{sj} - \overline{P}_j\right) \tag{28.3}$$

Then, using σ_j as a natural scale for P_j, we determine dimensionless parameters:

$$D_{sj} = \frac{P_{sj}(s) - \overline{P}_j}{\sigma_j}, \tag{28.4}$$

which all have mean values equal to zero and dispersions equal to unity. Then, to characterise each site, we determine integral, multi-component indexes as linear combinations of the above dimensionless parameters:

$$E_s = \sum_{j=1}^{j=N} \alpha_j D_{sj}, \tag{28.5}$$

where coefficients α_j ($\sum \alpha_j = 1$) give specific weights to each parameter P_{sj}. In this context, $\alpha_j = 0$ would mean that P_{sj} is excluded from the integral index; $\alpha_j = 0$, that P_{sj} is the only parameter composing the index; and equal weights: $\alpha_j = N^{-1}$, that all parameters are considered equally important. Generally, the choice of parameters, P_{sj}, and determination of their weight coefficients, α_j, are subjects of ecological expertise – according to expert appraisal of the importance of each parameter in particular problem under consideration. For convenience, we determine parameters of this type in such a way that higher values correspond to worse ecological situations [1, 2].

28.3 Application to River Biferno

The above MAC-based eco-indexes are used to assess the ecological and sanitary state of the River Biferno in the province Molose, Italy. The river basin is a complex region with concentrated human population and industry, which results in pronounced contamination. The industrial area is located in the city of Termoli, just near the river mouth. Owing to collecting channels and depuration plants, effluents coming from the industrial area do not enter the river, and arrive directly at the Adriatic Sea. Below we demonstrate application of the two MAC-based eco-indexes using data from the enhanced monitoring campaign of the year 2009, and from available retrospective data [10, 11]. We applied, in particular, (i) microbiological and sanitary-bacteriological eco-indexes to determine sanitary-bacteriological state of the river and to disclose zones hazardous for human health due to contamination by epidemiologically dangerous bacteria (total bacterial count, CBT 22°C; total bacterial count, CBT 37°C; Enterococci; and Escherichia coli); and (ii) heavy-metal eco-index to characterise concentration of heavy metals in water.

28.3.1 Microbiological and Sanitary-Bacteriological Indexes

The spatial distribution of the microbiological eco-index (MI) and sanitary Escherichia Coli (EC) eco-index shown in Fig. 28.1 demonstrates critical contamination of water by intestinal flora in the mouth of the river and strong maxima in

Station	Escherichia coli, UFS/100 ml	
	Italian	EU
A	3	3
B	7333	7333
C	3333	3333
D	2000	2000
E	8333	8333
F	2800	2800
G	3000	3000
H	3767	3767
I	2850	2850
M	9333	9333
N	2067	2067
O	22667	22667
Italian Classes		
I low	<100	<100
II moderate	100-1000	100-1000
III critical	1001-5000	1001-10000
IV strong	5001-20000	10001-100000
V exessive	>20000	>100000

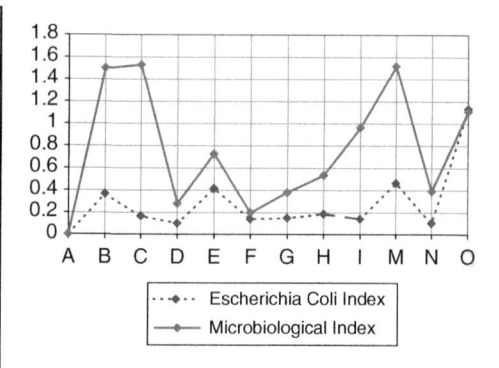

Fig. 28.1 Comparison of microbiological eco-indexes with European (EU) and Italian (National Decree 152/99) classifications for summer 2009

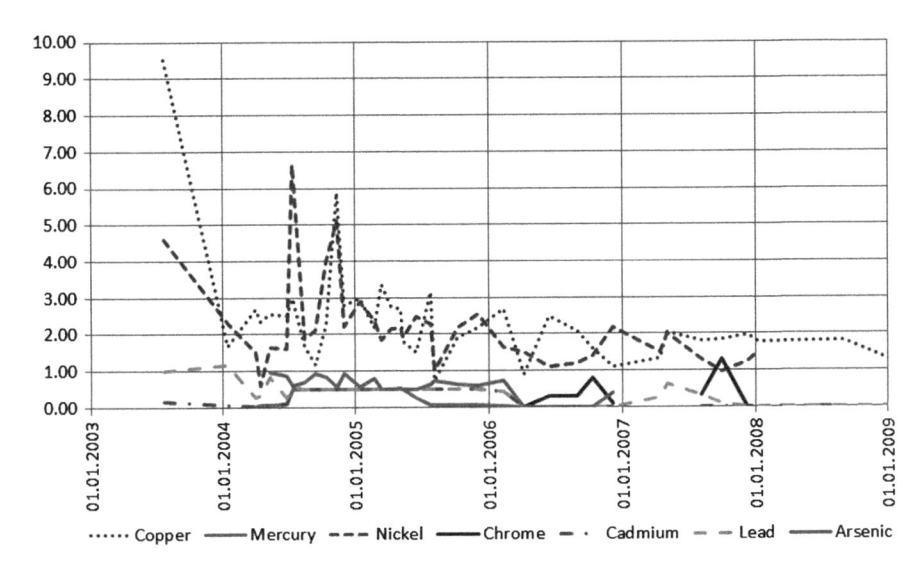

Fig. 28.2 Temporal distribution of the heavy-metal concentrations

bacterial pollution at the stations E, F and O. As seen from both Fig. 28.1 and the table, standard EU and Italian classifications do not allow for recognising these essential features.

28.3.2   Heavy-Metal (HM) Index

Heavy metals belong to the category of most toxic pollutants that are dangerous for living organisms. Our retrospective analysis of River Biferno for the last decade based on the HM-index revealed that the HM-loading generally did not exceed MAC, approaching a maximum in 2004–2006 and then decreasing substantially. This conclusion correlates quite well with the above analysis of microbiological regimes. The advantage of integral indexing is clearly seen from comparisons of temporal-variation curves for separate metals shown in Fig. 28.2 and the integral heavy-metal index in Fig. 28.3 (both calculated for the entire river).

Our analyses of ecological state of River Biferno using MAC-based integral eco-indexes disclosed the following essential features of river's ecosystem, which otherwise would be difficult to detect: (a) steady increase in the trophity level, indicative of increasing anthropogenic impact since 2006, (b) local points of pollution at stations C, E, BF6 and O, (c) deterioration of water by intestinal flora at stations B, E, M and O in the summer, up to the critical water quality class 5 at stations E, F and O in the winter-spring, and (d) generally allowable (<MAC) level of heavy-metal loading and its substantial decrease since 2000. As a result of the sustainability of self-purification processes during the past years, we conclude that River Biferno remains within the category of "oligotrophic-mesotrophic" water bodies [10, 11].

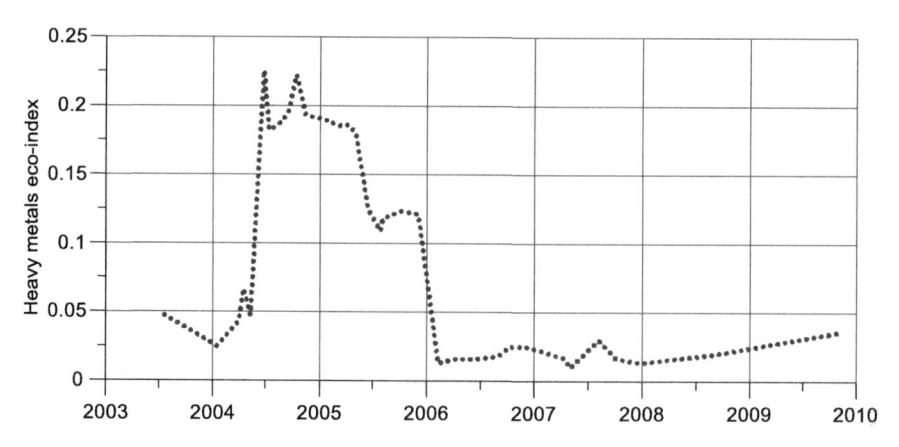

Fig. 28.3 Temporal distribution of the integral heavy-metal eco-index

28.4 Concluding Remarks

Eco-indexing is a convenient method for assessment of water quality, especially in the conditions of strong anthropogenic contamination. It offers the following:

- presenting essential results from monitoring in a clear and illustrative way
- quickly and reliably detecting deviations of water ecosystem from its acceptable state
- employing different systems of basic parameters and by this means better addressing different monitoring tasks
- essentially extending the list of indicators (including those that characterise functioning of hydrobionts, missed in standard methods
- illustratively comparing inter-annual observations and disclosing basic trends and relationships in water ecosystems and water quality
- efficiently calibrating numerical models of water ecosystems

Acknowledgements This work has been supported by EC FP7 projects MEGAPOLI (Contract No. 212520) and PBL-PMES (No. 227915; the Italian national project "Applied research in integrated information system for global management of air quality in industrial settlement of town Termoli and surrounding areas" (Molise, Italy, 2007–2010); and Federal Targeted Programme "Research and Educational Human Resources of Innovation Russia 2009–2013" (Contract No. 02.740.11.5225); and the Russian Federation Government Grant No. 11.G34.31.0048.

References

1. Bashmakova IKh (2004) Ecological indices for estimation of ecological state and water quality in estuarine zones of big rivers. Hydrobiol J 40(3):76–82
2. Bashmakova I (2010) Complementary instruments for assessment of ecological state and ecological risks of water ecosystems, IOP conference series. Earth Environ Sci 13:6 pp, doi:10.1088/1755-1315/13/1/012011

3. Bashmakova I, Levikov S (2003) Statistical analysis and modelling of hydrochemical, hydrobiological and ecological conditions in the Danube river basin. 35th international conference IAD, Apr 2003, Novi Sad, Yugoslavia
4. Bashmakova I, Levikov S (2007) Application of integral indices to the assessment of ecological risks and damages. In: Air, water and soil quality modelling for risk and impact assessment. Springer Verlag, New York
5. Baumert H, Levikov S, Pfeiffer KD, Goldman D (2002) The MARS data collection and analysis. Work reports, HYDROMOD Scientific Consulting, Wedel
6. Edberg SC, LeClerc H, Robertson J (1997) Natural protection of spring and well drinking water against surface microbial contamination. Pt. II: indicators and monitoring parameters for parasites. Crit Rev Microbiol 23:179–206
7. EU Bathing Water Quality Directive (1976) EU Bathing Water Quality Directive 76/160/EEC
8. EU Bathing Water Quality Directive (2006) EU Bathing Water Quality Directive 2006/7/EC
9. EU Water Framework Directive (2000) EU-Water Framework Directive 2000/60/EC Council Directive 2000/60/EC of 23 October 2000: Establishing a framework for community action in the field of water policy
10. Fernando HJ, Bashmakova I, Mammarella M-C, Grandoni G, Morgana JG (2010) Background scientifico. In: Mammarella MC et al (eds) La meteodiffusivita per una migliore gestione della qualità dell´aria. Amadore Editore, Roma, pp 35–39, 144 pp. ISBN 978-88-6081-794-5
11. Fernando HJ, Mammarella M, Grandoni G, Bashmakova I, Smirnov A (2010) Indicatori della qualita dell´ambiente. In: Mammarella MC et al (eds) La meteodiffusivita per una migliore gestione della qualità dell´aria. Amadore Editore, Roma, pp 105–119, 144 pp. ISBN 978-88-6081-794-5
12. Pollard P, Huxham M (1998) The European water framework directive: a new era in the management of aquatic ecosystem health? Aquat Conservat Mar Freshwat Ecosyst 8:773–792
13. Smith VH (2003) Eutrophication of freshwater and coastal marine ecosystems: a global problem. Environ Sci Pollut R 10:126–139
14. Sokal RR, Rohlf FJ (2000) Biometry: the principles and practice of statistics in biological research. Freeman, New York, 887 pp
15. Van der Oost R, Vindimian E, van den Brink PJ, Satumalay K, Heida H, Vermeulen NPE (1997) Biomonitoring aquatic pollution with feral eel (Anguilla anguilla. III. Statistical analysis of relationships between contaminant exposure and biomarkers). Aquat Toxicol 39:46–75

Chapter 29
On Some Issues of the Anthropogenic Transformation of Water Ecosystems (Case Study of Lake Sevan)

Trahel Vardanian

Abstract This paper discusses the influence of economic development in the basin of Lake Sevan. It shows that the decrease of lake's water level has led to changes in the hydrological regime. The development in the basin has caused disruption of thermal and hydro-chemical regimes of the lake, deteriorating the water quality and increasing the water turbidity. The circulation of biological and chemical species has changed as well.

Studies of the chemical composition of the water were launched at the end of the nineteenth century, and the first salt balance was determined in the 1930s. According to routine observations, 1 l of the lake's water contains about 0.7 g salt in ionic form, which has changed by about 5–10% as a result of the decrease of water level. There have been changes in the general mineralization that are related to the decrease of the water level and the magnitude of its flow.

In 1928–1930, well before artificial changes to the lake's water level were realized, the total mineralization was 718.4 mg/l, while today it is 673 mg/l (1999–2002). The decrease of total general mineralization of lake water is strongly related to the massive outflow of salty water, which removed salts that have been accumulating in the lake for centuries.

The studies of separate components of Lake Sevan balance show that under conditions of the global climate warming, evaporation from the lake's surface may reach up to 145×10^6 m^3/annum. Averaging of the consequences of possible climate change on river runoff points at the possible decrease of water resources of the lake's basin by −2.51% by the end of the first half of the twenty-first century.

T. Vardanian (✉)
Department of Physical Geography and Hydrometeorology, Yerevan State University,
1, Alek Manoukian Street, Yerevan 0025, Armenia
e-mail: tvardanian@ysu.am

H.J.S. Fernando et al. (eds.), *National Security and Human Health Implications of Climate Change*, NATO Science for Peace and Security Series C: Environmental Security, DOI 10.1007/978-94-007-2430-3_29, © Springer Science+Business Media B.V. 2012

It is hard to predict the future developments of these processes. However, the issue of Lake Sevan is not entirely settled as the ecosystem of the lake is damaged and is undergoing the process of eutrophication. The flora and fauna of the water and coast have undergone serious and irreversible changes.

Keywords The anthropogenic changes of lake Sevan level • Hydrochemical composition and regime of the lake • Mineralization of lake Sevan water • Ionic composition of the lake in different years • Hydrobiological regime • The thermal regime • Climate change • Ecosystem of Lake Sevan • Eutrophication of Lake Sevan

29.1   Introduction

Numerous rivers, lakes and, in general, water bodies have undergone considerable changes under the impact of the economic activity of man. As a result, hydrological, hydrochemical, and biological conditions, as well as the morphometrical elements of these bodies, have been destroyed. Lake Sevan and its basin may serve as a classical example in this respect (Fig. 29.1).

Lake Sevan is perceived as a miracle of nature, an ecologically and economically significant water body, and a national treasure for Armenia. At 1,900 m above sea level it is one of the highest lakes in the world. But it is especially unusual because it is a fresh water lake located in the dry subtropical climate belt. Other lakes in the same belt, such as Van and Urmia (in the Armenian Highland), the Dead Sea, Tuz (Middle East), Lobnor (Central Asia), and Issik-Kul (Central Asia), are all salty and their water is not fit for irrigation and drinking. The salinity of Lake Sevan water is

Fig. 29.1 Lake Sevan

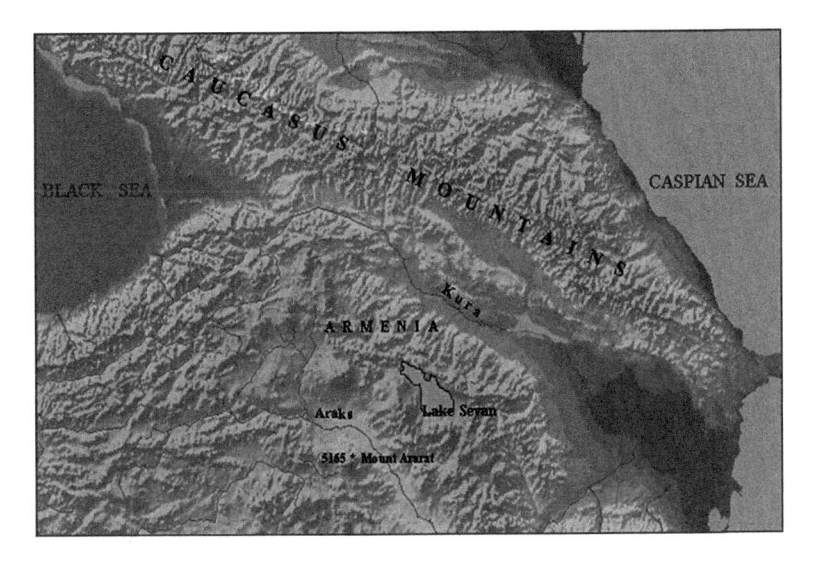

Fig. 29.2 The physical-geographic position of Lake Sevan

Table 29.1 Some hydrometric indices of Lake Sevan

Indices	Unit of measurement	Before the drop of the level	Present-day condition
Drop of lake level	m	0.0	17.4
Height above sea level	m	1,915.9	1,898.5
Watershed surface	km²	3,475	3,639
Lake surface	km²	1,416	1,252
Mean depth	m	41.3	27.8
Maximum depth	m	98.7	81.4
Water amount	km³	58.5	34.8

only 0.6–0.7 g/l, which makes it a major source of water for Armenia, and potentially for other countries in the region (Fig. 29.2). The lake is surrounded by mountain ranges exceeding 3,000 m in altitude which are the sources for the water for the numerous streams flowing into the lake.

Before 1930, the surface of Lake Sevan was 1,916 m above sea level. The surface of the drainage area of the lake before its artificial drop (1930s) was 3,475 km², which is larger than the surface of the lake by 2.5 times (1,416 km²), and the volume of lake water was 58 billion m³ (Table 29.1).

Lake Sevan is fed by water from inflowing rivers, precipitation falling on the surface and by groundwater inflow. Water is removed from the lake by evaporation, infiltration, and flow out through the Hrazdan river.

Rivers are the main water source for Lake Sevan (Fig. 29.3). There are 28 rivers and streams of 10 km and longer (Table 29.2). The river network is rather dense in

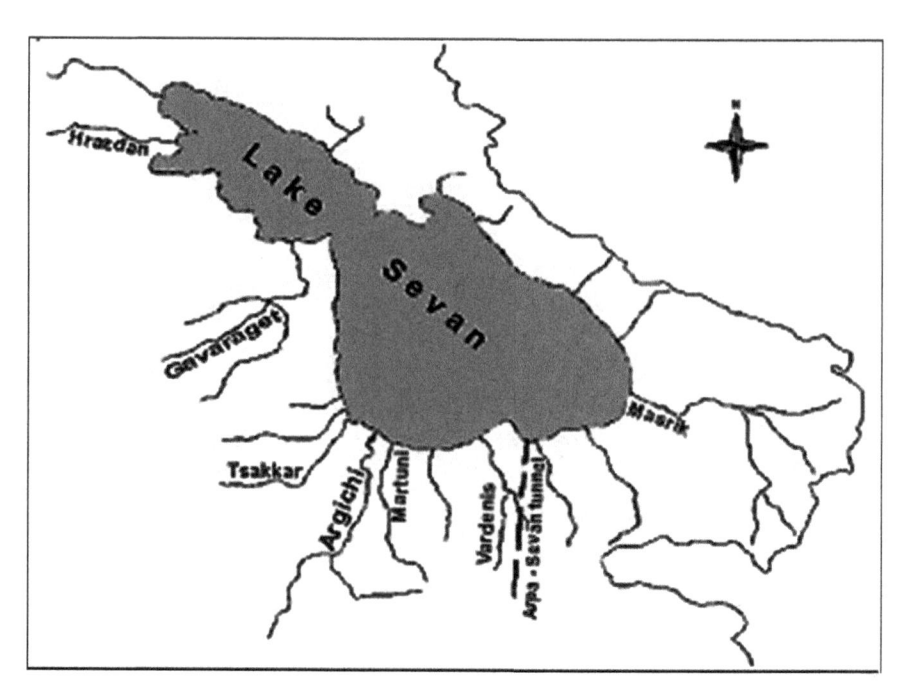

Fig. 29.3 The river network of Lake Sevan

Table 29.2 Some hydrometric and hydrological characteristics of relatively large rivers of the basin of Lake Sevan

River-observation post	River length	Size of watershed basin, km²	Mean height of watershed, m	Mean annual discharge, m³/s	Runoff, l/s km²
Argiji-V. Getashen	51	384	2,470	5.55	14.5
Gavaraget-Noraduz	41	467	2,430	3.51	7.5
Masrik-Torf	45	685	2,310	3.42	4.9
Vardenis-Vardenik	24	116	2,680	1.64	14.1
Karchaghbjur-Karchaghbjur	26	117	2,650	1.15	9.8
Martuni-Geghovit	20	85	2,760	1.41	16.6
Dzknaget-Tsovagjugh	21	85	2,220	1.06	12.5

the south and south-western parts of the basin, which has the largest rivers, the Argiji, Gavaraget, Masrik, and Vardenis.

The anthropogenic changes in the level of lake Sevan took place during the last 70 years after the decision to deepen the river bed of the Hrazdan River, the only river flowing out of the lake, with the aim of using the lake water for irrigation and energy generation.

The aim of this research is to explore the influence of the dynamics of the artificial lowering of the lake's level on the ecological and other relevant changes,

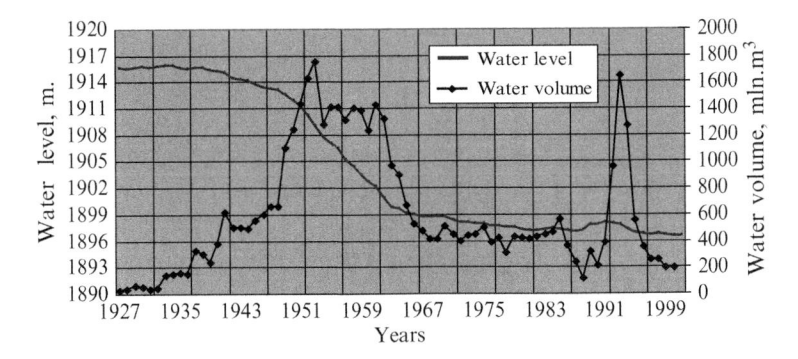

Fig. 29.4 The graph of perennial fluctuation of lake Sevan level: annual total water outflow volume and average lake level

as well as the regime of the lake (in the last 70 years), with the case of Lake Sevan.

29.2   Ecological and Associated Changes

In the world limnology, there has been no other case when a lake level was artificially lowered by 18 m within three to four decades (1930–1970), and by another 2 m in the last decade (1990–2000) (Figs. 29.4 and 29.5).

Sevan is the only lake in this respect, and is considered to be a large natural laboratory where one can observe all those processes connected with the decrease of erosion basis of flowing into the lake rivers, and which cannot be studied under laboratory conditions. Among these processes, the hydrological, thermal, hydrochemical, carbon regime of the lake, as well as biological conditions, which have served as a rich material for scientific researches, are rather important.

The decrease of erosion activated channel processes of the rivers flowing into the lake (Fig. 29.6). It brought about the violation of the balanced profiles of river valleys, formed within thousands of years. The active down-cutting erosion destroyed the foundations of bridges and caused their collapse.

29.3   The Hydrochemical Regime

Lake Sevan is one of the world's fresh-water lakes. Studies of the chemical composition of the water were launched at the end of nineteenth century, and the first salt balance was determined in the 1930s [1]. According to routine observations, 1 l of

Fig. 29.5 As a result of the decrease of the lake's water level, the Island became a Peninsula

the lake's water contains about 0.7 g salt in ionic form. The hydro-carbonate ion (HCO3) and chlorine (Cl), at 414.7 and 62.3 mg/l respectively, dominate in the ionic composition among non-metals. Among cations, magnesium (Mg), with 55.9 mg/l, and sodium plus potassium (Na + K), with 98.7 mg/l, are dominant (Table 29.3).

Observations of the chemical composition of Lake Sevan water in its natural state, i.e. before its artificial lowering (1930), and the changes that have taken place since then are shown in Table 29.3 and Fig. 29.7.

Almost all the ion concentrations have changed by about 5–10% as a result of the drop of the lake's water level. Calcium concentration has changed most of all. At the end of the nineteenth century, the calcium concentration was 38 mg/l (Stakhovskij 1893), while a century later it was only 21 mg/l [2]. This phenomenon has not yet

Fig. 29.6 The riverbed of the Dzknaget river, which [Riverbed] has deepened as a result of the drop of Lake Sevan level

Table 29.3 The mineralization of lake Sevan water and its ionic composition in different years (mg/l)

Ionic composition	Before the drop of the level (1928–1930)	Present-day condition
Ph	9.2	8.6
Ca	33.9	20.6
Mg	55.9	55.4
Na + K	98.7	92.8
HCO_3	414.7	373.0
CO_3	36.0	21.2
CL	62.3	68.0
SO_4	16.9	29.2
$\sum U$	**718.4**	**660.2**

been closely studied, but some opinions link the lack of calcium in Lake Sevan water to the intensification of plankton photosynthesis. The concentration decrease started in 1978. Simultaneously, lake productivity dropped and the growth of blue-green algae abruptly decreased. These changes were observed after an increase of turbidity. The current low level of calcium may lead to another change. The lake water is a rich solution containing Ca^{2+} and CO_3^{2-} which originate form a poorly soluble $CaCO_3$ carbonate, which settles on the lake bottom.

Writing the ions in the water in decreasing order, we have:

$$HCO_3 > Na + K > Cl > Mg > SO_4 > CO_3 > Ca,$$

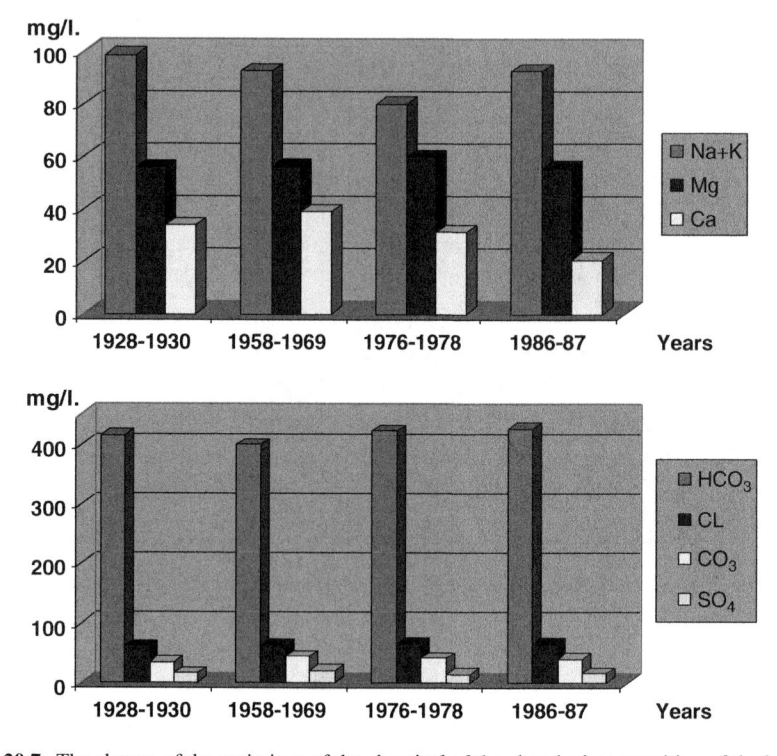

Fig. 29.7 The change of the main ions of the chemical of the chemical composition of the lake

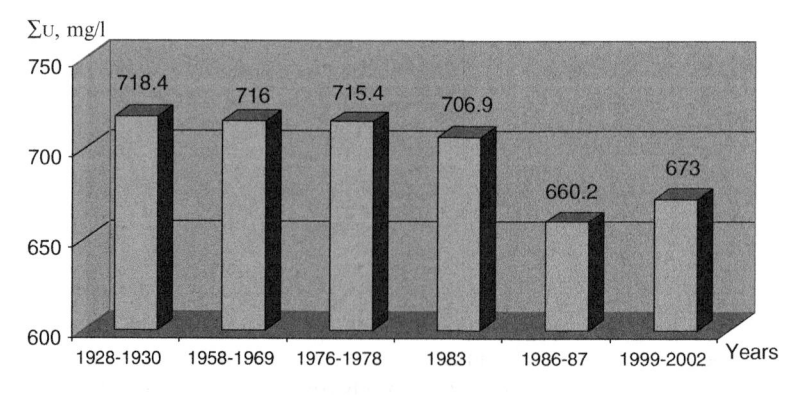

Fig. 29.8 The change of the total mineralization (∑U, mg/l)

Starting with **HCO₃ > Mg + Ca** indicates that according to Aliokhin's classification, Lake Sevan belongs to the magnesium group of the hydro-carbonate lake category.

There have also been changes in the general mineralization of the Lake that are related to the drop of the level of the Lake and the magnitude of its flow (Fig. 29.8). In 1928–1930, well before the artificial change in lake level, the total mineralization was 718.4 mg/l. Even during the rapid changes of 1958–1969 it changed little, being 716 mg/l. After 1980, the mineralization sharply decreased, becoming 706.9 mg/l as low as 660 mg/l in 1986–1987 and 673 mg/l at present (1999–2002).

The decrease of the total general mineralization of the lake water is strongly related to the massive outflow of salty water, removing the salts which had been accumulating in the lake for ages. For comparison, the total mean mineralization of the river waters flowing into the lake is 160–180 mg/l [3].

29.4 The Thermal Regime of the Lake

A few lake water temperature observations were made by Davidov in the 1920s, but no fixed observation stations were established until the end of the 1940s. The early data are sparse, but they suggest that the artificial drop of the water level seriously affected all the components of the thermal regime. As a consequence of the reduction in water mass, Lake Sevan now warms and cools very fast. In Big Sevan the mean monthly temperatures range averaged 16.8°C in 1951–1960 and 18.3°C in 1971–1980 [4]. Before the drop in water level, ice rarely covered the entire surface. From 1890 to 1960 the lake was covered with ice only 9 times [5], but now the ice covers it almost every year with a layer 20–30 cm thick. All this is explained by the quality of heat accumulation. Before the drop in water level, the heat content was 700*1,012 kcal, but today it is only 500*1,012 kcal [4].

According to the observations, the normal mean annual temperature on the lake surface is 9–10°C. The lowest temperature is in February, 1–2°C, and the highest, 17–18°C, in August. Usually the water temperature near the coast is 2–3°C is higher than that in the lake center, and the near-coast water temperature may be as high as 23–24°C.

29.5 The Hydro-Biological Regime and the Conservation Issues

In contrast to numerous other lakes in the world, Lake Sevan previously had a nutrient regime favorable for biologic activity. The artificial decrease of the lake level and the changes of the hydrological, thermal, saline, and gas regimes could not but affect the biological processes of the lake. Many biological indices have dramatically changed (Table 29.4). Most obvious are the changes which have led to more favorable conditions for algae growth and the first signs of eutrophication. From 1964 onwards, the lake started to 'bloom' with numerous algae, particularly the blue-green ones. The 'blooming' decreased the water quality and caused bacterial pollution. In the 1960s the number of bacteria per liter was more than twice the concentration found before the drop of the lake level, although the waters still had

Table 29.4 The change of some chemical and biological indicators of lake Sevan as a result of the drop of the water level [6]

Indicators	Measurement unit	Before drop of the level	Present condition
Permeability	M	14.3	4.5
PH hydrogen indicator	–	9.2	8.7
Sum of ions	g/m^3	720	680
Oxygen quantity in hypolymnion	g/m^3	8.0	2.0
Mineral nitrogen	g/m^3	0.003	0.16
General nitrogen	g/m^3	0.07	0.64
Mineral phosphorus	g/m^3	0.32	0.007
General phosphorus	g/m^3	0.37	0.08
Phytoplankton	g/m^3	0.32	2.4
Biomass		91.0	483.6
Primary product	$kcal/m^2$ p.a.	1000.0	5000.0
Macrophytes	$10^3 t/p.a.$	900.0	26.0
Zooplankton	g/m^3	0.45	0.70
Zoobenthosis	g/m^3	3.38	11.0
Fish	t/p.a.	1000.0	2400.0

the conditions of an oligotrophic lake. From the 1980s, however, the number of bacteria doubled again, giving a concentration which now corresponds to the conditions of a mesotrophic lake.

The drying of the coastal swamps and wetlands has affected about 100,000 ha. A direct consequence is that only 30 species of birds remain in the area, where recently there were as many as 160 different species [6].

29.6 The Influence of Climate Change on the Water Level of Lake Sevan

The studies of the separate components of Lake Sevan balance show that under conditions of global climate warming, evaporation from the lake's surface may reach up to 145 mln·m³/annum [7]. Averaging of the consequences of the possible climate change on river runoff points at the possible decrease of water resources of the lake's basin by −2.51% by the end of the first half of the twenty-first century [8]. This means that the water economy situation will seriously change, since the decision on the issue of keeping the ecosystem of the lake in a regular state will become complicated. To solve that issue, it will be necessary to increase the volume of the water taken into Lake Sevan via the tunnel Vorotan-Arpa-Sevan.

In case of various scenarios of climate change, the mean annual water level in the lake will decrease by 3–4 cm (without consideration of the artificial water provision via the aforementioned tunnel as well as the water draw-off through the derivational canal Geghamavan), and the free runoff will become twice as small [7]. In this case, the amount of generated energy will reduce. Climate warming will lead to more drawdowns from the lake, which will mean that the necessity of compensating the deficit drawdown from the lake will increase as much as the total of additional need for soil humidification and decrease of a local alimenting runoff. In this case, the amount of generated energy may grow in respect with present-day conditions. Thus, the increase of renewable water resources under condition of global climate warming as well as evaporation increase from the surface of the lake will bring about negative changes in the water basin of Lake Sevan.

29.7   Conclusions

In conclusion, it must be noted that the decrease of the lake's water level and the economic development in the basin have brought about the change in the ecosystem of the lake. The latter caused the disruption of the thermal and hydro-chemical regimes of the lake, the quality of the water deteriorated, and water turbidity increased. The inner circulation of the water substances, as well as the circulation of the biological substances altered as well.

In the next 20–30 years it is envisaged to increase the level of the lake by only 4–6 m, because if the level rises more than that, the recent coastal constructions (roads, railway, resort houses and others) as well as tree-shrub vegetation (artificially planted and grown after the drop of water level) will go under water. However, submergence is reality today. In the recent 5 years, the water level of the lake has increased by more than 2 m, which is the effect of the growth surface flow and decrease of outflow from the lake. As a result, a new issue emerged namely, the coastal green zone is under water (considerable water lever increase took place so fast and unexpectedly that there was no time to clear the coastal line from its green cover), and thus endangering the lake with eutrophication anew (Fig. 29.9).

Fig. 29.9 The coastal green zone is under water

It is hard to predict the future developments of these processes. However, the issue of Lake Sevan is not entirely settled, the ecosystem of the lake is damaged, undergoing the process of eutrophication. The flora and fauna of the water and coast underwent serious and irreversible changes. These are the old and new issues of Lake Sevan and its basin.

References

1. Liatti SJa (1932) The hydrochemical study of lake Sevan: materials on the researches of lake Sevan and its basin. Part IV, issue 2, Leningrad
2. Vardanian TG (1990) On several problems of mineralization and ion Structure of Lake Sevan water. Scientific notes of Yerevan State University, Natural Sciences, N 1(173), Yerevan, pp 140–145
3. Vardanian TG (1999) The condition of mineralization of the rivers of lake Sevan basin. In: Proceedings of the national academy of sciences of the RA, Earth Science, LII, N2–3, Yerevan, pp 97–100
4. Musaelian SM (1993) The ecology and economy of lake Sevan and its basin. Yerevan University Press, Yerevan 164 p (in Russian)
5. Gabrielian HK (1980) The pearl Sevan. "Hayastan" Press, Yerevan 135 p (in Armenian)
6. The biodiversity of Armenia, the first report. RA ministry of nature protection, Yerevan (1999), pp 13–15
7. Armenia: climate change problems/Collected articles, Issue I. Gabrielyan AH (ed). Yerevan: "Science" Press of National Academy of Science of RA (1999), 373 p
8. Vardanian TG (2009) Forecasting and evaluations of the runoff of rivers in lake Sevan basin in various scenarios of climate changes. Materials of all-Russian conference, dedicated to the 100-the anniversary of Professor O.A. Drozdov "Contemporary issues in climatology", St-Petersburg, Russia, pp 47–48

Chapter 30
Effects of Climate Change on Egypt's Water Supply

Gamal Elsaeed

Abstract Egypt is plagued by a water shortage as well as water resource management issues. Egypt, as a developing country, is at particular risk for being unable to provide clean drinking water and adequate sanitation systems for citizens, ensure sustainable irrigation, use hydropower to produce electricity, and maintain diverse ecosystems. The Egyptian Environmental Affairs Agency report notes that Egypt's fresh water budget runs a deficit: supply, which comes from the Nile (95%), precipitation (3.5%) and ground water (1.5%) is less than current demand. Egypt has available fresh water reserves of 58 billion m^3, but an annual water demand of about 77 billion m^3. This deficit is met through recycling treated sewage and industrial effluent (four billion m^3) and recycling used water, mainly from agriculture (eight billion m^3). An additional four billion m^3 is extracted from the shallow aquifer and three billion m^3 comes from the Al Salam Canal Project. Egypt is therefore in a situation where it must plan for several different future scenarios, mostly negative, if climate change results in increased temperatures and decreased precipitation levels. Even in the absence of any negative effects of climate change, Egypt is dealing with a steady growth in population, increased urbanization, and riparian neighbors with their own plans for securing future water needs. All of these will require Egypt to put water resource planning as a top national security priority.

Keywords Egypt • Climate change • River Nile • Water resources • Nile basin

G. Elsaeed (✉)
Department of Civil Engineering, Faculty of Engineering, Shobra, Banha University,
Cairo, Egypt
e-mail: gelsaeed@feng.bu.edu.eg

H.J.S. Fernando et al. (eds.), *National Security and Human Health Implications of Climate Change*, NATO Science for Peace and Security Series C: Environmental Security, DOI 10.1007/978-94-007-2430-3_30, © Springer Science+Business Media B.V. 2012

30.1 Introduction

Any assessment of Egypt's water resources recognizes the country's enormous reliance on the Nile, which makes up about 95% of Egypt's water budget. Other sources of Egypt's water budget, precipitation and ground water, do not make up more than 5% of the available supply, although the effect of increases or decreases in precipitation near the sources of the Nile can have a larger than expected effect on Nile flows.

Egypt's total water budget is produced by a combination of three variables: the Nile (95%), precipitation (3.5%) and ground water (1.5%). The Nile produces 55.5 billion m^3, while the latter two variables combine to form safely about 2.2 billion m^3 of fresh water. In total, Egypt has available fresh water reserves of 58 billion m^3.

Egypt's annual water demand is about 77 billion m^3. The deficit between Egypt's water supply and demand must be met through recycling. The 19 billion m^3 deficit is filled by a combination of treated sewage and industrial effluent (four billion m^3) and recycling used water, mainly from agriculture (eight billion m^3). An additional four billion m^3 is extracted from the shallow aquifer and three billion m^3 comes from the Al Salam Canal Project.

Recycling is partly natural and partly intentional. Water reclaimed from agriculture is a natural process of drainage waters returning to the Nile. The remaining two sources of recycled water, the Al Salam Canal and extraction from the shallow aquifer, are manmade solutions to the deficit.

Consumption of the 77 billion m^3 in annual water demand in Egypt is mainly from agriculture (62 billion m^3). An additional 10% (eight billion m^3) is used as drinking water. Approximately 95% of the population relies on this water for drinking purposes. The remaining demand comes from industry (7.5 billion m^3).

The following paper will focus on the impact of climate change on water supply and the potential challenges Egypt will face in the future if the balance between water supply and demand is altered.

30.2 Vulnerability of Water Resources to Climate Change

Managing water resources will become a more complex endeavor with climate change. Analysis predicts that climate change will intensify and accelerate the hydrological cycle, which will result in more water being available in some parts of the world and less water being available in other parts of the world (most of the developing world). Weather patterns are predicted to be more extreme. Those regions adversely affected will experience droughts and/or possible flooding.

Is Egypt vulnerable? The answer is yes. The Nile waters are highly sensitive to climate change, both in amount of rainfall and variations in temperature. And since these two factors are also interrelated, i.e., temperature changes affecting rainfall, it can be expected that climate change will take the form of changes in levels of

Table 30.1 Change of flow corresponding to uniform change in rainfall for Nile sub-basins

	Change in rainfall					
	−50	−25	−10	+10	+25	+50
Sub basin	Corresponding change in water flow (%)					
Atbarra (Atbara)	−93	−60	−24	+34	*84	+187
Blue Nile (Diem)	−92	−62	−24	+32	+78	+165
Blue Nile (Khartoum)	−98	−77	−31	+36	+89	+149
Lake Victoria (Jinja)	−20	−11	−4	+6	+14	+33
White Nile (Malakal)	−41	−28	−11	+19	+48	+63
Main Nile (Dongla)	−85	−63	−25	+30	+74	+130

Table 30.2 Nile flows under sensitivity analysis

Precipitation	−20%	−20%	−20%	0%	0%	+20%	+20%	+20%
Temperature	0	2	4	2	4	0	2	4
Flow (BCM)	32	10	2	39	8	147	87	27
% of base	37	12	2	46	10	171	101	32

precipitation as a result of changes in temperature, or other factors, and that the resulting effect on the Nile flows will be from moderate to extreme, with the latter scenario most likely in the long term.

In terms of levels of sensitivity of Nile flows, the Nile waters are separated into three areas, containing sub-basins: the Eastern Nile, comprised of the Atbara and Blue Nile, the Equatorial Nile, which is Lake Victoria at Jinja, and the Baba El Ghazal Basin, which is the White Nile at Malakal.

The range of sensitivity to rainfall differs from one sub-basin to another. The Eastern Nile is extremely sensitive, the Baba El Ghazal Basin moderately sensitive, and the Equatorial Nile only minimally sensitive. Table 30.1 summarizes the levels of sensitivity to rainfall in the different Nile sub-basins.

Nile water flows are also sensitive to temperature changes. The EEAA* report cites Hulme et al. [4] who argue that changes in temperature affect evaporation and evapotranspiration correspondingly. Increases in evaporation and evapotranspiration as a result of increases in temperature could reduce the levels of water flows in some Nile sub-basins by double or triple the percentage of evapotranspiration. Table 30.2 displays the results in the EEAA* report of a study by Strezpek et al. [7] on Nile sensitivity to temperature changes.

The sensitivity of Nile flow to climate change is strongly supported by the above data. For example, an increase in temperature of 4°C coupled with a 20% decrease in precipitation could decrease the flow in Nile by 98%. A slightly smaller increase in temperature (2°C) with the same reduction in precipitation could result in an 88% decrease in Nile flows. Thus climate changes have a potentially dramatic effect on Nile flows and thus on water resources for Egypt, which is heavily dependent on the Nile for its water supply.

30.3 Scenarios of the Effect of Climate Change on Nile Flows

Table 30.3 shows the results of three Global Circulation Models (GCMs) used in 1996 to estimate future Nile flows and cited in the EEAA* report. Variables in the studies include precipitation, temperature, increases in CO_2, and flow rate in Nile. An assumption in the models is that increases in CO_2 concentrations would result in increases in temperatures. Results indicate that even with increases in the amount of precipitation, Nile flows would decrease in two of the three scenarios as a result of rises in temperature.

The conclusions drawn from the above calculations are as follows:

- Contributions from the Equatorial Nile to downstream Nile flows can be reduced to zero by only minor increases (2.7–4.8°C) in temperature or minimal decreases (15–17%) in precipitation
- Water loss from evaporation and evapotranspiration, currently occurring to a large extent on the water surfaces and from vegetation on the Bahr el Ghazal basin would be even greater with any increases in temperature
- Climate changes would most likely result in the Eastern Nile retaining its essential role in preserving Nile flows

The EEAA* report cites a research model by Strezpek et al. [8], who developed ten different scenarios for Nile flows. Nine of the ten predict reductions in Nile flows from 10% to 90% by the year 2095. Even in the short term, by 2025 losses are estimated at 5–50%. Figure 30.1 shows the Strezpek models of changes.

The question of the vulnerability of the Nile flows to amounts of rainfall upstream was demonstrated in a 2005 United Nations Environment Program (UNEP) (2005) [9] study cited in the EEAA* report. Gauging stations along the Nile measure water levels. Figure 30.2 displays a stream hydrograph charting the levels taken at the Atbara gauging station, established in the early twentieth century. Monitoring at the Atbara station was recorded for a 90-year period: (1907–1997). Over this period, water levels rose and fell but can be divided into three recognizable periods: rising slightly from 1907 to 1961; dropping sharply from 1962 to 1984, and recovering from 1987 to 1997 to pre-drop levels. What is significant is that these increases and decreases in water levels coincided with increases and decreases in the amount of flows from the Ethiopian highlands as a result of rainfall. This is strong evidence of the effect of upstream rainfall on downstream water levels.

A 2007 study by Bergen University in Norway under the Nile Basin Research Program (cited in the EEAA report), focuses on three sub-catchments of the Nile basin: the Atbara in Ethiopia near the border with Eritrea, the Kagera on the Uganda-Rwanda border, forming the Southwest of the Lake Victoria Basin, and the Gilgel Abbay in the Blue Nile Basin, which is the main feeder of Lake Tana in Ethiopia. Figure 30.3 shows the location of three catchments in red in relation to the Nile river, in blue.

Research conducted during the period August-December 2007 on these three sub-basins supported other research, such as that of Kite and Waitutu (1981),

Table 30.3 Nile flows under GCM scenarios

	Base	UKMO	GISS	GFDL
Precipitation	100	122	131	105
Temperature	0	4.7	3.5	3.2
Flow (billions m^3)	84	76	112	20
% of base	100	91	133	24

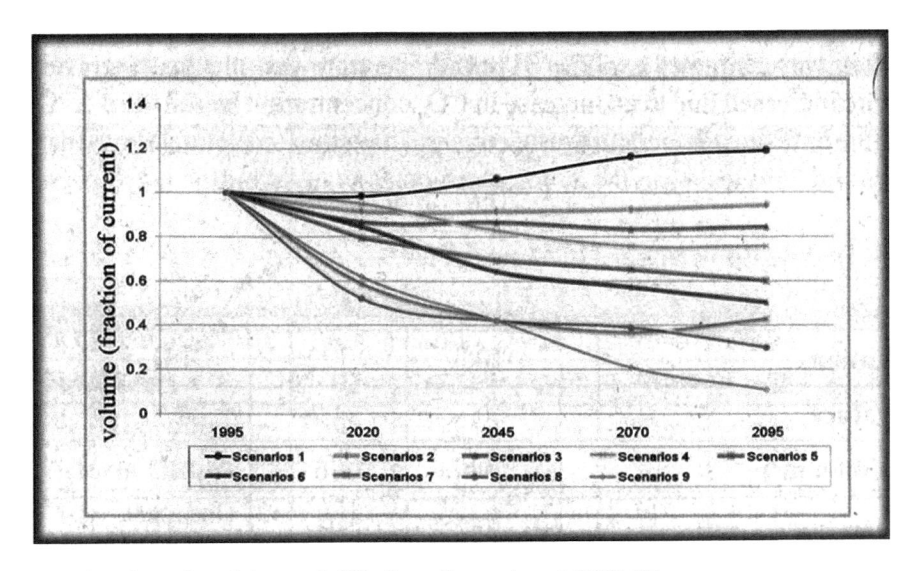

Fig. 30.1 Scenarios of changes in Nile flows (Strepzek et al. 2001) [8]

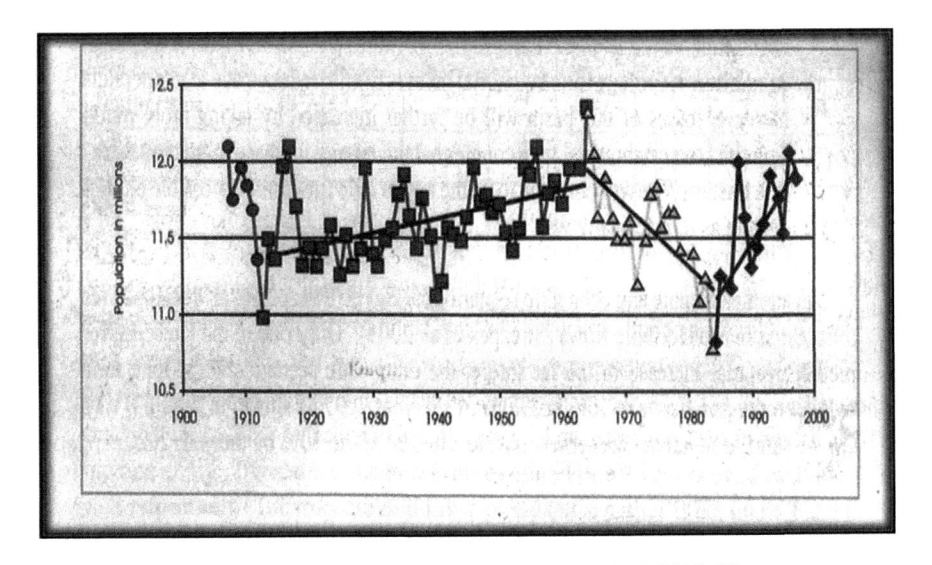

Fig. 30.2 Annual average stream levels on the Nile at Atbara (UNEP 2005) [9]

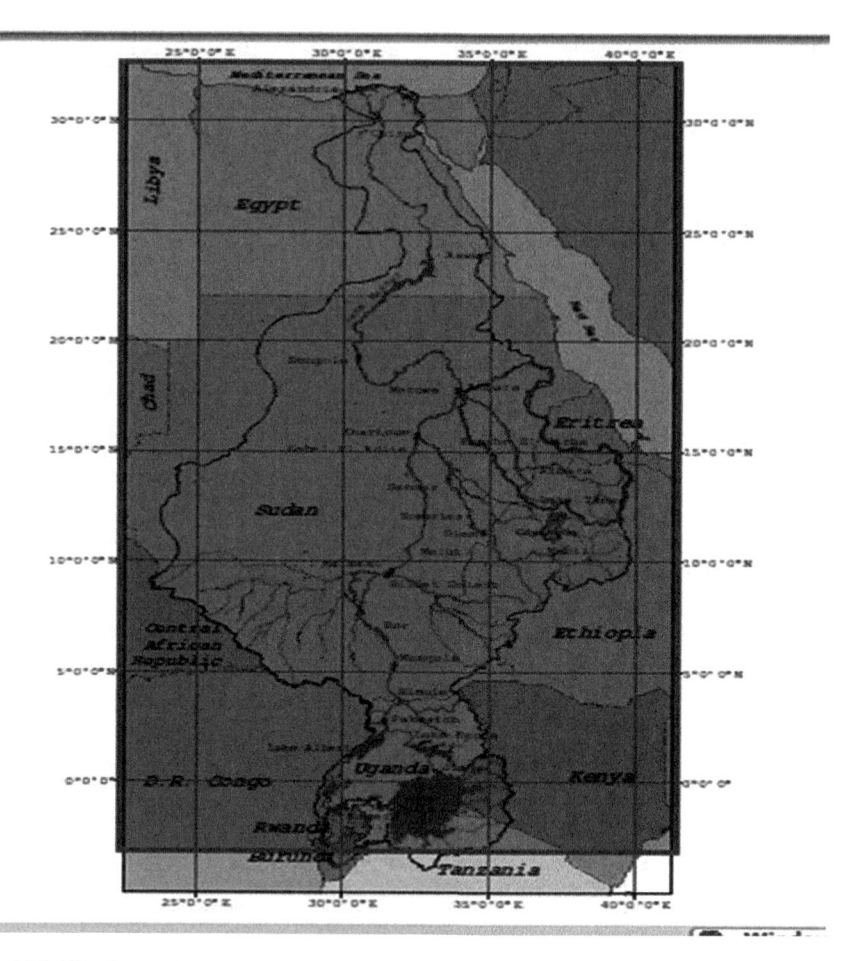

Fig. 30.3 The three sub-catchments of the Bergen University study [4]

studying the Nzoia River, a tributary of Lake Victoria. The results show that runoff
is more sensitive to precipitation changes than to evapotranspiration. The Kagera
catchment has a large base flow due to the regulating effect of the lakes and swamps
in the sub-basin. Of the three catchments, the Kagera is the most sensitive to evapotranspiration and this may be due to its higher aridity [4]. The significance of this
study is twofold: it is necessary to understand both the main features of each catchment and the sensitivity of each catchment to changes in rainfall and evapotranspiration because this assists countries in the Nile region to take more proactive
measures in light of fluctuations in Nile flows as a result of fluctuations in rainfall.

The EEAA* report cites Agrawala et al. [1], who concluded in an Organization
for Economic Cooperation and Development study that Egypt's vulnerability in
terms of water resource dependence on the Nile is tied to trends in population

growth, land use, and agriculture and economic activity being almost exclusively focused along the Nile Valley and Delta. As demand increases, due to growth in population and any increases in temperature leading to greater evaporative losses of the country's allocation of Nile water, Nile water availability is likely to be increasingly stressed. Any activity upstream that added to the diminishing of available water resources in Egypt, whether manmade by upstream riparian countries or otherwise unaccounted for could seriously exacerbate this stress on Egypt. Countries downstream in the Nile basin, of which Egypt is by far the most populous and the most dependent on the Nile for its water needs, are sensitive to the variability of the runoff from the Ethiopian part of the basin, according to the International Water Management Institute (IWMI) together with Utah State University (Kim et al. [6]) and cited in the EEAA* report.

The report notes that future hydropower dam operation in the upstream part of the Nile basin may have an impact on the Nile basin if future climate scenarios materialize. They are summarized as follows:

- Climate changes in the form of more precipitation and higher temperatures in most of the Upper Blue Nile River Basin
- Higher and more severe low flows, but droughts of less frequency over the mid to longer term
- Minimal or negligible effect of dam operations on water availability to Sudan and Egypt

The results are, however, uncertain with existing accuracy of climate models. In other words, there is no clear indication that suggests any one specific scenario, and the region could take actions to produce hydropower, increase flow duration and increase water storage capacity without affecting outflows to riparian countries into the 2050s.

In summary, all studies clearly show that the results obtained of the impact of climate change in the Nile Basin are strongly dependent on the choice of the climate scenario and the underlying GCM experiment.

30.4   Other Vulnerability Indicators

It is easy to overlook factors beyond climate change that make Egypt vulnerable to water shortages in the future: The following looks at other consideration for Egypt and the Nile Basin.

30.4.1   Population Growth and Urbanization

Population growth and extension of inhabited areas increase wastewater disposal causing deterioration of river quality from upstream to downstream. Upstream excessive urbanization may result in increased flooding downstream due to the reduction in infiltration and evapotranspiration from natural vegetation and more runoff downstream.

30.4.2 *Water Related Conflicts*

Upstream Nile countries depend more directly on rain, which sustains forestry, wildlife, wetlands, rain-fed agriculture, fishing and groundwater recharge. For tail end countries, Nile water is the only source of irrigated agriculture and drinking water supply. The per capita share of water in the Nile Basin stands now at 1,000 m^3 per capita per year. This is expected to drop by 50% by the year 2050. Water problems upstream are related to drainage, flood protection occasional drought, infrastructure, while water problems downstream are mainly related to scarcity.

There is high potential for trans-boundary cooperation rather than conflict. Projects for decreasing losses and preventing flood hazards upstream could be developed to generate additional river flows for downstream countries.

30.5 Conclusions

The above discussion reveals the following important points:

- Natural flows in the River Nile Basin as a whole and in separate sub-basins are extremely sensitive to changes in precipitation and temperature increase
- Estimates of the order of magnitude of the effect of GHG emissions on temperature and precipitation rate are extremely uncertain
- Both high and low natural flows of Nile water have positive and negative impacts on the water system in Egypt. Higher flows require bigger storage capacity and a larger conveyance and distribution network. Reduced rates of natural flows limit the ability of the economy to cope with all development activities, especially agriculture, industry, tourism, hydropower, generation, navigation, fish farming and environment required for providing the ever growing population with potable and domestic requirements.
- Little has been published internationally on the effect of climate change on precipitation on the coastal strips running parallel to the Mediterranean and the Red Sea, except that stated by the IPCC (2007) [5] on the prediction of movement of the rain belt form north to south.
- Sea level rise will certainly affect groundwater aquifers in the Nile delta, in particular those close to the northern strip. These aquifers, although brackish, were considered future hope. However, increased salinity may cause them to be unusable.

30.6   Adaptation

Egypt is a developing country, with a majority of its population dependent on government subsidies and other low-income support. This, coupled with higher than desired population growth, presents the government with immediate challenges that make it more difficult to justify placing long term water management needs at the top of their list of national priorities.

Research on adaptation to climate change has produced suggestions for initiatives, some structural, some soft, some on the local level, and others regional, requiring a consensus of some or all of the ten countries sharing the Nile basin. Following are some of the policies collected from the different sources of information of this document:

30.6.1   Adapt to Uncertainty

Maintain storage at Aswan High Dam at lower elevations and allocate other storage, to receive or absorb surplus water in the event of emergencies. Examples of other storage areas include:

- the Toshka and Qantara Depressions, currently dry
- the Qaroun and Wadi Natroun areas, where limited ground waters are collected
- the coastal lakes of Manzala, Borroulas, Edko and Mariout, which are salty,
- cultivated areas, particularly in highly elevated lands

30.6.2   Adapt to Increase of Inflow

Revive the plan to store in upstream lakes in light of the present development of the Nile Basin Initiative.

30.6.3   Adapt to Inflow Reduction

Egypt's per capita share of water will be reduced by half by 2050 even in the absence of climate change. Some of the measures that need to be taken according to the National Water Resources Plan (NWRP) developed by the Ministry of Water Resources and Irrigation are the following:

- Physical improvement of the irrigation system
- More efficient and reliable water delivery

- Better control on water usage
- Augmented farm productivity and raised farmers income
- Empowerment and participation of stakeholders
- Quick resolution of conflicts between users
- Use of new technologies of weed control
- Redesign of canal cross sections to reduce evaporation losses
- Cost recovery systems
- Improvements to drainage
- Change of cropping patterns and on farm irrigation systems

30.6.4 Develop New Water Resources

- Reevaluate in light of impacts of climate change previous upper Nile conservation projects to increase Nile flows
- Explore deep groundwater reservoirs in the Sinai Peninsula and the Western desert as potential sources of water if needed
- Promote rain harvesting as a possible solution to destructive Red Sea area flash floods
- Desalinate brackish groundwater
- Increase recycling of treated wastewater (both domestic and industrial)
- Increase reuse of land drainage water

30.6.5 Soft Interventions

- Promote public awareness
- Develop circulation models
- Increase research in all fields of climate change and its impact on water systems
- Encourage exchange of data and information between Nile Basin countries
- Enhance precipitation measurement networks in upstream countries of the Nile Basin

References

1. Agrawala, Shardul, Moehner A, El Raey, My Conway D, Van Aalst M, Hagenstad M, Smith J (2004) Development and Climate Change in Egypt, Focus on Coastal Resources and the Nile, Working Party on Global and Structural Policies, Organization for Economic Cooperation and Development, (OECD)
2. Conway D (2005), from Headwater Tributaries to International River: Observing and Adapting to Climate Variability and Change in the Nile basin. Glob Environ Chang 15:99–114

3. Egypt Second National Communication Report by the Egyptian Environmental Affairs Agency (EEAA), May 2010
4. Hulme M, Conway D, Kelly PM, Subak S, Downing TE (1995) The Impacts of Climate Change on Africa, SEI, Stockholm, Sweden
5. IPCC (2007) Climate Change 2007, Impacts, Adaptation and Vulnerability, Contribution of Working Group II to the Fourth Assessment Report of the Intergovernmental Panel on Climate Change, Cambridge University Press, Cambridge, UK
6. Kim U, kaluarachchi JJ, Smakhtin VU (2008) Climate Change Impacts on Hydrology and Water Resources of the Upper Blue Nile River Basin, Ethiopia, Research Report 126, International Water Management Institute
7. Strezpek KM, Yates DN, El Quosy DED (1996) Vulnerability Assessment of Water Resources in Egypt to Climate Change in the Nile Basin, Climate Research 6: 89–95
8. Strezpek KM, Yates DN, Yohe G, Tol RJS, Mader N (2001) Constructing "Not Implausible" Climate and Economic Scenarios for Egypt, Integrated Assessment 2: 139–157, Integrated Assessment Society
9. UNEP (2005), Facing the Facts: Assessing the Vulnerability of Africa's Water Resources to Environmental Change

Chapter 31
Relative Impacts of Climate Change on Water Resources in Jordan

Ibrahim M. Oroud

Abstract The relative impacts of anthropogenic forcings and climate change on water stress in Jordan during the period 2030–2050 are investigated. The more likely figure for the population of Jordan based on natural growth only would be between 13 and 15 million people by 2050. Given this conservative projection, annual water needs for domestic purposes alone would be between 700 and 800 million m³, with the current level of water consumption. This quantity is equivalent to the total renewable water resources of the entire country even without a climate change. A rise in temperature and a drop in total precipitation or both as suggested by Global Climate Models would add another dimension to the water crisis in Jordan. A climate change will lead to a reduction in renewable water resources by 20–40%. Thus, there is a composed freshwater shortage risk caused by population growth and climate change. The outcome would be a serious water deficit risk that produces a permanent water supply crisis in this politically volatile and environmentally fragile region. Alternative freshwater sources must be sought (e.g., Red Sea-Dead Sea conveyance project; sharing freshwater resources) to meet the growing freshwater demands due to population growth and the anticipated blue water decline caused by warmer and drier climatic conditions.

Keywords Climate change • Eastern Mediterranean • Jordan water resources • Water resource vulnerability in Jordan

I.M. Oroud (✉)
Prince Faisal Center for Dead Sea, Environmental and Energy Research,
Mu'tah University, Karak, Jordan
e-mail: ioroud@mutah.edu.jo

H.J.S. Fernando et al. (eds.), *National Security and Human Health Implications of Climate Change*, NATO Science for Peace and Security Series C: Environmental Security, DOI 10.1007/978-94-007-2430-3_31, © Springer Science+Business Media B.V. 2012

31.1 Introduction

Greenhouse gases (GHG) have been building up in the atmosphere since the turn of the twentieth century. For instance, carbon dioxide increased from about 280 ± 5 ppm in 1880 to a current value close to 390 ppm. Other radiatively active atmospheric constituents such as methane and nitrous oxides are also increasing at a similar or greater rate as well. The build-up of GHG enhances atmospheric blanketing, thereby increasing near surface air temperature [1]. The eastern Mediterranean is located in a transitional climate zone between a vast subtropical desert to the south and south-west and a more humid environment to the northwest (Europe). Long-term observations in the eastern Mediterranean indicate that this area is experiencing a temperature rise [2]. There is almost a consensus among most global climate models (GCM) that a global warming will affect the eastern Mediterranean adversely [1]. Model results suggest that the near surface air temperature will increase by 1–3°C following an equivalent doubling of greenhouse gases in the atmosphere. The projections for precipitation amount, its temporal distribution and variability are not as certain. Due to the northward displacement of the polar front during the winter months, however, it is expected that the eastern Mediterranean will experience less cyclogenic activities, and as such fewer winter storms. This means that precipitation in this region will decline following the proposed climate change. Jordan receives ~75% of its precipitation in the winter months, December through February, due mainly to cyclogenic activities. During transitional periods (Spring and Fall), precipitation is associated with the passage of cyclones and also due to thunderstorm activities triggered by dynamic instability which is linked to upper air cold intrusions along with a Red Sea surface trough. During the rainfall season of this year (2010/2011), very few winter storms were recorded in this area, with a long drought event that extended until the end of January.

Globally, this area is probably the least fortunate in its water resources, and thus a climate change towards drier and warmer conditions will cause a significant drop in blue and green water availability in this already water-stressed environment. Additionally, this area has been experiencing a rapid population increase due to natural growth and influx of immigrants.

It is evident that anthropogenic activities and natural forcings work hand-in-hand to adversely impact Jordan's limited water resources in the very near future. As such, two questions of operational importance need to be addressed: (1) what would be the near future water needs in the country for the various sectors (domestic, agricultural and industrial)? and (2) what would be the water status (availability) for the near future following a climate change? Answering these two questions properly is essential for a better assessment of the relative impacts of climate change and population growth on water availability and water stress during the period 2030–2050. This assessment is quite operational for adequate planning of water resources and in the country and its allocation for the various sectors. This paper is within the framework of the Glowa Jordan River Project phase III. This contribution is geared as a risk assessment measure of water resource availability in Jordan and the associated hazard levels.

31.1.1 Current Water Status in Jordan

Currently, Jordan ranks the third poorest country in the world in its water resources. The average annual renewable water resources in the country are estimated at about 800 million cubic meters (M m³) (Ministry of Water and Irrigation, Jordan, 2010). This figure experiences substantial interannual fluctuations due to natural rainfall variability plaguing the eastern Mediterranean. Anthropogenic factors along with natural forcings have been working hand-in-hand in exacerbating the water status in the country. The massive population growth, the increased demands for agricultural products, and the establishment of more industrial compounds have put further strains on the very limited water resources in the country. Natural population growth during the past 50 years was about 3.6% with a doubling period of less than 20 years. The population of Jordan, however, swelled by about 11 times during the past half century [3] which gives a virtual population growth of ~4.8%. More recently, the invasion of Iraq caused a mass movement of Iraqis towards Jordan

Official figures provided by the Ministry of Water and Irrigation, Jordan (2010), indicated that current domestic freshwater supply is ~150 l per person/ day, which gives a total annual freshwater need of 330 M m³. Nowadays, with only six million inhabitants (excluding recent refugees), most households in Jordan receive a specified amount of water during the summer months, and domestic water is supplied for 24 h per week. Renewable water resources in the country were not enough to meet the water demands, and as such non-renewable fossil freshwater resources have been intensively exploited during the past several decades. These measures have caused a steady drop in the level of underground aquifers and have resulted in poor water quality (high salinity).

The population growth was paralleled by a similar increase in irrigated agriculture. The area of irrigated lands in the Jordan Valley increased from ~15 thousand hectares in the early 1960s to ~37 thousand hectares in 2003. Likewise, irrigated agriculture in the desert region increased from virtually nil in the early 1970s to ~17 thousand hectares in 2008 (Ministry of Water and Irrigation, 2010). The substantial increase in the irrigated agricultural land caused further demands on freshwater. The availability of irrigation water will shrink in the near future, however, because of growing demands on this resource from other sectors, mainly domestic. The future water status in the country would indeed look quite bleak should population growth continue unabated and a climate change towards warmer and/or drier conditions prevail in the near future.

31.2 Future Water Needs

There are several factors that will influence future water needs in a country, which include: (1) population growth, (2) land use changes, (3) economic development, (4) climatic changes, and (5) technological developments. Figure 31.1 shows the

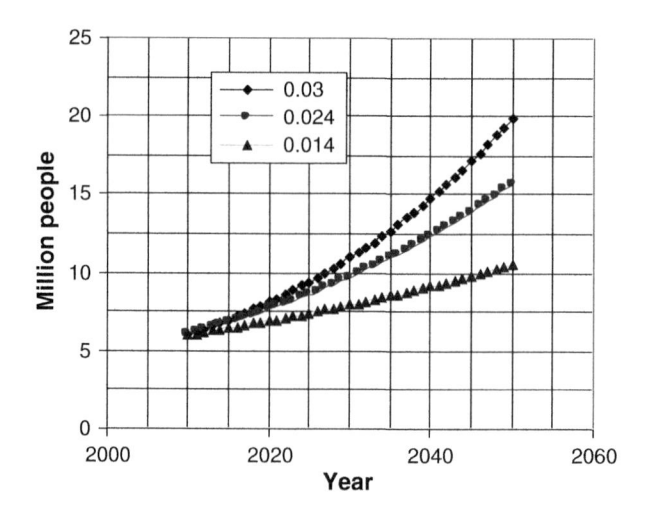

Fig. 31.1 Population growth scenarios in Jordan between 2010 and 2050

projected population of Jordan up to 2050 using three population growth scenarios (3.0%, 2.4%, 1.4%) and assuming that geopolitics in the area remains "quasi-stationary!!" (i.e., no sudden influxes of refugees, stable political systems!, no epidemics of infectious diseases, non-conventional wars, etc.). The first growth rate represents a relatively large figure, although it is not unreasonable; the second figure represents the current population growth, and the third one represents a restricted growth rate caused by economic hardships and stringent official policies. According to these scenarios, by 2050, the population of Jordan is expected to be between 10 and 20 million people. A more likely figure for the population of Jordan by 2050 would probably be around 13–15 million people.

With no technological breakthroughs such as desalination of sea water at a reasonable cost, or large scale atmospheric moisture condensing with relatively little cost, the future water needs will progressively worsen as time passes even without a climate change. Assuming that water supply stays at this precarious edge, projections based on future population growth scenarios indicate that annual domestic water needs alone will range from a minimum of 550 Mm^3 to a value close to 1,100 Mm^3 by 2050. A more likely figure would probably be between 700 and 800 Mm^3. Thus, renewable freshwater resources of the entire country will barely meet domestic water needs even without a climate change.

Currently, the agricultural sector accounts for about 65% of total freshwater consumed in Jordan (Ministry of Water and Irrigation, 2010). With the current level of population growth and the parallel hike of food commodities world wide, irrigation water needs in Jordan are expected to rise in the future. Paradoxically, the amount of water allocated for irrigation must drop in the near future because of demands by relatively more needy sectors, the domestic one in particular. Industrial water needs are expected to be close to 200 Mm^3. Although treated sewage /grey effluents are expected to be intensively utilized in the irrigation of certain restricted crops, it is very unlikely that tangential, yet costly, uncertain, and potentially hazardous measures such as this will make any difference in this marginal and volatile environment.

Based on population growth alone, it is obvious that the future water status in the country looks quite bleak even without a climate change. Should the climate of the eastern Mediterranean become warmer and/or drier or both, the country will face a serious freshwater dilemma in the very near future which will ultimately lead to social and political unrest. The anticipated climate change adds another unpleasant dimension to the future water crisis in Jordan.

31.3 Climate Change Impact on Water Resources

A climate change will impact water resources in at least three ways: (1) affect water availability, (2) influence irrigation water demands, and (3) increase evaporation losses from dams and open water conveyance canals. The impact of climate change on the available water resources in Jordan is investigated using a water balance model with a temporal resolution of 1 day. The model is run over the mountainous areas of Jordan. Figure 31.2 shows the study area along with the average annual precipitation and average annual temperature. A more detailed description of the model is presented elsewhere [4, 5]. The water balance of a soil column may be expressed using the following form,

$$P = \delta s + ET + Ro + D_r \qquad (31.1)$$

where P is daily precipitation, δ_s is change in soil water storage, ET, Ro and D_r are evapotranspiration, runoff and deep percolation beyond the root zone (all terms are mm/day). The soil column is divided into four layers, 0.25 m each. Surface albedo responded dynamically to foliage coverage and moisture status in the top soil layer. Direct evaporation and transpiration were calculated separately. Evaporation from the soil proceeds at its potential rate when moisture with the skin soil layer exceeds a threshold value (6 mm for clay-like soils). Transpiration is determined by vertical root extension and distribution, and proceeds at its potential rate when moisture within a given layer exceeds a critical threshold level. When soil moisture within the root zone declines below the critical level, transpiration proceeds proportional to the available plant water. In this model, surface runoff occurs when the top soil layer reaches saturation. Underground recharge occurs when the entire soil column reaches the field capacity. The model was run for current conditions and results appear to be commensurate with field observations and data collected by the Ministry of Water and Irrigation (Jordan).

The effect of climate change can be identified using a *space-for-time approach*. Thus, climate gradient can be used as a surrogate for that purpose. Figure 31.3 shows the linkage between annual blue water availability and annual precipitation as calculated by the model. The model was run for climate change scenarios assuming

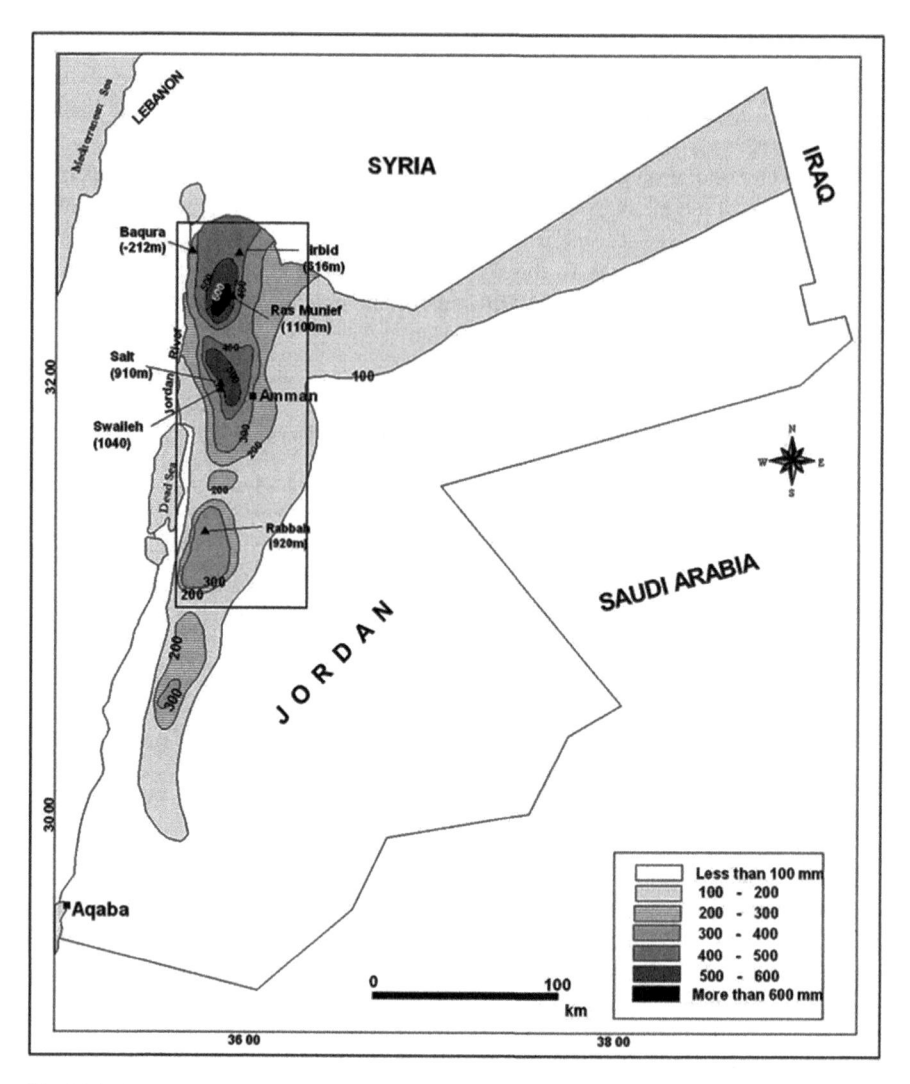

Fig. 31.2 Study area

an increase in air temperature and a reduction in precipitation. A climate change towards warmer and drier conditions causes a reduction in blue water availability. Precipitation intensity is expected to increase, however, leading to greater probability of runoff, which will partially compensate for precipitation reduction [6]. Of course, underground recharge will be seriously reduced. Conservative calculations show that a 2°C temperature increase along with a 15% reduction in precipitation decreases water availability, on average, by ~25–40% depending on the level of aridity. This means that renewable water resources in the country following a

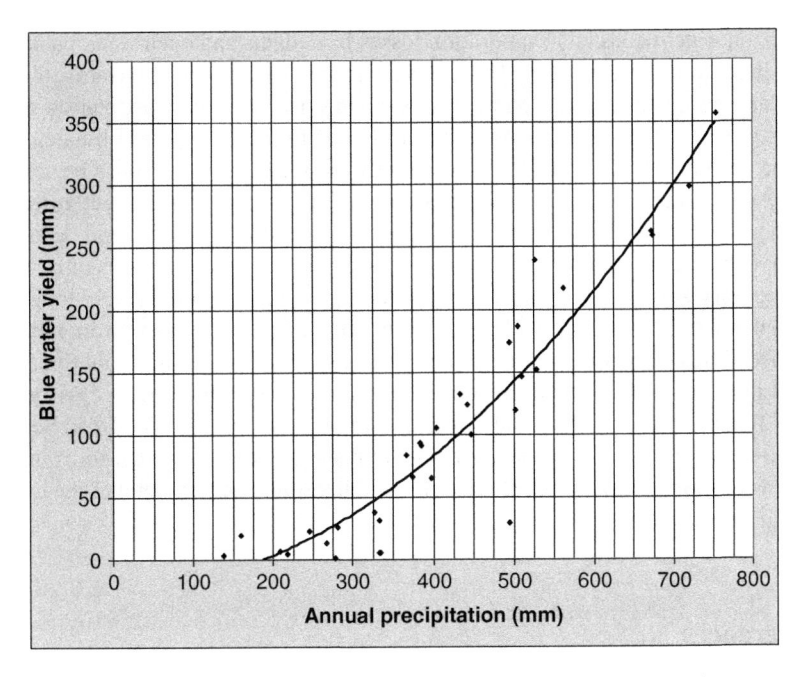

Fig. 31.3 Linkage between annual precipitation and annual water yield in the study area

warmer, drier climate will probably be between 500–650 M m3 by 2050. This conclusion is commensurate with those presented in Margane et al. (2008) and Suppan et al. (2008) [7, 8].

Irrigation water needs under current climate conditions and following a climate change were calculated for the Jordan Valley using a simulation model. Following a climate change, the irrigation water needs will increase by around 15%. This is equivalent to 40–50 M m^3 of extra irrigation water needed to maintain the irrigation water needs in the Jordan Valley at the current land use regime. Evaporation from dams is relatively small compared to other changes, but will contribute to further deterioration of the water crisis in the future.

31.4 Discussion and Conclusion

Population growth and climate change towards warmer and drier conditions go hand-in-hand in exacerbating the water situation in Jordan. Water demands due to population growth are estimated to increase between 350 and 450 M m^3 annually by 2050. A climate change towards warmer and drier conditions will decrease blue water availability in Jordan by about 200–350 M m^3/year by 2050. The increased irrigation water needs and evaporation losses from dams and open water canals will contribute more water stress. Thus, there will be two combined hazards, a demographic one resulting from population swelling due to natural population growth and regional politics, and a natural hazard linked to the anticipated climate change in the eastern Mediterranean. Subsequently, serious measures must be taken to address the anticipated severe water shortages in the country. Currently, the Disi project, which will bring around 80–100 M m^3 of freshwater for several decades, is under construction. The Disi water will be used for domestic purposes only. This is a temporary solution to a long lasting water problem. The more likely long- term sustainable option would be a large scale desalination of sea water through the Red Sea-Dead Sea conveyance project. This project will serve two important purposes: (1) it provides additional freshwater for domestic use in this fragile environment, and (2) it will preserve the Dead Sea from further deterioration. The Dead Sea level is currently falling at a rate greater than 1 m per year. Additional measures include: increasing water conveyance efficiencies and introduce and implement the concept of water saving measures.

References

1. Inter Governmental Panel on Climate Change (2007) The physical science basis. IPCC, Geneva
2. Zhang X et al (2005) Trends in Middle East climate extremes from 1950 to 2003. J Geophys Res 110:D22104. doi:10.1029/2005JD006181
3. Oroud IM, Alrousan N (2004) Urban encroachment on agricultural lands in Jordan during the second half of the twentieth century. Arab World Geogr 7:165–180
4. Oroud IM (2008) The impact of climate change on water resources in Jordan. In: Zereini F, Hotzl H (eds) Climatic changes and water resources in the Middle East and North Africa. Springer, New York, pp 109–123
5. Oroud IM (2011) Climate change impact on green water fluxes in the eastern Mediterranean. In: Leal Filho W (ed) Climate change and the sustainable management of water resources. Springer, Berlin
6. Schulz O, Busche H, Benbouziane A (2008) Decadal precipitation variances and reservoir inflow in the semi-arid upper Draa basin. In: Zereini F, Hotzl H (eds) Climatic changes and water resources in the Middle East and North Africa. Springer, New York, pp 165–178
7. Margane A, Borgstedt A, Subah A (2008) Water resources protection efforts in Jordan and their contribution to a sustainable water resources management. In: Zereini F, Hotzl H (eds) Climatic changes and water resources in the Middle East and North Africa. Springer, New York, pp 325–345
8. Suppan P, Kunstmann H, Heck A, Rimmer A (2008) Impact of climate change on water availability in the near East. In: Zereini F, Hotzl H (eds) Climatic changes and water resources in the Middle East and North Africa. Springer, New York, pp 47–58

Appendix

Group Photo

H.J.S. Fernando et al. (eds.), *National Security and Human Health Implications of Climate Change*, NATO Science for Peace and Security Series C: Environmental Security, DOI 10.1007/978-94-007-2430-3, © Springer Science+Business Media B.V. 2012

List of Participants

Dr. Ana Alebić-Juretić
Associate Professor
University of Rijeka
Air Pollution Division
Teaching Institute of Public Health/School of Medicine
Kresimirova 52a
HR-51000 Rijeka
Croatia
Tel: 385 51 358 752   Fax: 385 51 358 753
alebic@rijeka.riteh.hr

Prof. Pinhas Alpert
Professor & Head
Tel-Aviv University
Porter School of Environmental Studies
Department of Geophysics and Planetary Sciences
Tel-Aviv 69978
Israel
Tel: 972 3 6407380   Fax: 972 3 6409282
pinhas@post.tau.ac.il

Dr. Alexander Baklanov
Vice Director
Danish Meteorological Institute
Center for Energy, Environment and Health
Lyngbyvej 100
DK-2100 Copenhagen
Denmark
Tel: 45 39147441   Fax: 45 39147400
alb@dmi.dk

Dr. Iryna Kh. Bashmakova
University Researcher
University of Helsinki
Department of Physics
Division of Atmospheric Sciences
PL 48 (Erik Palménin aukio 1), 1D17b
HELSINGIN YLIOPISTO
Finland
Tel: 09 19151651
iryna.bashmakova@helsinki.fi

Dr. Anne-Lise Beaulant
Meteo France CNRM/GMME/TURBAU
42 av. G. Coriolis
31057 Toulouse Cedex
France
Tel: 05 61 07 93 05
anne-lise.beaulant@meteo.fr

Dr. Robert Bornstein
Professor
San Jose State University
Department of Meteorology & Climate Science
One Washington Square
San Jose, CA 95192–0104
USA
Tel: 1 (408) 924–5205   Fax: 1 (408) 924–5191
pblmodel@hotmail.com

Dr. Antonio Cenedese
Professor
Sapienza University of Rome
Dipartimento di Ingegneria Civile edile e ambientale
Facoltà di Ingegneria
via Eudossiana, 18
00184 ROMA
Italy
Tel: 39 06 44585 095   Fax: 39 06 44585 094
antonio.cenedese@uniroma1.it

Dr. Vincenzo Costigliola
President
European Medical Agency
Belgium
Vincenzo@EMAnet.org

Dr. Gamal Elsaeed
Associate Professor
Civil Engineering Department
Faculty of Engineering
Shobra
Banha University
Egypt
g.elsaeed@feng.bu.edu.eg

Dr. Roger A. Falconer
Halcrow Professor of Water Management
Cardiff University
Director Hydro-environmental Research Centre
School of Engineering
The Parade
Cardiff CF24 3AA
United Kingdom
Tel: 44 29 2087 4280
FalconerRA@cf.ac.uk

Dr. Harindra Joseph S. Fernando
Wayne & Diana Murdy Professor
University of Notre Dame
Department of Civil Engineering & Geological Sciences
156 Fitzpatrick Hall
Notre Dame, IN 46556
USA
Tel: 1 (574) 631 9346   Fax: 1 (574) 631 1063
Fernando.10@nd.edu

Dr. Jacques Ganoulis
UNESCO Chair and Network INWEB
Aristotle University of Thessaloniki
International Network of Water/Environment Centres for the Balkans
Department of Civil Engineering
Thessaloniki 54124
Greece
Tel: 30 2310 99 56 82   Fax: 30 2310 99 56 81
iganouli@civil.auth.gr

Dr. Giovanni Grandoni
Senior Scientist
Enea, Casaccia
Department of Environmental Technologies
Enea Atmosphere Research Team
Via Anguillarese, 301
00123 Rome
ITALY
Tel: +39 (0)6 30483625   Fax: +39 (0)6 30486571
giovanni.grandoni@enea.it

Dr. Ivana Herceg Bulić
University of Zagreb
Andrija Mohorovičić Geophysical Institute
Faculty of Science
Horvatovac 95
10000 Zagreb
Croatia
Tel: 385 1 46 05 930   Fax: 385 1 46 80 331
iherceg@mail.irb.hr

Dr. Julian C.R. Hunt
Professor of Climate Modeling
University College London
Department of Earth Science
Gower Street, London
WC1E 6BT
United Kingdom
Tel: 44 020 7679 7743
jcrh@cpom.ucl.ac.uk

Dr. Amela Jericevic
Head
Meteorological and Hydrological Service
Air Quality Research and Application Department
Grič 3 HR
10000 Zagreb
Croatia
Tel: Fax: 385 (0)1 4565 630
jcriccvic@cirus.dhz.hr

Prof. Harry D. Kambezidis
Director
National Observatory of Athens
Research & Sustainable Development
Lofos Nymphon
PO Box 20048
GR-11810 Athens
Greece
Tel: 30 210 3490119   Fax: 30 210 3490113
harry@meteo.noa.gr

Dr. Adnan Kaplan
Professor
Ege University
Institute of Natural and Applied Sciences
Department of Landscape Architecture
35040 Bornova
İzmir
Türkiye (Turkey)
Tel: +90 232 3111419   Fax: +90 232 3881864
adnankaplan@gmail.com; adnan.kaplan@ege.ed.tr

Dr. Marcus DuBois King
CNA
Environment & Energy Research Team
4825 Mark Center Drive
Alexandria, VA 22311–1850
USA
Tel: 1 (703) 824 2505   Fax: 1 (703) 933 6247
kingm@cna.org; mdk7@georgetown.edu
(currently: Research Director, Georgetown University)

Dr. Zvjezdana Bencetić Klaić
Professor
University of Zagreb
Andrija Mohorovičić Geophysical Institute
Department of Geophysics
Horvatovac 95
10000 Zagreb
Croatia
Tel: 385 1 4605929   Fax: 385 1 4680331
zklaic@gfz.hr

Dr. Hrvoje Kozmar
Assistant Professor
University of Zagreb
Faculty of Mechanical Engineering and Naval Architecture
Ivana Lučića 5
10000 Zagreb
Croatia
Tel: 385 1 6168162   Fax: 385 1 6156940
hrvoje.kozmar@fsb.hr

Dr. Nancy Lewis
Director
East-West Center
Research Program
1601 East-West Road
Honolulu, Hawai'i 96848–1601
USA
Tel: 1 (808) 944 7245   Fax: 1 (808) 944 7399
lewisn@eastwestcenter.org

Dr. Maria Christina Mammarela
Senior Scientist
ENEA – (Italian National Agency for New Technologies, Energy and Sustainable
Economic Development)
Department of Environmental Technologies
EART – Enea Atmosphere Research Team
LungoTevere Thaon di Revel n°76
00196-Rome
ITALY
Tel: +39(0)636272925   Fax: +39(0)636272691
mariacristina.mammarella@enea.it

Ms. Jennifer L. McCulley
Workshop Coordinator
Faculty Research Associate
Arizona State University
Center for Environmental Fluid Dynamics
P.O. Box 9809
Tempe, Arizona 85287–9809
USA
Tel: 1 (480) 965 9108   Fax: 1 (480) 965 8746
jenm@asu.edu

Prof. Ibrahim Oroud
Professor
Mu'tah University
Physical Climatology
P.O. Box 7
Karak
Jordan
Tel: 962 79 528 9874   Fax: 962 32372 380 ext. 4298
ioroud@mutah.edu.jo; ibrahimoroud@yahoo.com

Dr. Shlomit Paz
University of Haifa
Department of Geography and Environmental Studies
Mt. Carmel
Haifa
Israel
Tel: 972 4 8249617   Fax: 972 4 8249605
shlomit@geo.haifa.ac.il

Dr. Chava Peretz
Tel Aviv University
School of Public Health
Faculty of Medicine
Israel
Tel: 972 3 6405653
cperetz@post.tau.ac.il

Dr. Vesela Radovic
Assistant Professor
EDUCONS University
Faculty of Environmental Governance and Corporate Responsibility
Serbia
veselaradovic@yahoo.com

Dr. Kjeld Rasmussen
Assistant Professor
University of Copenhagen
Department of Geography and Geology
Øster Voldgade 10
DK-1350 Copenhagen K
Denmark
Tel: 45 3532 2563
kr@geo.ku.dk

Mr. Lukas Rüttinger
Project Manager
Adelphi Research
Caspar-Theyss-Strasse 14a
D - 14193 Berlin
Germany
Tel: 49 (0)30 89 000 68 37   Fax: 49 (0)30 89 000 68 10
ruettinger@adelphi.de

Mr. Stipo Sentic
Workshop Assistant
Graduate Student
University of Notre Dame
Department of Civil Engineering & Geological Sciences
156 Fitzpatrick Hall
Notre Dame, IN 46556
USA
Tel: 1 (574) 6315380   Fax: 1 (574) 631 1063
ssentic@nd.edu

Dr. Yves M. Tourre
Adjunct Senior Research Scientist
LDEO of Columbia University & Meteo France
Ocean and Climate Physics
25 W 54th Street
New York, New York 10019
USA
Tel: 1 (212) 489 1222   Fax: 1 (845) 365–8157
yvestourre@aol.com

Dr. Trahel Vardanian
Head
Department of Physical Geography and Hydrometeorology
Yerevan State University
Faculty of Geography and Geology
1, Alek Manoukian Street
Yerevan, 0025
Republic of Armenia
tvardanian@ysu.am

Dr. Peter O. Zavialov
Deputy Director
P.P. Shirshov Institute of Oceanology
Physical Sector
36, Nahimovski prospect
Moscow 117997
Russia
Tel: 7 (495) 1245994
peter@ocean.ru

Dr. Sergej S. Zilitinkevich
Professor
Finnish Meteorological Institute
PO Box 503 (Erik Palmenin aukio 1)
00101 Helsinki
Finland
Tel: 358 9 1929 4678   Fax: 358 9 1929 4103
Sergej.Zilitinkevich@fmi.fi

Index

H.J.S. Fernando et al. (eds.), *National Security and Human Health Implications of Climate Change*, NATO Science for Peace and Security Series C: Environmental Security, DOI 10.1007/978-94-007-2430-3, © Springer Science+Business Media B.V. 2012